Chemistry and Chemical Technology in XXI Century (CCT 2016)

MATEC Web of Conferences Volume 85 (2016)

Tomsk, Russia
17 - 20 May 2016

Editors:

Alexey Pestryakov
Elena Korotkova
Anatoly Vorobev

ISBN: 978-1-5108-3279-4

Printed from e-media with permission by:

Curran Associates, Inc.
57 Morehouse Lane
Red Hook, NY 12571

Some format issues inherent in the e-media version may also appear in this print version.

Printed by Curran Associates, Inc. (2017)

For additional information, please contact EDP Sciences – Web of Conferences at the address below.

EDP Sciences – Web of Conferences
17, Avenue du Hoggar
Parc d'Activité de Courtabœuf
BP 112
F-91944 Les Ulis Cedex A
France

Phone: +33 (0) 1 69 18 75 75
Fax: +33 (0) 1 69 28 84 91

contact-edps@webofconferences.org

Additional copies of this publication are available from:

Curran Associates, Inc.
57 Morehouse Lane
Red Hook, NY 12571 USA
Phone: 845-758-0400
Fax: 845-758-2633
Email: curran@proceedings.com
Web: www.proceedings.com

TABLE OF CONTENTS

INVESTIGATION OF RUTILIZING ADDITIVE INFLUENCE IN HYDRATED TITANIUM DIOXIDE AT THERMAL PROCESSING 1
Kantaev A.S, Lashtur A.L, Kraydenko R.I., Dyachenko A.N.

STUDY OF O-FLUORBENZONAL ELECTROCHEMICAL BEHAVIOR WITH THE CARBON ELECTRODE USING VOLTAMMETRY 5
Slepchenko Galina, Mikheeva Elena, Mezentseva Olga, Zaytsev Nikolay

SYNTHESIS AND THERMAL STUDY OF DUAL COORDINATION COMPOUND OF GADOLINIUM (III) AND CHROMIUM (III) 9
Bobrovnikova A., Cherkasova T., Shirokolobova A., Syskina A.

ALTERATION IN ELEMENTAL AND FUNCTIONAL COMPOSITION OF HEATED PEAT HUMIC ACIDS 15
Chukhareva Natalya, Sartakov Mikhail, Korotchenko Tatiana

COMPUTER MODELING AND SOFTWARE DEVELOPMENT FOR UNSTEADY CHEMICAL TECHNOLOGICAL SYSTEMS 23
Dolganov Igor, Dolganova Irena, Ivashkina Elena, Pisarev Mikhail

THE COMPOSITION OF BIOLOGICALLY ACTIVE COMPOUNDS IN PEAT FROM DIFFERENT CLIMATIC ZONES OF WESTERN SIBERIA 29
Serebrennikova Olga, Strelnikova Eugenia, Preis Yulia, Duchko Maria, Rousskikh Irina, Kaduchagov Petr, Peresedova Alexandra

DETERMINATION OF RHODIUM CONTENT BY THE METHOD OF STRIPPING VOLTAMMETRY IN ORES AND TECHNOGENIC RAW MATERIALS 33
Kolpakova N.A., Dyachenko E.N.

THE RESOURCE-SAVING TECHNOLOGY OF ALUMINUM NITRIDE OBTAINING DURING COMBUSTION OF ALUMINUM NANOPOWDER IN AIR 40
Ilyin Alexander, Mostovshchikov Andrei, Root Liudmila

THE STUDY OF THE PROCESS OF ALKALINE PRECIPITATION PURIFICATION OF SOLUTIONS FROM SILICON MACROSCALES IN THE AMMONIUM-FLUORIDE PROCESSING OF BERYLLIUM-CONTAINING MATERIALS 45
Dyachenko Alexander, Kraydenko Roman, Lesnikova Margarita, Malyutin Lev, Petlin Ilya

CALCULATION METHOD TO DETERMINE THE GROUP COMPOSITION OF VACUUM DISTILLATE WITH HIGH CONTENT OF SATURATED HYDROCARBONS 51
Nazarova Galina, Ivashkina Elena, Shafran Tatyana, Belinskaya Nataliya

ENGINEERING-GEOLOGICAL CONDITIONS OF SURFACE MINING IN AINSKOE GOLD FIELD AND PIT WALL STABILITY 56
Ol'khovatenko V.E., Pryanishnikov E.G.

DETERMINATION OF AU, PT, PD IN GOLD ORE MINERAL RAW MATERIALS BY STRIPPING VOLTAMMETRY 60
Kolpakova N.A., Oskina Y. A., Panova S.M., Pshenichkin A. Ya.

RECYCLING OF IRON-CONTAINING DEPOSITS – THE MAIN WAY TO INCREASE THE EFFICIENCY OF WATER-PROTECTIVE MEASURES ON THE TERRITORY OF THE TOM LOWER COURSE 63
Popov Viktor K., Pasechnik Elena Yu., Karmanova Anna

POTENTIAL USES OF NATURAL ZEOLITES FOR THE DEVELOPMENT OF NEW MATERIALS: SHORT REVIEW 69
Petranovskii V., Chaves-Rivas F., Hernandez Espinoza M.A., Pestryakov A., Kolobova E.

APPLICATION OF COMBINED SORBENT FOR PROCESS OF WATER PURIFICATION FROM MICROBIAL CONTAMINATION 74
Martemianova Irina, Zhuravkov Sergey, Plotnikov Evgenii, Martemianov Dmitrii, Korotkova Elena, Kratochvil Bohumil, Voronova Olesya, Antonova Svetlana

TECHNOLOGICAL CHARACTERISTICS OF GLAUCONITE ROCKS IN BAKCHAR DEPOSIT (WESTERN SIBERIA) 79
Rudmin M.A., Mazurov A.K., Makarov B.I., Reva I.V.

ADSORPTION OF INORGANIC IONS FROM AQUEOUS SOLUTIONS USING MINERAL SORBENT - TRIPOLI 86
Yurmazova Tatyana, Shakhova Nina, Tuan Hoang Tran

INFLUENCE OF HOMOGENIZATION OF CLAY BATCH FOR LABORATORY TEST ON THE STRUCTURE OF THE CERAMIC SAMPLES 93
Toropkov Nikita, Abakumov Alexander
ANALYSIS OF PRESSURE REDUCING EFFECTIVENESS IN THE REFORMING INDUSTRIAL UNIT WITH MATHEMATICAL MODELLING METHOD USING 98
Yakupova Inna, Chernyakova Ekaterina, Koksharov Anton
RHEOLOGICAL PROPERTIES OF INVERSE EMULSIONS STABILIZED BY ETHANOLAMIDES OF TALL OIL FATTY ACIDS 103
Yanovsky V. A., Andropov M. O., Fakhrislamova R. S., Churkin R. A., Minaev K. M., Ulyanova O. S.
DEVELOPMENT OF A PURIFICATION TECHNOLOGY FOR TREATMENT OF MEDIUM- AND LOW-ACTIVITY RADIOACTIVE WASTE OF RADIOCHEMICAL PRODUCTION FROM CO-60 AND CS-137 110
Apalkov Gleb, Dyachenko Alexander, Zhabin Andrey, Smirnov Sergey
MICROBIOLOGICAL SYNTHESIS OF METHYL ETHYL KETONE 116
Astashkina Anna, Kolbysheva Yuliya, Nikiforova Alena, Bakibayev Abdigali
DEVELOPMENT OF THE MATHEMATICAL MODEL OF DIESEL FUEL CATALYTIC DEWAXING PROCESS TAKING INTO ACCOUNT FACTORS OF NONSTATIONARITY 125
Frantsina Evgeniya, Belinskaya Nataliya, Popova Natalya
UP-DATED ORE COMPOSITION DATA (CENTRAL ORE-FIELD, KUZNETSK ALATAU) 130
Bushmanov A., Voroshilov V., Reva I., Rudmin M., Chernyshov A., Abramova R.
FEATURES OF CHEMICAL-MINERALOGICAL COMPOSITION AND PROCESSING PROPERTIES FOR MEDIUM-FERROUS BAUXITES 137
Govorova L.P., Vakalova T.V., Shvagrukova E.V., Reshetova A.A., Potapova A.P.
THE RESEARCH OF FINELY DISPERSED IRON POWDER MOISTENING APPLYING THE PH–METRY METHOD 143
Ikonnikova Kseniya, Ikonnikova Lyubov, Koltunova Ekaterina, Murashkin Maxim
APPLYING PALEOMAGNETIC ORIENTED CORING IN INVESTIGATING THE LITHOFACIES HETEROGENEITY OF TERRIGENOUS SEDIMENTS 150
Krasnoshchekova Lyubov, Cherdantseva Daria, Yurkova Maria
MODELLING OF THE KINETICS OF SULFURE COMPOUNDS IN DESULFURISATION PROCESSES BASED ON INDUSTRY DATA OF PLANT 154
Krivtcova Nadezhda, Tataurshikov Anton, Kotkova Elena
DEVELOPMENT AND EXPLOITATION FEATURES OF METHANE-COAL WELLS USING DOWNHOLE PUMPING EQUIPMENT 159
Kurganova Elena, Makarovskikh Alexandra, Chernykh Tatyana
MODIFICATION OF NATURAL SORBENT FOR PROVIDING IT WITH BACTERICIDAL AND BACTERIOSTATIC PROPERTIES 163
Martemianova Irina, Nadeina Luiza, Plotnikov Evgenii, Martemianov Dmitrii
CHEMICAL COMPOSITION OF BOTTOM SEDIMENTS IN SMALL MOUNTAIN RIVERS (ALTAI, THE RUSSIAN FEDERATION) 170
Savichev Oleg, SMatveenko Irina, Paromov Vladimir
TYPOMORPHIC CHARACTERISTIC FEATURES OF ACCESSORY ILMENITE IN GRANITOIDS OF THE POLYPHASE ALEISK-ZMEINOGORSK COMPLEX (N-W RUDNY ALTAI AREA) 177
Novoselov K.L., Gavrilova K.A.
VOLTAMMETRIC ESTIMATION OF THE CONTENT OF ANTIBIOTICS IN VETERINARY PREPARATIONS 181
Slepchenko Galina, Gindullina Tatyana, Deryabina Valentine, Shchukina Tatyana, Okanov Aydin, Kabieva Saule
THE DETERMINATION OF SORPTION MATERIAL PROPERTIES FOR WATER SOFTENING IN THE DYNAMIC CONDITIONS 185
Somin Vladimir, Kurtukova Lubov, Komarova Larisa
TUNGSTATE AND CARBONATE IONS SORPTION USING ANION EXCHANGERS AV-17-8 AND PUROLITE A400 190
Chegrintsev S., Kraydenko R., Dyachenko A., Kiselev A.
ANALYSIS OF OPTIMAL PROCESS FLOW DIAGRAMS OF LIGHT NAPHTHA ISOMERIZATION PROCESS BY MATHEMATIC MODELLING METHOD 193
Chuzlov Vjacheslav, Molotov Konstantin
Author Index

MATEC Web of Conferences
Open access conference proceedings in Materials science, Engineering and Chemistry
A flexible, fast and efficient publication service, for an open access recording of your event and a wide dissemination of your proceedings
www.matec-conferences.org
edp open Open Access Publishing from EDP Sciences

Statement of Peer review

In submitting conference proceedings to MATEC Web of Conferences, certify to the Publisher that I adhere to the Policy on Publishing Integrity of the journal in order to safeguard good scientific practice in publishing.

1. All articles have been subjected to peer review administered by the proceedings editors.
2. Reviews have been conducted by expert referees, who have been requested to provide unbiased and constructive comments aimed, whenever possible, at improving the work.
3. Proceedings editors have taken all reasonable steps to ensure the quality of the materials they publish and their decision to accept or reject a paper for publication has been based only on the merits of the work and the relevance to the journal.

Title, date and place of the conference

Chemistry and Chemical Technology in XXI Century,
May, 17-20, 2016
Tomsk, Russia

Proceedings editor(s):

Alexey Pestryakov,

Elena Korotkova

Anatoly Vorobev

Date and editor's signature

17.10.2016 A.Pestryakov

E.Korotkova

MATEC Web of Conferences 85, 01001 (2016) DOI: 10.1051/matecconf/20168501001
CCT 2016

Investigation of rutilizing additive influence in hydrated titanium dioxide at thermal processing

A.S Kantaev[1,a], A.L Lashtur[1], R.I. Kraydenko[1] and A.N. Dyachenko[1]

[1]National Research Tomsk Polytechnic University, 634050,Tomsk, Russia

Abstract. The influence of a rutilizing additive in hydrated titanium dioxide is studied at its dehydrationwithin the range of temperaturesfrom 100to 230° C. The quantityof major impurities isdetermined using an atom-emission spectrometer. Also, the X-ray diffraction analysis of obtained samplesis performed.

1. Introduction

Heterogeneous reactions occupy a dominant position in the technologies of rare, scattered, and radioactive elements. The majority of processes belong to this type of reactions. The processes of dehydration and calcinationare accompanied by boiling and vaporization.The multistage sequential flowof a process and the product formationare the main features of heterogeneous reactions.The main objective of heterogeneous kinetics is to determine arate-limiting stage of a technological process and the rate of the whole process. All the processes have a multistage character, and it is impossible to separate individually each stage, therefore, the total energy of activation is determined, which consists of elementary steps [1]. The activation energy is a criterion of the rate-limiting stage. The higher activation energy is, the more difficult formolecules to interact, and the slower the whole process runs. The activation energy is associated with a rate constant. It is possible to determine the rate constant experimentally [2]. Determination of kinetic parameters is confined to measuring the rate of reaction at varioustemperatures; it is definedas mass-change of one of reacting substances in time [2].

Hydrated titanium dioxide (HTD) has a complex molecular structure, and it could contain a variable amount of hydroxyl groupstied with titanium. At thermal processing of HTD the processes of dehydration and polymorphic transformation of anataseinto rutile occur. In this case, the temperature range, in which these processes are implemented, depends,to a large extent, on conditions of HTD obtainment, the size of its crystals, the presence of impurities, and other parameters.

2. Experimental

At carrying out the work kinetic research of HTD thermal processing has been conducted with a rutilizing additive and without it at various temperatures. The installation for kinetic researchis shown in Figure 1.

[1]Corresponding author: akantaev@tpu.ru

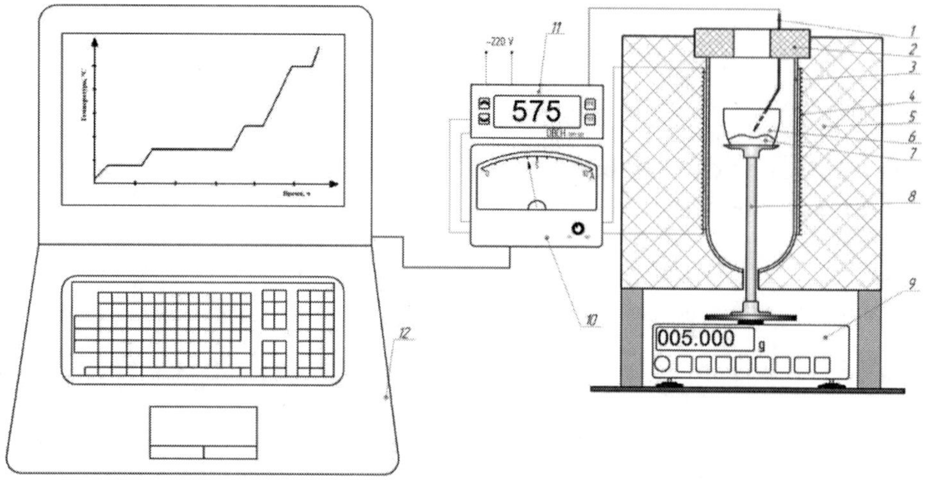

Figure 1. Scheme of a kinetic installation:1 –a thermocouple; 2 – a cover; 3 – a quartz tube; 4 –an electric heating tube; 5 –heat insulation;6 – a platinum crucible; 7 –a weighted amount of the studied substance; 8 –a crucible stood; 9 –an electronic weighing machine; 10 –an ammeter; 11 –a programmable thermostat controller "OVEN" TPM 501;12 – a personal computer.

The weighted amounts of hydrated titanium dioxide (11 g) were dissolved in chlorohydric acid at pH = 2 with the addition of zinc metal (Zn), or aluminum (Al) as a rutilizing additive. After 3 hours of reaction the unreacted metal was removed from the solution, and then the 25% aqua-ammonia solution was added to pH = 8. The resulting precipitate was separated from the solution, and it was washed with distilled water. The resulting precipitate was studied usingthe kinetic installation, the process of dehydration was examined according to weight loss, which was, mainly, caused by H_2O evaporation and NH_3 excess. The drying temperature varied from 100° C to 230° C, dehydration time - from 1 to 3 hours[3].

The structure of acrystal lattice in the samplesobtained after thermal processingwas determinedusing the X-ray diffraction analysis (XRD)atthe X-ray diffractometer XRD-7000S of Shimadzu Company. The content of major impurity elements was defined at the atomic emission spectrometer iCAP 6300 Duo of Thermo Electron Corporation Company [4].

3. Resultsand discussion

The experimental data for constructingdependence diagramsof mass-change on time were obtainedat various temperatures, using the installation at HTD thermal processing with a rutilizing additive and without it. The results are shown in Figure 2.

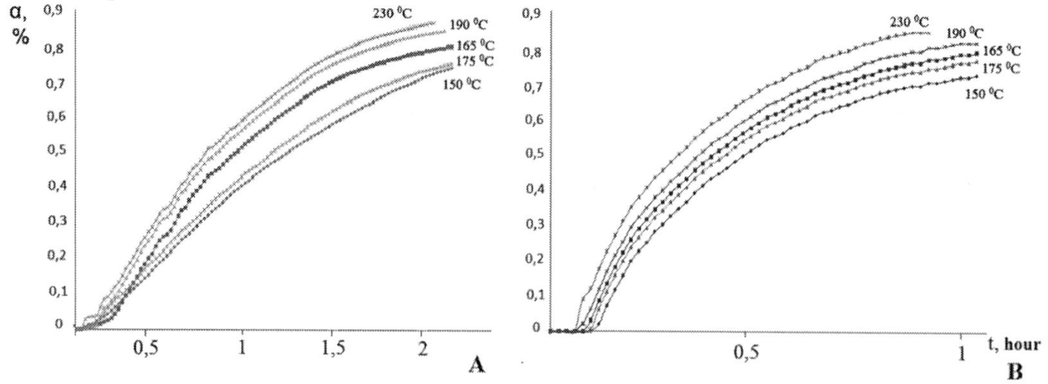

Figure 2. Dependence of water removal rate fromhydrated titanium dioxide on time at various temperatures: A –without adding a rutilizing additive; B - with adding a rutilizing additive.

Initially, a dehydration process runs intensively with kinking at $\alpha \sim 0.8$, the temperature of 190 - 230° C (after, approximately, 40 minutes).The processes proceed evenly at the temperature below 190° C, similar to the processes with addition of therutilizing additive - zinc metal.

The process of HTD dehydration proceeds with evenlydependence of mass-change on time (Figure 2).Several sections with different rates of evaporation are traced in all dependencies: the first section –HTD heating for 10 minutes, the second section - intensive water evaporation (generation and volatilization of gases), and thenthe processes achieve a plateau, where dehydration occurs. The rate of water evaporation increases with the increase of a temperature.Application of a rutilizing additive influencesthermal decomposition, and the process runs faster than without rutilization.

The activation energy of the process is of 2356.3 J/mol within the range of temperatures from 150° C to 230° C (Figure 2 A).

For this process the dependence of waterremoval rate from the precipitate on the time and temperature is as follows:

$$\alpha = 1 - \left(1 - 2{,}4 \cdot exp\left(-\frac{2356{,}3}{R \cdot T}\right) \cdot t\right)^3$$

The process runs in an intra-diffusion reaction area.It is determined by a dense structure of hydrated titanium dioxide and also by the complexity of water evaporation through a HTD isolation layer. The process can be intensified using mechanical mixing.

The activation energy of the process is of1107.8 J/mol within the range of temperatures from 150° C to 230° C at HTD drying with the use of rutilizing additives (Figure 2 B).

Water removal rate in the given range is calculated as follows:

$$\alpha = 1 - \left(1 - 2{,}4 \cdot exp\left(-\frac{1107{,}8}{R \cdot T}\right) \cdot t\right)^3$$

Consequently, the process of dehydration runs in the intra-diffusionreaction area,and it is limited by the diffusion of resultants in pores.The method used to stimulate the process is mechanical mixing.

According to the results of the samples studied using the X-ray diffraction analysis, it is observed that recrystallization of an anatase formintoa rutile oneoccurs as a result of thermal processing.

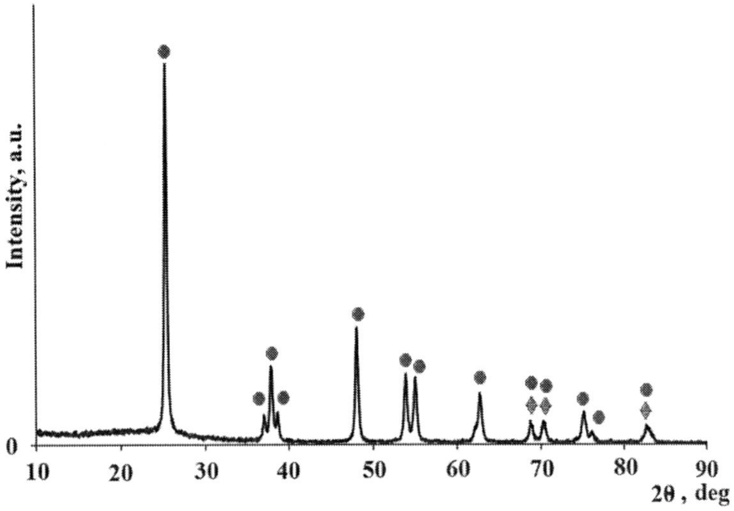

Figure 3. Diffraction patterns of samples without a rutilizing additive: ◆ - TiO_2 (rutile), ● - TiO_2 (anatase)

 DOI: 10.1051/matecconf/20168501001

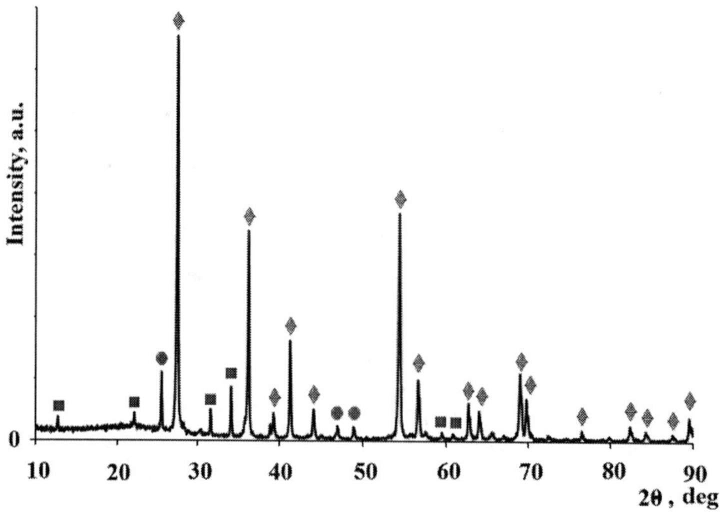

Figure 4. Diffraction patterns of samples with a rutilizing additive: ◆ - TiO_2 (rutile), ● - TiO_2 (anatase), ■ - Zn_2SiO_4 (villemit).

The results of the analysis forSample №1 (fig.3) - without addition of a rutilizingadditive, and Sample №2 (fig.4) - with addition of it analyzed at the atomic emission spectrometer iCAP 6300 Duo are shown in Table 1.

Table 1. Results of element analysis at the atomic emission spectrometer iCAP 6300 Duo

Denotation	Content of main components, %	Content of main impurities, %	
	Ti	Fe	Zn
Sample 1	58,21	0,29	0,094
Sample 2	59,75	0,078	0,23

The results of the element analysis show that Zn metal addition decreases the content of Fe by substitution. TiO_2 has a sorption activity, and, apparently, Zn^{2+} displaces Fe^{3+}.

4. Conclusion

Based on experimental data dehydration of hydrated titanium dioxide with addition of a rutilizing additive and without it is calculated using mathematical processingwithin the range of temperatures from 100 ° C to 230° C in time.The results of a structural analysis show as follows: addition of the rutilizing additive in the form of Znmetal decreases the temperature of thermal processing and facilitates formation of crystallization rutile centers. For speeding up the processes, which run in the intra-diffusion area, it is necessary to remove the layer inhibiting a chemical reaction fromreacting particles. It could be arranged in rotating machines.

References

1. V.A. Borisov, A.N. Dyachenko, R.I. Kraidenko,Mechanism of reaction between cobalt(II) oxide and ammonium chloride. Russian Journal of Inorganic Chemistry, Vol. 57, Issue: 4, pp 499-501, (2012)
2. G.G. Andreev, A.N. Dyachenko, O.E. Permyakov, Lectures on heterogeneous chemical kinetics, Tomsk:TPU Publishing, p 120 (2008)
3. A.S. Kantaev, A.L. Lashtur, Obtaining Titanium Dioxide of Rutile Form of Crystal Lattice from The Solutions Contained Fluorammonium Titanium Complex Salts // International MultiConference of Engineers and Computer Scientists: Proceedings, Hong Kong, March 16-18, 2016. - London: International Association of Engineers, - Vol. 2 - pp 719-720 (2016)
4. V.V. Smirnova // Modern problems of science and education. №5(2012)

Study of o-fluorbenzonal electrochemical behavior with the carbon electrode using voltammetry

Galina Slepchenko[1], *Elena* Mikheeva[1], *Olga* Mezentseva[1] and *Nikolay* Zaytsev[2]

[1] National Research Tomsk Polytechnic University, Institute of nature resources, 634050 Tomsk, Russia
[2] Moscow Technological University, 119454 Moscow, Russia

Abstract. A high sensitive stripping voltammetric method was for the first time proposed for determination of o-fluorbenzonal. This method is based on the ability to be electrochemically reduced with the carbon electrodes of different types. The working conditions of o-fluorbenzonal voltammetric determination are: pH of the background electrolyte, electrolysis potential and time scan rate. The effective dissociation constant of various forms of o-fluorbenzonal, number of electrons involved in the electrode process were calculated.

1.Inroduction

To solve the problem of drug provision of the citizens with modern Russian drugs domestic drugs need to be developped. At the Department of Biotechnology and Organic Chemistry of National Research Tomsk Polytechnic University the drug of o-fluorbenzonal (galonal) was synthesized, it can be used as an antiepileptic drug, like other barbiturates. Figure 1 shows its chemical structure.

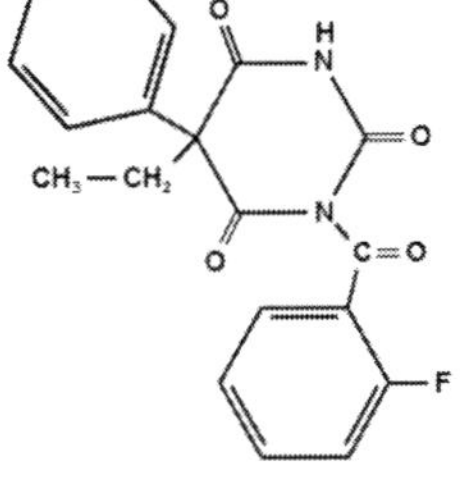

Fig1. Structure of o-fluorbenzonal

Along with this, the present substance has an enzyme-inducing activity, like phenobarbital, and therefore, may be recommended to enhance the antitoxic function of the liver [1] including for treatment of extrahepatic cholestasis [2]. Since the drug belongs to the group of barbiturates with the ability to cumulate, dose adjustments depending on the patient organism for the prevention of toxic effects of overdose and accompanied by inhibition of the respiratory center is nessesary [3].

Due to the high requirements to the quality of drugs, the importance of the use of modern standardized methods for analysis increases. The most versatile method is the voltammetry. It is widely used for determination of trace amounts of substances, in particular organic ones.

The aim of the work was to study the electrochemical behavior of o-fluorbenzonal on the carbon electrodes by the differential voltammetry.

2. Experiment

The measurements were carried out using the carbon-coated (CCE) and glassy carbon electrodes (GCE), and the saturated silver chloride electrode (SCE) was a reference electrode. Oxygen from the solution was removed by a stream of nitrogen with oxygen mass fraction not more than 0.01% (GOST 9293-74). The voltammograms were recorded on the voltammetric complex STA (TU4215-001-20694097-98) in differential pulse mode.

PH was determined using the portable pH-meter (pH-673). The error in determining solution pH did not exceed ± 0.1%.

The initial solution containing 1000.0 mg/dm^3 of o-fluorbenzonal was prepared by dissolving the sample in dymethylformamide (DMF).

Corresponding author: slepchenkogb@mail.ru

DOI: 10.1051/matecconf/20168501002

3. Results and discussion

To determine o-fluorbenzonal (galonal), a highly sensitive method of differential voltammetry was first proposed, it is based on its ability to be electrochemically reduced with the carbon-containing electrodes of different types.

The graphite (GE) and glassy carbon (GCE) electrodes were used as working, the first one is impregnated with polyethylene and paraffin in a vacuum.

These electrodes were used due to their high chemical and electrochemical stability, a wide area of the working potentials, and mechanical surface renewal simplicity. The potential for reduction of organic matter is determined by the composition, structure and degree of adsorption on the graphite hexagons having the π-band structure. The structural similarity of the electrode material and stereospecific adsorption, apparently favored by the transfer of electrons and reversible reduction of o-fluorbenzonalon the graphite electrodes. The greatest value of the analytical signal, least residual current, and better reproducibility in the voltammogram signals were observed on the glassy carbon electrode, which was selected as working.

The advantages of the glassy carbon electrode can also include high chemical resistance in corrosive environments, non-porousity, and high hardness. Also, the glassy carbon electrodes are commercially available, so it is not required for the researchers to produce their own ones. The potential of the peak of o-fluorbenzonal on the GCE is -1.38 ± 0.05V.

The voltammogram of o-fluorbenzonal electroreduction on the GCE is shown in Figure 2.

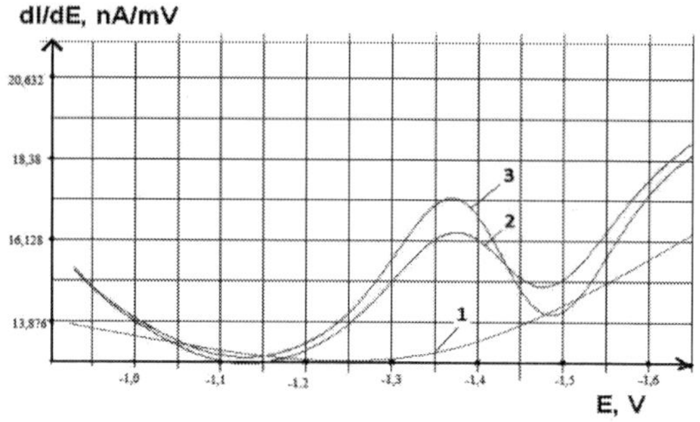

Fig2. The voltammogram of o-fluorbenzonal elecroreduction on the GCE,w=50 mV/sec, E_e = -0.7V,τ_e = 30 sec: 1 – the background electrolyte 0.1 MNaH_2PO_4; 2 – $C_{o\text{-fluorbenzonal}}$ = 0.002mg/ml; 3 – $C_{o\text{-fluorbenzonal}}$ = 0.004mg/ml.

To select the background electrolyte,electroreduction of galonal was investigated in the solutions: 0.1M NaH_2PO_4, 0.01M $Na_2B_4O_7$, 0.1M Na_2CO_3, 0.1M KCl, 0.1M Na_3PO_4. All of these electrolytes can be used for the voltammetric determination of o-fluorbenzonal but the working background electrolyte was chosen 0.1M Na_3PO_4, with help of which the distinct peaks of electroreduction are recorded with good reproducibility; the relative standard deviation Sr is less than 0.05 (Figure 3).

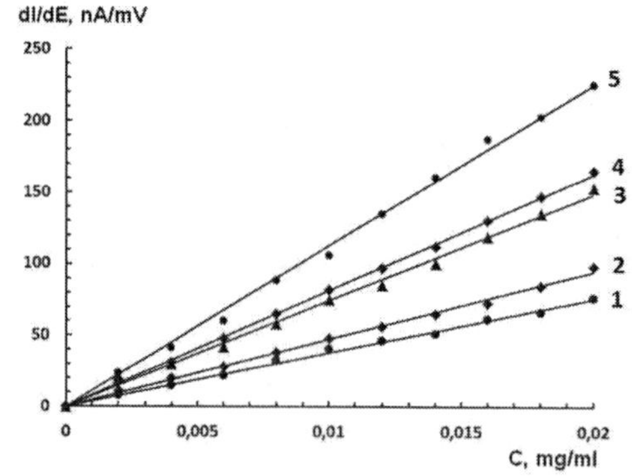

Fig3. The calibration curve of the o-fluorbenzonal electroreduction current dependence in different background electrolytes of the GCE, w=50mV/sec: 1 –0.1 M NaH_2PO_4; 2 -0.01M $Na_2B_4O_7$; 3 –0.1 M Na_2CO_3; 4– 0.1 M KCl; 5 –0.1 M Na_3PO_4.

It is known that the organic substance can be adsorbed on the carbon-coated electrodes due to the structural interaction of adsorbate (organic matter-molecule) and the adsorbent (the electrode). The dependence of the current value of o-fluorbenzonal electrochemical reduction is shown in Figure 4. Figure shows that the optimum value of the adsorption concentration potential for galonal on the glassy carbon electrode surface is - (0.5 - 0.6) V.

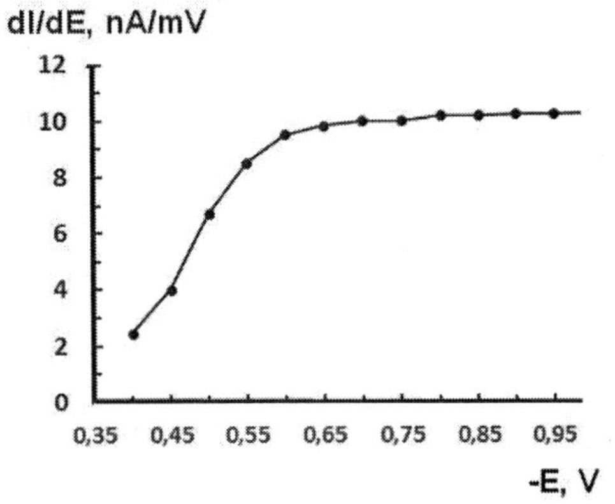

Fig4. The dependence of elecroreduction current of o-fluorbenzonal on electrolysis potential of the GCE: the background electrolyte – 0.1 M Na_3PO_4, $C_{o\text{-fluorbenzonal}}$=2 mg/l in DMF, t_e = 30 sec, w= 50 mV/sec.

The dependence of the current value of galonal electroreduction on the electrolysis time of the electrode in the solution demonstrates the role of adsorption processes. The dependence of the electroreduction current of o-fluorbenzonal on pre-adsorption accumulation time is shown in Figure 5. Figure shows that the pre-accumulation time of galonal should not exceed 30 seconds.

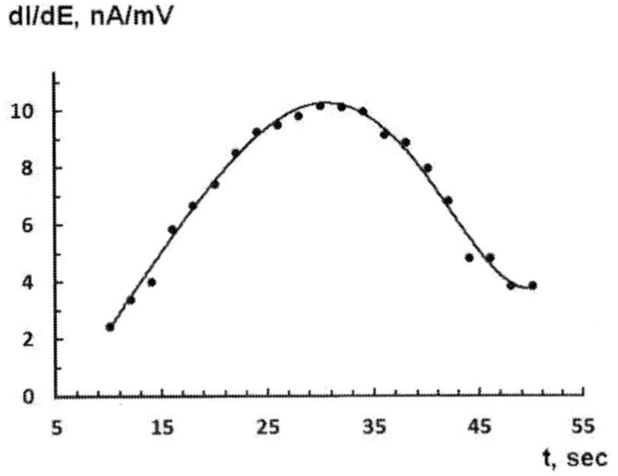

Fig5. The dependence of elecroreduction current of o-fluorbenzonal on electrolysis time on the GCE: the background electrolyte – 0.1 M Na_3PO_4, $C_{o\text{-fluorbenzonal}}$ = 2 mg/lin DMF, E_e = -0.7 V, w= 50 mV/sec.

To increase the detection sensitivity, the differential voltammograms shooting mode was applied. The advantage of this type of shooting is the possibility of obtaining a distinct peak reducing the value of the measurement error of the analytical signal. When registering voltammograms, the scan rate of polarizing voltage of 50 mV/sec was selected. At higher speeds, detection sensitivity increased but the residual current increases at the same time. Using slower speeds significantly reduces the magnitude of the analytical signal of o-fluorbenzonal electroreduction.

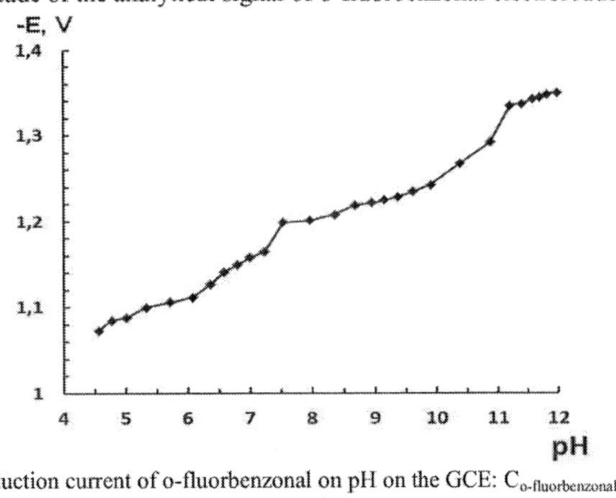

Fig6. The dependence of elecroreduction current of o-fluorbenzonal on pH on the GCE: $C_{o\text{-fluorbenzonal}}$= 2 mg/l in DMF, E_e = -0.7 V, w= 50 mV/sec.

The redox properties of o-fluorbenzonal (galonal) have not been previously studied. The paper presents the research of the electrochemical reduction process of o-fluorbenzonal depending on pH, which is an important factor in the voltammetric determination of organic substances. The dependence of the electrochemical reduction potential on pH on the CGE is shown in Figure 6. The dependence of the maximum potential of the cathodic peak on pH is complex. With an increase in concentration of hydroxide-ion, the peak shifts to the cathode region, it means to the difficulty in the recovery process of the organic matter, which appears to be linked to the preceding protolytic deprotonation reaction of the galonal molecular form. Deprotonation step may precede the electron attachment of the depolarizer molecule to the electrode or occur simultaneously. The effective values of the dissociation constants of different forms of o-fluorbenzonal were calculated, which are $3.16 \cdot 10^{-8}$ and $6.31 \cdot 10^{-12}$ respectively.

With an increase in pH of the background electrolyte, the organic matter changes its form in the solution, it is determined by the protolytic balance. Since an electron breaking from a neutral particle requires less energy than from the cationic potential, the electroreduction potential of o-fluorbenzonal in alkaline solution (where the neutral form of the galonal molecule dominates) is lower than in acidic.

The peak potential of the organic compound in the solution is not determined by the ratio of the concentrations of oxidized and reduced forms in the Nernst equation, it is caused by the transfer kinetics of the electrons, and, perhaps, protons dependent on adsorption of organic matter and structure of double electric layer. In this case, the pH dependence of E on pH may be expressed by:

$$E½ = E^{\circ} + \frac{RT}{zF} \ln \alpha_{H+}^{\upsilon} = E^{\circ} - \frac{RT}{zF} pH \qquad (1)$$

where ν - the number of protons involved in the electrode process.

The slope angle tangent of E depending on pH is 0.0589, indicating participation of one electron and one proton in the electrode processes.

The background electrolytes of 0.1 M Na_3PO_4 (pH 4-7) recommended for voltammetric determination of galonal have a pH optimum when o-fluorbenzonal is in the stable protonated form.

4. Conclusion

Thus, the operating conditions for the voltammetric determination of o-fluorbenzonal are determined on the basis of this research namely, background electrolyte - a solution of 0.1M Na_3PO_4, electrolysis time - 30 seconds, electrolysis potential - -0,6V. Subsequently, using these conditions, a quantitative determination method will be developed.

References

[1] A.Saratikov Regulators detoxification enzyme systems include nitrogen-containing compounds 264 p (2002)
[2] T. Novozheeva, I. Chursina, A. Novozheeva, A. Saratikov 2004 Pharm. Chem. J. **38, 1** 3-4 (2004)
[3] T.Vergeichik Toxicological chemistry 400 p (2009)

Conflict of interest: The authors declare that they have no conflict of interest.

Synthesis and thermal study of dual coordination compound of gadolinium (III) and chromium (III)

A. Bobrovnikova[1,a], **T. Cherkasova**[1,b], **A. Shirokolobova**[1,c], **A. Syskina**[2,d]

[1]T.F. Gorbachev Kuzbass State Technical University, Kemerovo, Russia
[2]National Research Tomsk Polytechnic University, Tomsk, Russia

[a]bobrownickowa.al@yandex.ru, [b]ctg.htnv@kuzstu.ru, [c]nastja_shirokolo@rambler.ru, [d]annesyskina@yahoo.com

Abstract. The paper presents experimental data of thermal studies of coordination compound based on gadolinium nitrate with the DSCmethod; the study was performed in the temperature range from 298 to 1273 C. The thermal analysis has several advantages over the other methods: experimental set up flexibility, fast data receive, automatic data processing, small substance amount. The process of thermal decomposition of the received compound was carried out in both inert and oxidizing environments. During the experiment the temperature was defined, above which heating causes decomposition of the investigated substance and weight loss of the sample.

1. Introduction

In modern industry the development of energy - and resource efficient technologies, such as nuclear power engineering industry, space systems, advanced medical equipment and weapons, military and special equipment, and other trends in the development of science are practically impossible without the use of rare earth metals and their compounds. Even small amounts of these substances in industrial production allow producing technical products, unique in properties and quality, and their derivatives [1].

Gadolinium compounds are intensively used in a number of existing and innovative areas of production, they are used for image visualization in medicine, for optical recording, in glass and ceramics production, in laser technology Gadolinium compounds are included in crystal scintillators and magnets, which are very demanded in advanced technology production [2].

Gadolinium opens a subgroup of heavy lanthanides and has a half-filled f-shell, which has high energy stability. For this reason, the gadolinium ions, unlike some other lanthanides, are always trivalent Gd^{+3}. Gadolinium contains the so-called "gadolinium break", the violation of the monotonic change of properties of lanthanide compounds according to the atomic number.

At present, the Cambridge structural database contains more than 2,000 (2537) crystal compounds with gadolinium as a complex former, among them only one substance has hexamethylphosphorotriamide (HMPA) in its composition [3] (Fig. 1.). A small number of structurally characterized compounds with Gd and HMPA, available in the Cambridge structural database, indirectly indicates the lack of study on hexamethylphosphorotriamide complex formation.

 DOI: 10.1051/matecconf/20168501003

Previously [4-6], ammonium tetra(isothiocynato)diaminechromate(III) hemihydrade (Reinecke's salt) was used as a ligand. The compounds with gadolinium (III), Reinecke's salt, and HMPA were not found in the structure.

The purpose of this paper is thermal study of synthesized thiocyanate complex based on gadolinium and hexamethylphosphorotriamide in a wide temperature range via the method of differential scanning calorimetry (DSC).

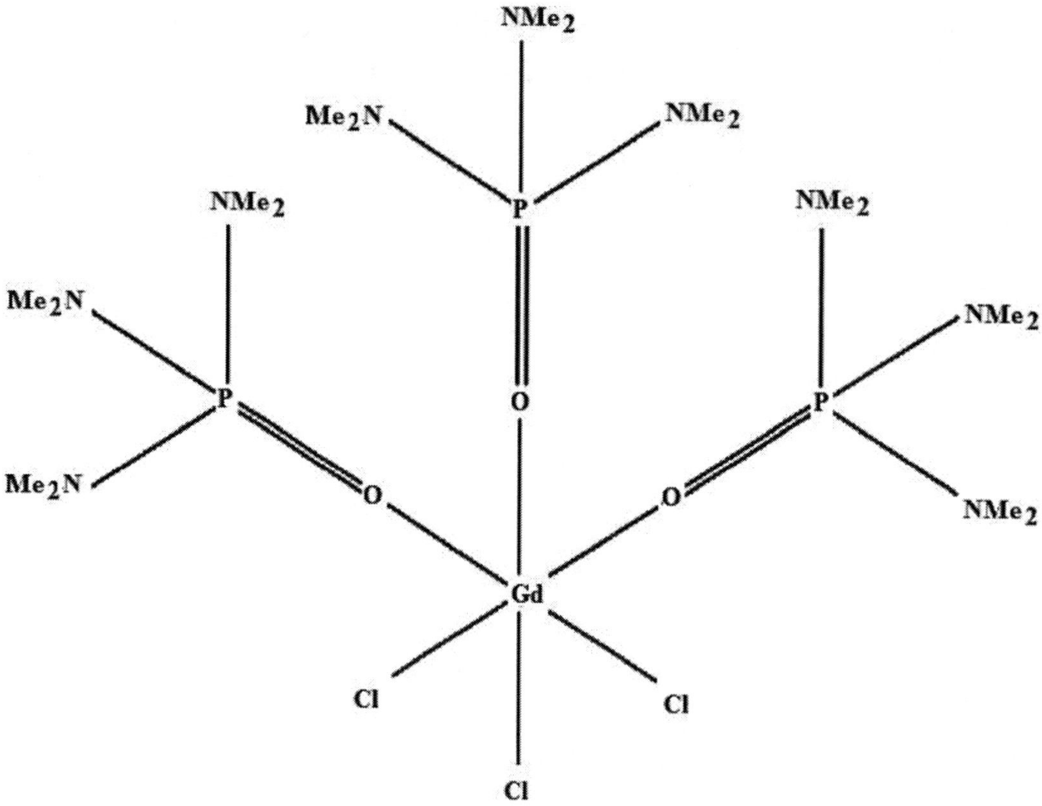

Figure 1. Complex structure fac-$[GdCl_3(HMPA)_3]$

This method is used as a type of thermal analysis in the study of thermal properties of substances. Combined DSC-TG analysis (i.e. registration of thermal effects with simultaneous current change of weight of the sample) allows controlling the processes of dehydration, removing volatile compounds, as well as binder burnout. The obtained data make it possible to determine the enthalpy of phase transitions with exact reference to the temperatures of reactions [7-10].

2. Experiment

Ammonoium tetra(isothiocynato)diaminechromate(III) hemihydrates $NH_4[Cr(NH_3)_2(NCS)_4]{\cdot}0,5H_2O$, hexamethylphosphorotriamide $((CH_3)_2N)_3PO$ and $Gd(NO_3)_3{\cdot}H_2O$ line «C.P.» are chosen as initial substances for synthesis of coordinate substances.

While mixing aqueous solutions of gadolinium nitrate and Reinecke's salt, in molecular ratio 1:1, with later dropwise addition of organic ligand – HMPA, at pH 6-7 and the solution concentration is 0.05-0.5 mole/l fine-crystalline precipitates settle out, the composition $[Gd(HMPA)_4(NO_3)_2][Cr(NH_3)_2(NCS)_4]$ (I).

To carry out the synchronous thermal analysis, NETZSCH STA 449 C Jupiter analyzer was used, it allows simultaneous registration of thermogravimetry (TG) and differential scanning calorimetry

DOI: 10.1051/matecconf/20168501003

(DSC) curves under programmable non-isothermal heating conditions, with the standard α-Al_2O_3 at the heating rate of 10 K/min in the temperature range from 298 to 1273 K.

The results of the study on the thermal behavior of the complex in air are shown in Fig. 2, in an inert atmosphere – in Fig. 3.

3. Results and discussion

The present study is a continuation of the thermolysis processes research in air and in an inert atmosphere of double complex compounds (DCC) with cations – lanthanide complexes with neutral organic ligands and inert complex rhodanate anions of chromium (III); the research is devoted to the study of thermal decomposition processes of a new compound.

Thermogram analysis shows that the complex compound is thermally stable up to the temperature of 385 K. Ammonia molecules split off and burn, and the molecules of HMPA remove when it is over 385 K (in the temperature range from 385 to 413 K).

The main weight loss occurs in the range from 473 to 873K (Fig. 2). In the DSC curve, the discussed range is presented by a number of exoeffects with extremes caused by sample decomposition and decomposition products combustion.

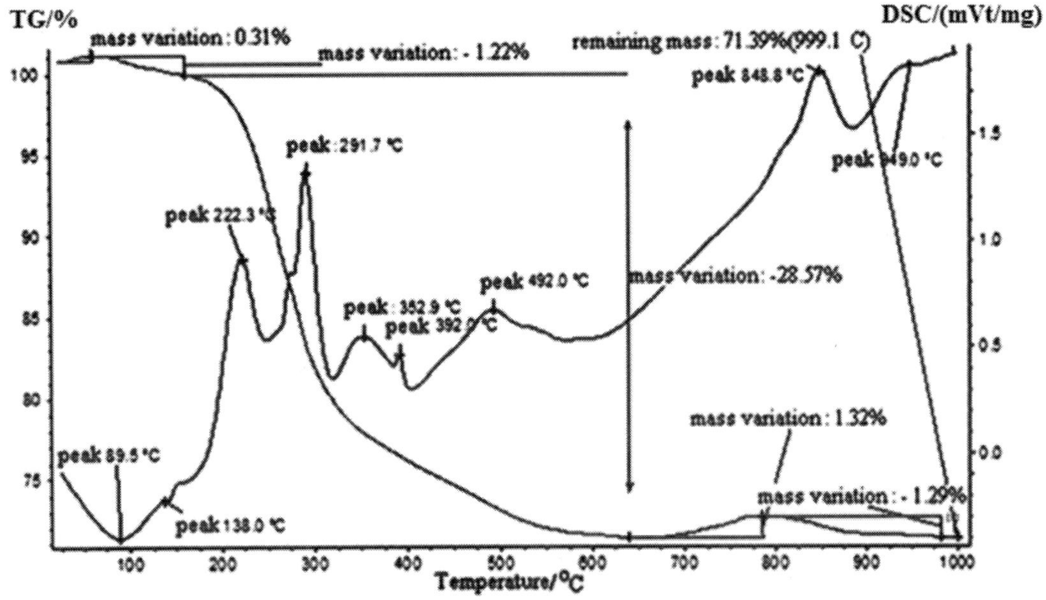

Figure 2. The thermogram of the compound composition $Gd(HMPA)_4(NO_3)_2][Cr(NH_3)_2(NCS)_4]$in air

The DSC curve of the sample I shown in Fig. 2 demonstrates six expressed thermal effects: the first two are endothermic effects starting at 363 and 433 K, the following two are exothermic effects starting at 483 and 533 K, they are followed by endoeffect at 583 K and exoeffect at 713 K. Exothermic effects correspond to a significant loss of the sample weight.

To interpret the processes of thermal decomposition of the obtained complex in inert argon atmosphere, the curves of thermal decomposition of HMPA and Reinecke salt were taken.

The thermolysis processes $NH_4[Cr(NH_3)_2(NCS)_4]{\cdot}0{,}5H_2O$ are characterized by a number of effects. In the temperature range from 340 to 353 K there is a gradual removal of one water molecule and ammonium cation; the sample weight loss is 10.10%. Then the decomposition of the anionic part of the complex begins (the sample weight loss is 43.55 % at a maximum rate). Three endoeffects were recorded at the temperature of 461, 544, 581 K corresponding to the removal of two ammonia molecules and two isothiocyanate ligands.

To confirm the correctness of the thermolysis process interpretation, the IR spectra of the solid decomposition products of the complex I were taken.

There are no absorption bands of the valence vibrations of ammonia NH-groups (ν(NH) = 3366, 3310, 3165 cm^{-1}), and the main absorption bands of isothiocyanate groups (ν(CN), ν(CS), δ(NCS)) and group P-O remain in the IR spectrum of the decomposition product I at the temperature of 473 K. The decomposition of HMPA molecules and anionic parts of the complex (Δm = 28.46%) can be seen in the temperature range from 490 to 715 K. There are no major absorption bands of isothiocyanate groups in the IR spectrum of the decomposition product I at the temperature 673 K.

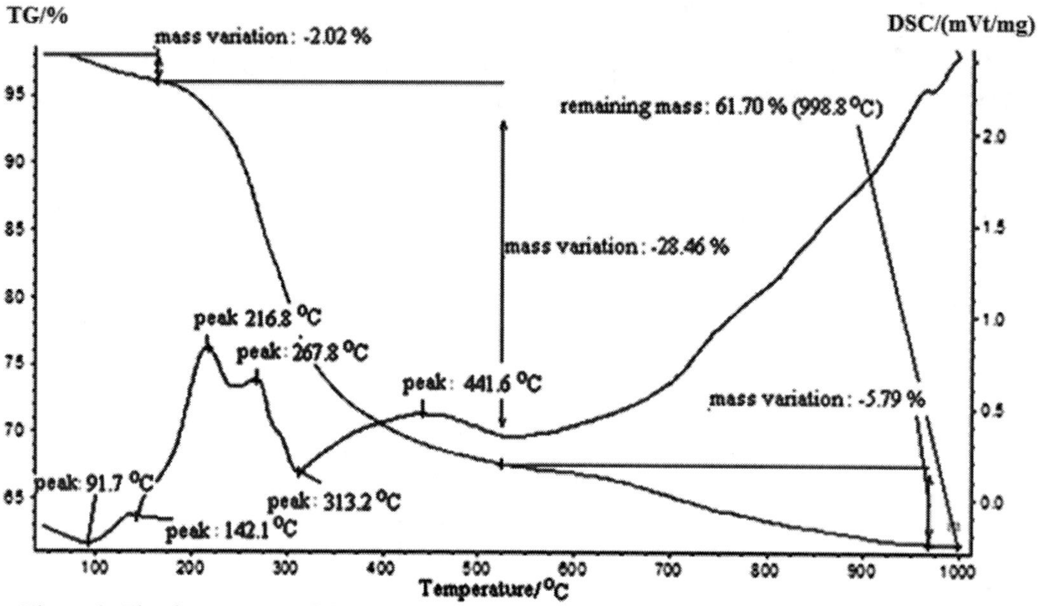

Figure 3. The thermogram of the compound composition $[Gd(HMPA)_4(NO_3)_2][Cr(NH_3)_2(NCS)_4]$ in argon atmosphere

Thus, the behaviour modes of samples in air and argon atmosphere are similar (Fig. 4).

The data of the sample weight loss make it possible to determine the number of HMPA moles per formula unit of the compound (about 2 moles).

 DOI: 10.1051/matecconf/20168501003

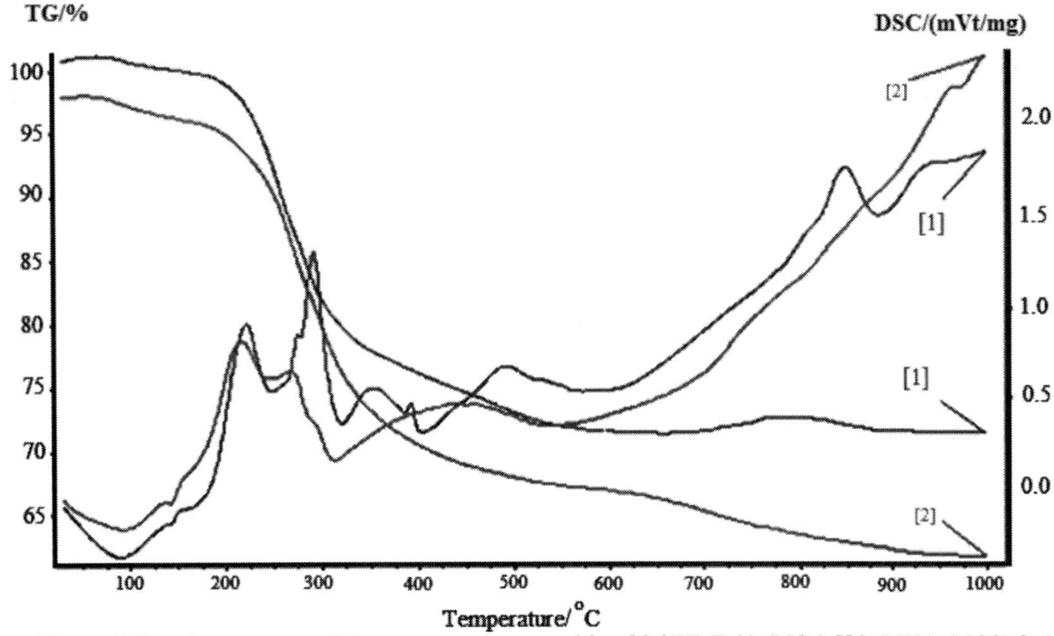

Figure 4. The thermogram of the compound composition $[Gd(HMPA)_4(NO_3)_2][Cr(NH_3)_2(NCS)_4]$: 1 – study in air, 2 – study in argon atmosphere

The double complex compound is less stable to heat due to the presence of organic molecules, as well as the coordinated ammonia in the anions of Reinecke salt. The ligands nature influences the thermal behavior of substances.

4. Conclusion

The technique to obtain a new dual complex compound of gadolinium (III) with Reinecke salt and hexamethylphosphorictriamide was developed.

The application of modern simultaneous thermal analysis allowed studying the behavior of the new compoundwith high accuracy at low sample consumption. The thermolysis process of the sample is complex and multistageous. The main part of the organic component is removed in the temperature range from 473 to 873 C.

The coordination compound with the inert anion of chromium (III) is suitable for obtaining the mixed oxide systems by thermal decomposition of the double complex compound in air at relatively low temperatures.

The work was performed within the project of the Ministry of Education of Russia (project number 10.782.2014K).

References

[1] Pavlishchuk A, Kolotilov S, Zeller M, Thompson L and Addison A 2014 *Inorganic Chemistry* **53** 3 pp 1320-1330
[2] Babailov S, Chuikov I and Kruppa A 2016 *Inorganica Chimica Acta* **439** pp 117-122
[3] Gavriluta A, Tommasino J, Iasco O, Druta V, Luneau D, Kuhn P, Arion V, Claiser N and Novitchi G 2015 *European Journal of Inorganic Chemistry* **9** pp 1616-1624
[4] Cherkasova T, Tatarinova E, Cherkasova E 2014 *Polzunovskiy Bulletin* **3** pp 98-101
[5] Isakova I, Cherkasova T, Peresypkina E and others 2012 *Russian Journal of Inorganic Chemistry* **57** 9 pp 1221-1226

[6] Cherkasova E, Cherkasova T, Peresypkina E and Virovets A 2015 *Russian Journal of Inorganic Chemistry* **60** 4 pp 493-499
[7] Bobrovnikova A, Cherkasova T, Shirokolobova A and Bolsunovskaia L 2014 *Procedia Chemistry* [Electronic resource] – Mode of access: http://www.sciencedirect.com/science/article/pii/S1876619614001351 **10** pp 530-534
[8] El-Shenawy A, Atta A and Refat M 2014 *International Journal of Electrochemical Science* **9** pp 5187-5203
[9] Molero-Sánchez B, Chen M, Birss V, Prado-Gonjal J, Ávila-Brande D and Morán E 2015 *International Journal of Hydrogen Energy* **40** 4 pp 1902-1910
[10] Zhai Y, Huang Z, Zhong Z and Coombs TA 2015 *IEEE Transactions on Applied Superconductivity* **25** 3 pp 697-505

MATEC Web of Conferences **85**, 01004 (2016) DOI: 10.1051/matecconf/20168501004
CCT 2016

Alteration in elemental and functional composition of heated peat humic acids

Natalya Chukhareva [1,*], *Mikhail* Sartakov [2] , and *Tatiana* Korotchenko [1]

[1]Tomsk Polytechnic University, Institute of Natural Resources, 634050, Lenin Avenue, 30,Tomsk, Russia
[2] Yugra State University, Khanty-Mansiysk, Russia

Abstract. The article analyzes the effect of thermal modification of different-type peat on the alteration of elemental and functional composition of peat humic acids. Based on the data of IR-spectra and readings of electron paramagnetic resonance, structural alterations are identified. It is shown that the impact of peat characteristics on humic acids is preserved after thermal modification. It is revealed that the strongest alteration of humic acid composition and properties caused by peat heating are typical to humic acid samples extracted from the peat with low decomposition degree.

1 Introduction

Humic acids (HAs) are polydisperse components of soils, peat, and brown coal which are unique as they are characterized by a number of specific features. They are commonly used in the agricultural, environmental, and industrial sectors.

The latest studies have shown that HAs are heterpolycondensates carrying a great variety of functional groups which make macromolecules of HA different in composition, which, in its turn, contributes to the diversity of HA samples. This does not permit a holistic description of HA chemical structure. By now, it has been possible to propose some models and approximate formulas of HA molecule chemical structure [1].

In order to obtain more detailed description of HA chemical structure, it is proposed to use the methods of thermal modification, which makes it possible to reduce the effect of unstable peripheral structural groups on HA properties.

Thermal peat treatment at 250 °C in its decomposing gases is one of the above-mentioned methods which is actively employed in the experiment to yield required humic acids from peat [2, 3].

The main objective of the study is to identify the effect of thermal modification of different-type peat on the structural alteration of HA.

2 Research technique

For the research, humic acids extracted from 0.25 mm fraction of air-dried terrestrial, transition, and fen peat samples heated in decomposing gases according to technique [3] have been chosen. The peat samples are characterized by low and medium decomposition degree (R=5…45 %), the data being provided in table 1.

Table 1. Different-type peat characteristics.

Peat type	Group	R,	T,	Technical	Elemental analysis, %	Code HA

* Corresponding author: Natasha@tpu.ru

DOI: 10.1051/matecconf/20168501004

		%	°C	analysis *, %		daf**				
				W^a	A^d	C	H	O+S	N	
terrestrial peat										
Sphagnum	moss peat	5	-	8.4	1.9	52.1	6.3	40.7	0.9	HA(1Ph)
			250	1.2	2.1	55.1	5.5	37.9	1.5	HA($1Ph_{250}$)
Fuskum-peat	moss peat	10	-	6.2	1.3	52.2	5.4	41.7	0.7	HA(3Ph)
			250	1.3	2.4	60.3	5.0	33.0	1.7	HA($3Ph_{250}$)
Megallanicum	moss peat	10	-	7.8	2.9	54.3	4.2	39.8	1.7	HA(2Ph)
			250	1.5	5.1	59.9	4.0	33.8	2.3	HA($2Ph_{250}$)
Cottongrass-sphagnum	grass-moss	25	-	8.2	1.9	55.8	6.0	36.7	1.5	HA(4Ph)
			250	0.7	2.5	58.7	5.8	33.2	2.3	HA($4Ph_{250}$)
Cottongrass-sphagnum	grass	35	-	6.6	2.3	57.1	6.1	35.5	1.3	HA(5Ph)
			250	1.8	3.5	62.2	6.0	29.9	1.9	HA($5Ph_{250}$)
Scheuchzeria	grass	40	-	11.1	7.8	57.9	5.9	33.7	2.5	HA(6Ph)
			250	0.8	13.7	59.8	5.7	31.4	3.1	HA($6Ph_{250}$)
transition peat										
Scheuchzeria	grass	20	-	7.4	8.4	53.8	6.1	37.2	2.9	HA(1Pm)
			250	1.0	11.7	66.4	5.3	24.0	4.3	HA($1Pm_{250}$)
Sedge-sphagnum	grass-moss	25	-	6.7	6.6	55.9	5.2	36.9	2.0	HA(2Pm)
			250	1.3	8.2	58.1	4.9	33.3	3.7	HA($2Pm_{250}$)
Sedge-sphagnum	grass-moss	35	-	7.2	4.4	56.9	5.9	35.1	2.1	HA(3Pm)
			250	1.2	5.2	59.7	5.7	31.9	2.7	HA($3Pm_{250}$)
fen peat										
Sedge-hypnum	grass-moss	25	-	8.2	8.9	43.4	5.6	45.1	5.9	HA(1Pl)
			250	0.5	18.2	54.6	5.3	33.8	6.3	HA($1Pl_{250}$)
Woody-sedge peat	woody	30	-	7.1	10.3	56.0	6.0	34.9	3.1	HA(2Pl)
			250	0.9	10.7	58.1	5.6	32.8	3.5	HA($2Pl_{250}$)
Sedge peat	grass	35	-	7.6	6.1	52.8	6.3	36.4	4.5	HA(3Pl)
			250	1.2	6.9	58.1	5.8	31.5	4.6	HA($3Pl_{250}$)

* W a – moisture of analysis sample, Ad – ash content on air-dried basis, Vodaf – yield of volatile matter on dry ash-free basis, **daf – on air-dried and ash free basis; O+S is obtained by difference, with S content being not more than 2 %

The humic acids (HA и HA_{250}) were extracted in accordance with the following technique [4]: 1) bitumen was extracted by the use of benzene; 2) water-soluble and easy-hydrolysable substances were extracted from bitumen-free peat samples (extraction of 4%-HCℓ); 3) humic acids were extracted from peat samples free from bitumen, water-soluble and easy-hydrolysable substances (treatment of 0,1H by solution NaOH and further precipitation of HA by 10 %-HCℓ).

The functional acid groups (carboxy group and phenolic hydroxyl) in macromolecule HA and HA_{250} were identified by defining their total content by means of Barium oxydatum, followed by identification of COOH group by Calcium acetate, and OH group identification by the difference (OH+COOH) – (COOH) according to technique [4].

All samples HA and HA_{250} were examined for C, H, N, O+S content (hereinafter oxygen) by means of automatic elemental analyzer EURO EA3000 in accordance with technique [5]. The discordance between the data met the following conditions: $n \geq 3$, when $P = 0.95$.

The conditions of the experiment aimed at identifying paramagnetic properties of HA and HA_{250} (readings of electron paramagnetic resonance) were as follows: electron

DOI: 10.1051/matecconf/20168501004

paramagnetic resonance (EPR) –spectrometer Bruker ELEXSYS E-540 (Novosibirsk Institute of Organic Chemistry of the Siberian Branch of Russian Academy of Science); X-band at 10 GHz; g-factor was identified by inside frequency meter spectra, with gauge correction to g-factor DPPH; measurements were carried out on a Bruker ELEXSYS E-540 in EPR ampoules, MW power = 10 dB. The number of electron spins was calculated in accordance with the method developed by «Bruker» - Absolute spins [7]. The accuracy of spin number determination was ±15 %.

The optical properties of HA and HA_{250} were studied using the Nicolet iS10 FT-IR spectrometer, Termo Fisher Scientific. The IR spectrometer was fitted with an add-on device containing ZnSe crystal for frustrated internal reflectance. The experiment parameters were as follows: resolution – 4 cm^{-1}, the number of sample scans and comparison spectrum – 128, scan range was from 4000 up to 650 cm^{-1}. The spectra were registered and processed using Omnic 8.3 [6].

3 Results and discussion

The obtained data on elemental and functional composition of HA and HA_{250} are given in table 2.

Table 2. The characteristics of HA elemental and functional composition.

Code HA	W^a, %	A^d, %	Elemental composition, % daf				Functional com position, mg-eq/g	
			C	H	O	N	COOH	OH
humic acids of terrestrial peat, decomposition degree 5-10 %								
HA(1Ph)/ $HA(1Ph_{250})$	2.1/ 3.1	0.1/ 0.1	56.2/ 62.0	6.1/ 4.7	35.8/ 31.5	1.9/ 2.1	3.5/ 3.9	4.0/ 3.9
HA(2Ph)/ $HA(2Ph_{250})$	2.1/ 3.7	0.1/ 0.4	57.8/ 64.7	5.9/ 5.4	34.7/ 28.1	1.6/ 1.8	3.6/ 4.1	3.8/ 3.8
HA(3Ph)/ $HA(3Ph_{250})$	5.5/ 4.1	0.5/ 0.3	58.4/ 62.6	5.7/ 5.1	34.2/ 30.3	1.7/ 2.0	3.3/ 3.8	3.5/ 3.5
alteration of elemental composition and functional groups in HA, % respectively								
			+ΔC	-ΔH	-ΔO	+ΔN	+ΔCOOH	-ΔOH
$HA(1Ph_{250})$/HA(1Ph)			10.3	22.9	12.0	10.5	11.4	2.5
$HA(2Ph_{250})$/HA(2Ph)			12.0	8.5	19.0	12.5	13.9	-
$HA(3Ph_{250})$/HA(3Ph)			7.2	10.5	11.4	17.6	15.2	-
humic acids of terrestrial peat, decomposition degree 25-40 %								
HA(4Ph)/ $HA(4Ph_{250})$	5.7/ 3.1	0.8/ 0.2	60.0/ 60.7	5.2/ 4.9	32.2/ 31.7	2.6/ 2.7	3.4/ 3.6	3.1/ 3.1
HA(5Ph)/ $HA(5Ph_{250})$	9.2/ 3.6	0.7/ 0.1	58.9/ 60.7	5.3/ 4.9	33.7/ 32.2	2.1/ 2.2	3.5/ 3.7	3.2/ 3.2
HA(6Ph)/ $HA(6Ph_{250})$	6.4/ 8.2	0.1/ 0.2	61.5/ 62.0	5.3/ 4.9	30.8/ 30.3	2.4/ 2.8	3.7/ 3.9	3.3/ 3.1
alteration of elemental composition and functional groups in HA, % respectively								
			+ΔC	-ΔH	-ΔO	+ΔN	+ΔCOOH	-ΔOH
$HA(4Ph_{250})$/HA(4Ph)			1.2	5.8	1.6	3.8	5.9	-
$HA(5Ph_{250})$/HA(5Ph)			3.1	7.5	4.5	4.8	5.7	-
$HA(6Ph_{250})$/HA(6Ph)			0.8	7.5	1.6	16.7	5.4	6.1
humic acids of transition peat, decomposition degree 20-35 %								
HA(1Pm)/ $HA(1Pm_{250})$	8.1/ 4.9	0.6/ 0.1	60.0/ 62.0	5.3/ 4.8	31.7/ 30.0	3.0/ 3.2	3.1/ 3.3	3.3/ 3.1
HA(2Pm)/ $HA(2Pm_{250})$	5.0/ 5.1	0.8/ 0.4	58.8/ 61.4	5.1/ 4.5	32.7/ 30.6	3.4/ 3.5	2.7/ 3.0	3.6/ 3.4
HA(3Pm)/ $HA(3Pm_{250})$	6.9/ 3.5	0.1/ 0.2	60.1/ 61.5	4.9/ 4.2	32.0/ 31.1	3.0/ 3.2	2.1/ 2.4	4.1/ 4.1

alteration of elemental composition and functional groups in HA, % respectively								
			+ΔC	-ΔH	-ΔO	+ΔN	+ΔCOOH	-ΔOH
$HA(1Pm_{250})/HA(1Pm)$			3.3	9.4	5.4	6.7	6.5	6.1
$HA(2Pm_{250})/HA(2Pm)$			4.4	11.8	6.4	2.9	11.1	5.6
$HA(3Pm_{250})/HA(3Pm)$			2.3	14.3	2.8	6.7	14.3	-
humic acids of fen peat, decomposition degree 25-35 %								
$HA(1Pl)/$ $HA(1Pl_{250})$	8.1/ 4.1	1.1/ 0.2	59.8/ 62.0	4.8/ 4.3	32.6/ 30.5	2.8/3.2	3.3/ 3.5	3.2/ 3.1
$HA(2Pl)/$ $HA(2Pl_{250})$	9.8/ 3.4	0.3/ 0.4	61.3/ 63.7	5.0/ 4.6	30.1/ 28.1	3.6/3.6	3.1/ 3.3	3.3/ 3.4
$HA(3Pl)/$ $HA(3Pl_{250})$	9.4/ 5.1	0.4/ 0.1	61.2/ 63.6	4.7/ 4.8	30.9/ 28.2	3.2/3.4	2.9/ 3.1	3.4/ 3.4
alteration of elemental composition and functional groups in HA, % respectively								
			+ΔC	-ΔH	-ΔO	+ΔN	+ΔCOOH	-ΔOH
$HA(1Pl_{250})/ HA(1Pl)$			3.7	10.4	6.4	14.3	6.1	3.1
$HA(2Pl_{250})/ HA(2Pl)$			3.9	8.0	6.6	-	6.5	3.0
$HA(3Pl_{250})/ HA(3Pl)$			3.9	2.1	8.7	6.3	6.9	-

The data suggest that the average content of C, H, O, N elements is within the range specific to peat humic acids [8, 9]. Based on peat decomposition degree, certain regularities in elemental composition of HA samples were identified. High values of C (58.8÷61.5 % daf) and N (2.1÷3.6) are characteristic feature of terrestrial, transition, and fen peat HAs with the average decomposition degree (R = 20-40 %). High values of H (5.7÷6.1 % daf) and O (34.2÷35.8 % daf) are characteristic feature of terrestrial peat HAs with low decomposition degree (R = 5-10 %).

The preliminary heat treatment of peat HA_{250} led to carbon increase, oxygen and hydrogen decrease. The results demonstrated minimum changes in nitrogen content as compared to alteration degree of other elements.

The above-mentioned regularities are not preserved after peat heat treatment. At the same time, when comparing the relative alteration of element content (+ΔC, -ΔH, -ΔO, +ΔN) in samples HA_{250}, the effect of peat R finds its maximum.

The maximum relative alterations of element content are obtained for $HA(1Ph_{250})$, $HA(2Ph_{250})$, $HA(2Ph_{250})$ of thermally treated terrestrial peat with low decomposition degree R (5 – 10 %). They are as follows: +ΔC = 7,2÷12,0 % respectively; -ΔH = 8,5÷22,9 % respectively; -ΔO = 11,4÷19,0 %; +ΔN = 10,5÷17,6 % respectively.

The rest samples $HA(4Ph_{250})$, $HA(5Ph_{250})$, $HA(6Ph_{250})$, $HA(1Pm_{250})$, $HA(2Pm_{250})$, $HA(3Pm_{250})$, $HA(1Pl_{250})$, $HA(2Pl_{250})$, $HA(3Pl_{250})$ are characterized by lower values: +ΔC = 0,8÷4,4 % respectively; -ΔH = 2,1÷14,3 % respectively; -ΔO = 1,6÷8,7 %; +ΔN = 2,9÷14,3 % respectively.

Thus, it has been stated that the higher peat R is, the lower is the effect of peat heat treatment on relative alterations of C, H, O N content in the organic macromolecule of humic acids.

The data on functional composition indicate that the number of oxygen-containing functional groups is within the range typical for peat humic acids in all samples HA and HA_{250}[10].

Peat heat treatment leads to the increase in COOH-group content in modified humic acids. The number of hydroxyl OH-groups decreases or remains unchanged (table 2).

The increase in carboxyl groups in HA_{250} composition can stipulate their reaction capacity [11]. At the same time, it is necessary to consider not only the number of functional groups, but also their differentiation based on acidity level (difference in molecular environment within the macromolecule).

 DOI: 10.1051/matecconf/20168501004

When comparing infrared spectra of HA and HA_{250}, it has been revealed that all HA samples are characterized by uniformity of functional groups.

The absorption bands that are typical for HAs are observed in infrared spectra of HA and HA_{250} in the OH-group stretch (3400 $см^{-1}$). The bands are linked by hydrogen bonds. The absorption bands are also observed in CH_2-group stretch, methyl CH_3-group (2920 cm^{-}1 and 2850 cm^{-1}), carboxyl C=O-group (1725–1700 cm^{-1}), C=C polycoupled aromatic systems and monoaromatic structures (1620–1600 cm^{-1} and 1520–1500 cm^{-1}). Faint absorption bands are defined in vibration bending of C–CH_3-groups (1390–1370 cm^{-1}) and C–O stretch of acids and phenols (1250–1200 cm^{-1}). The band 1100–1000 cm^{-1} corresponds to C–O-group stretch of alcohols (figure 1).

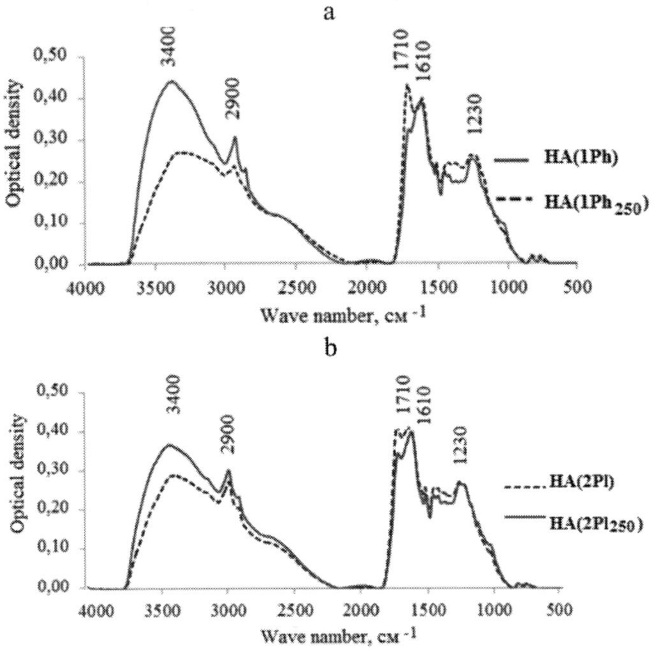

Fig. 1. IR-spectra of humic acids in air-dried and heat treated terrestrial peat, R = 5 % (a) and fen woody peat, R = 30 % (b).

Peat heat treatment in HA_{250} IR-spectra decreases the intensity of OH-groups and C-H-groups stretch at 3400 cm^{-1} and 2920 cm^{-1}, which, in its turn, indicates the presence of dehydration reactions at low-temperature thermal decomposition at the stage of peat heating.

The intensity alteration of absorption bands of C=O-groups at 1725–1700 cm^{-1} and C–O-groups of acids and phenols at 1250–1200 cm^{-1} has been determined. This can be caused by interaction of HA thermal decomposition products (alcohol ester, alcohol oxide, and ketone) with water forming new carboxyl groups [10, 11]. Besides, carboxyl groups can also be formed due to hydrolyze ester bonds that bind other peat components. This fact has been proved by analyzing kinetic parameters of peat components thermal decomposition in [12].

To obtain the data on HA structure, characterized by hetero-polycondensates of irregular composition (methyl-bearing hydroxycarboxylic acids with fused rings including O, N, S-bearing heterocycles and side chains, as well as other functional groups in nucleus and peripheral parts [1, 4, 10]), atomic ratios H:C and O:C have been calculated on the basis of elemental composition.

In order to obtain the characteristics of HA and HA_{250} aromaticity using the Van-Krevelen's equation [11], aromaticity index (A) and the number of aromatic rings (B) for the statistical carbon nucleus of a molecule have been defined.

The obtained number of spins in the sample s = ½ was recalculated for 1g of the sample size ($I \cdot 10^{17}$). The results are given in table 3.

Table 3. Humic acid aromaticity.

HA code	Atomic relations		the Van-Krevelen's equation parameters		Readings of electron paramagnetic resonance	
	H:C	*O:C*	*A*	*B*	Spin for g, $I \cdot 10^{17}$	g-factor
humic acids of terrestrial peat, decomposition degree 5-10 %						
HA(1Ph)/ HA($1Ph_{250}$)	1.25/ 1.13	0.48/ 0.38	3.11/ 4.40	0.48/ 1.33	0.97/ 1.80	2.0039/2.0034
HA(2Ph)/ HA($2Ph_{250}$)	1.21/ 0.99	0.45/ 0.33	3.30/ 4.02	0.57/ 1.02	1.78/ 1.81	2.0034/2.0034
HA(3Ph)/ HA($3Ph_{250}$)	0.17/ 1.14	0.44/ 0.36	3.42/ 4.16	0.63/ 1.12	1.85/ 3.09	2.0030/2.0038
humic acids of terrestrial peat, decomposition degree 25-40 %						
HA(4Ph)/ HA($4Ph_{250}$)	1.04/ 0.97	0.40/ 0.39	3.86/ 4.14	0.90/ 1.11	0.0/ 1.77	2.0038/2.0038
HA(5Ph)/ HA($5Ph_{250}$)	1.07/ 0.96	0.43/ 0.40	3.73/ 4.17	0.81/ 1.13	2.02/ 2.29	2.0031/2.0035
HA(6Ph)/ HA($6Ph_{250}$)	1.02/ 0.94	0.38/ 0.37	3.91/ 4.28	0.94/ 1.22	0.92/ 3.57	2.0038/2.0032
humic acids of transition peat, decomposition degree 20-35 %						
HA(1Pm)/HA($1Pm_{250}$)	1.05/ 0.92	0.40/ 0.36	3.79/ 4.34	0.85/ 1.28	1.82/ 2.61	2.0036/2.0034
HA(2Pm)/ HA($2Pm_{250}$)	1.03/ 0.88	0.42/ 0.37	3.89/ 4.56	0.92/ 1.48	1.42/ 1.80	2.0035/2.0037
HA(3Pm)/ HA($3Pm_{250}$)	0.96/ 0.82	0.40/ 0.38	4.15/ 4.89	1.12/ 1.82	1.17/ 1.98	2.0035/2.0034
humic acids of fen peat, decomposition degree 25-35 %						
HA(1Pl)/ HA($1Pl_{250}$)	0.95/ 0.81	0.41/ 0.37	4.22/ 4.91	1.17/ 1.84	2.18/ 2.89	2.0033/2.0035
HA(2Pl)/ HA($2Pl_{250}$)	0.97/ 0.86	0.37/ 0.33	4.14/ 4.67	1.11/ 1.59	1.41/ 1.53	2.0039/2.0034
HA(3Pl)/ HA($3Pl_{250}$)	0.91/ 0.89	0.38/ 0.33	4.38/ 4.50	1.31/ 1.43	1.85/ 2.30	2.0035/2.0036

The evidence presented suggests that the values of H:C and O:C atomic ratios for HA/HA_{250} (humic acids of air-dried peat/humic acids of heat treated peat) are within the range 0.91 ÷1.25/ 0.81÷1.14 and 0.37 ÷ 0.48/0.33÷0.39.

The alteration of H:C and O:C values is the evidence of change in the content of macromolecule cyclic components and oxidation degree. The obtained results show that preliminary peat heat treatment decreases the values of H:C and O:C atomic ratios for HA/HA_{250}. This fact indicates that the content of cyclic components can increase in macromolecule. Moreover, it presents the evidence of higher modification degree of modified HA_{250} initial substance compared with HA samples. This is proved by the parameters which were obtained using the Van-Krevelen's equation: higher values of A, B were obtained for HA_{250} samples compared with HA.

The impact of peat decomposition degree (R) on the calculated atomic relations and the Van-Krevelen's equation parameters is defined: R growth leads to the decrease in H:C and O:C atomic ratios and increase in aromaticity index and the number of aromatic rings of the statistical carbon nucleus of HA and HA_{250} macromolecule.

The atomic ratios H:C and the value of the Van-Krevelen's equation parameters reveal the difference between peripheral and nucleus parts of HA and HA_{250} macromolecules. At the same time, it should be noted that modified HAs demonstrate aromaticity increase under the action of preliminary heat treatment which stipulates chemical transformations in peat when it is heated up to 250 °C: 1) organic matter decomposition and emission of gaseous products (CO, CO_2, H_2O) during thermal decomposition; 2) interaction between peat group components and/or their residues in gaseous decomposition environment; 3) synthesis of new HAs [3, 6, 10-12].

The above is proved by the data obtained using EPR-spectrometer. The data show that peat heat treatment increases EPR signal for HA_{250} samples. The concentration of paramagnetic centers also increases: $I \cdot 10^{17}$ = 0.90÷2.18 (HA) and $I \cdot 10^{17}$ = 1.53÷3.57 (HA_{250}). According to [14], such results can be explained by the increase in the degree of aromatic polyconjugation of HA initial matter due to formation of new additional systems of polyconjugation caused by condensation reactions. This fact is in accord with carbon content in HA_{250} elemental composition and increasing number of active acidic COOH-groups, which, in its turn, contributes to forming additional paramagnetic centers (presence of hydrogen bonds formed by functional groups) and growing the Van-Krevelen's equation parameters.

The peat heat treatment does not lead to significant change in g-factor. For all examined HA and HA_{250} samples, g-factor ranges within 2.0030÷2.0039 and approximates the value of unbound electron g-factor (2.0023).

4 Conclusion

1. Heat treatment of different-type peat increases the carbon content and de-creases the content of hydrogen and oxygen in the elemental composition of HAs.
2. The more peat decomposition degree is, the more indifferent peat is to relative alteration of element content C, H, O N in organic part of modified HA macromolecule.
3. The absorption bands typical for HAs are preserved in IR-spectra of modified HAs.
4. As the result of peat heat treatment, IR-spectra of modified HA demonstrate the decrease in the intensity of OH-groups and C-H-groups stretch at 3400 cm^{-1} and 2920 cm^{-1} and increase in the intensity of absorption bands of C=O-groups at 1725–1700 cm^{-1} and C–O-groups of acids and phenols at 1250–1200 cm^{-1}. This is the evidence of dehydration reaction and further interaction of pyrogenic water with the products of peat thermal decomposition accompanied by the formation of new carboxyl groups.
5. Peat heat treatment increases aromaticity and condensity of HA macro-molecule nucleus part.
6. Peat heat treatment also contributes to the concentration of paramagnetic centers.

Thus, it can be stated that preliminary peat heat treatment affects the alteration of HA composition and increases the stable nucleus part of HA macromolecule. The preceding observations lead to the conclusion that peat thermal modification has a positive impact on HAs, as it allows eliminating peripheral structures and obtaining new properties.

References

1. I. I. Lishtvan *Fiziko-khimicheskaya mekhanika guminovykh veshchestv* (*Nauka i tekhnika*,1976) [in Russian]
2. N.V. Chukhareva, A.G. Zarubin, O.L. Bulgakova The Use of Principle Component Analysis in Type Classification of Air-dry Peat. IOP Conf. Ser.: Earth Environ. Sci. **21**, 1-5 (2014)

DOI: 10.1051/matecconf/20168501004

3. N.V. Chukhareva, T. V. Korotchenko, D.S. Rozhkova Impact of Heat Treatment on the Structure and Properties of Tomsk Region Peat. Procedia Chemistry, **10**, 535-540 (2014)
4. I. I. Lishtvan, N.T. *Korol' Osnovnye svoystva torfa i metody ikh opredeleniya* (*Nauka i tekhnika*, 1975) [in Russian]
5. V. P. Fadeeva Elemental Analysis of Organic Compounds with the Use of Automated CHNS Analyzers Journal of analytical chemistry **63** (11) 1094-1106 (2008)
6. N.V. Chukhareva, V. D. Tikhova, O. N. Zarubina, L. V. Shishmina The change of the peat element composition and optical properties under the thermal modification influence, the Bulletin of KrasGAU **11** 85-92 (2013)
7. P. Hofer, P. Carl Patent number US 7586305 B2. Method for determining the absolute number of electron spins in a sample of extended size (2009)
8. D.S. Orlov, L. A. Grishina *Praktikum po khimii gumusa* (MSU, 1981)
9. I. V. Perminova Analysis, classification, and prospect of humic acids properties (2000)
10. N.V. Chukhareva Issledovanie kinetiki termicheski aktivirovannykh iz-meneniy sostava i svoystv torfyanykh guminovykh kislot: extended abstract of Cand. Sci. (Chemistry) Dissertation (2003) [in Russian]
11. L. I. Tarnovskaya Zakonomernosti izmeneniya gruppovogo sostava torfa v protsesse termoliza: extended abstract of Cand. Sci. (Eng.) Dissertation (1985) [in Russian]
12. D. A. Bazhenov Modelirovanie fiziko-khimicheskikh zakonomernostey nizkotemperaturnogo razlozheniya torfa: extended abstract of Cand. Sci. (Chemistry) Dissertation (2000)
13. D.V. Van Krevelen *Nauka ob ugle* (GNT, 1960) [in Russian]
14. S. N. Chukov Paramagnitnaya aktivnost' pochvennogo gumusa i ee iz-menenie pri antropogennom vozdeystvii (na primere serykh lesnykh pochv): extended abstract of Cand. Sci. (Biology) Dissertation (2006) [in Russian]

DOI: 10.1051/matecconf/20168501005

Computer modeling and software development for unsteady chemical technological systems

Igor Dolganov[1], *Irena* Dolganova[1,*], *Elena* Ivashkina[1], and *Mikhail* Pisarev[1]

[1]Institute of Natural Resources, Tomsk Polytechnic University, Tomsk, Russia

Abstract. The paper deals with mathematical modeling in transient conditions to create a computer system that can reflect the behavior of real industrial plants. Such systems can respond to complex and pressing questions about the stability of the industrial facilities and the time spent on transients passing through the unstable regimes. In addition, such systems have a kind of intelligence and predictive ability, as they consider partial integral and differential systems of equations that are based on physical and chemical nature of the processes occurring in devices of technological systems.

1 Introduction

Currently, computer-modeling systems for calculation of the operating parameters of devices for complex petrochemical plants are widespread [1-4]. Such systems allow to develop chemical and mathematical models of processes as a set of systems of differential, integral, and algebraic equations, and then to solve them by various methods, including numerical. However several problems are hardly solvable for most systems. These problems are related with the modeling of non-stationary processes, depending on the astronomical time, such as changing system settings at occurrence of transient modes. The geometric dimensions of the elements affect primarily the time and nature of processes, and this determines their inertia in responding to changes in external conditions.

The article sets out the principle that is used for simulation of unsteady systems, considering the apparatuses parameters change through time at transition regime. As an example of chemical-technological system for modeling, we selected the technology of low-temperature gas separation. The choice is made in favor of this technology, as it is one of the most common and popular technologies of gas and gas condensate treatment [5-8].

This is due to the relative simplicity in hardware design and management of the process At that, it allows ensuring the necessary quality of products obtained from raw materials with a wide range of parameters variation.

2 Technological scheme description

The working range of pressures in process scheme is from 2.50 to 7.35 MPa, the temperature ranges from minus 45 to plus 30 ° C, the amount of incoming feedstock to the installation varies from 60 to 210 nm^3/h.

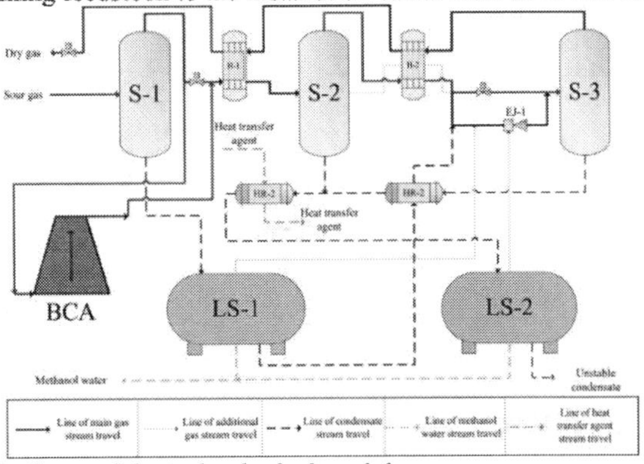

Fig. 1. Structure scheme of main flows of the technological module

* Corresponding author: dolganovaio@tpu.ru

S1, S2, S3 – separators; H-1, H-2, HR-1,HR-2 – heatexchangers;LS-1, LS-2 – liquid separators; EJ-1 – ejector; BCA - boosting compressor station.

In such a process scheme, there are three consecutive separation stages: at the first stage, the feed pressure is in the range of 6.0 to 7.0 MPa, a temperature of 5 to 25 °C. Then after passing through a heat exchanger and a control valve in the second stage separation, we observe the temperature decrease from minus 10 to plus 3 ° C.

In the third stage separation, we observe the decrease of the temperature of minus 35-minus 30 °C and a pressure to 3.7-5.5 MPa. In each liquid separator condensation followed by separation, with the greatest number of the gas stream withdrawn from the first stage. To make the further condensation possible is necessary to change the working conditions in the separators [9-12].

3 Mathematical description of technological system performance

We found out that the rate of materials accumulation in the apparatus can be represented by the following differential equation:

$$\frac{dN_{ij}}{dt} = \sum_k {}^{in}F_{ij}^k - \sum_k {}^{out}F_{ij}^k, \tag{1}$$

Initial conditions: $t = 0, N_{ij} = N_{ij}^0$;

here N_{ij} – amount of j-*th* substance in the i-*th* apparatus, mole;

${}^{in}F_{ij}^k$ –k-*th* flow rate of the j-*th* substance at the entrance to the i-*th* apparatus, mole/sec;

${}^{out}F_{ij}^k$ – k- *th* flow rate of the j-*th* substance from the i-*th* apparatus, mole/sec;

The material balance equation supplemented by the heat balance equation in a stationary form:

$$\frac{dQ_{ij}}{dt} = \sum_k {}^{in}F_{ij}^k \cdot H_{ij}^{T^{in}} - \sum_k {}^{out}F_{ij}^k \cdot H_{ij}^{T^{out}} + \Delta Q_{ij}^*, \tag{2}$$

Initial conditions: $t = 0, Q_{ij} = Q_{ij}^0$;

here Q_{ij} – amount of the heat of the j-*th* substance in the i-*th* apparatus, J; $H_{ij}^{T_{in}}$ – enthalpy of the j-*th* substance in the i-*th* apparatus at the input temperature, J/mole; $H_{ij}^{T^{out}}$ – enthalpy of the j-*th* substance in the i-*th* apparatus at the output temperature, J/mole; ΔQ_{ij}^* – energy, accepted by j-*th* substance in the i-*th* apparatus, J/sec.

It is shown that for the simulation of chemical-technological system consisting of a plurality of elements, mutually influencing each other, it is necessary to establish the nature of the connection between devices, which can be described by the system of equations:

$$\begin{cases} {}^{in}F_{ij}^k = f({}^{out}F_{mj}^n) \\ {}^{in}F_{ij}^k \cdot H_{ij}^{T^{in}} = \lambda({}^{out}F_{mj}^n \cdot H_{mj}^{T^{out}}) \end{cases}, \tag{3}$$

where ${}^{in}F_{mj}^n$ – n- *th* flow rate of j-*th* substance at the entrance to the m-*th* apparatus, mole/sec; ${}^{out}F_{mj}^n$ – n- *th* flow rate of the j-*th* substance from the m-*th* apparatus, mole/sec; $H_{mj}^{T^{in}}$ – enthalpy of the j-*th* substance in the m-*th* apparatus at the input temperature, J/mole; $H_{mj}^{T^{out}}$ – enthalpy of the j-*th* substance in the m-*th* apparatus at the output temperature, J/mole;

Type of the functional dependence of the material and heat flows depends on the type and design of the equipment.

Combining the presented Expressions, we obtained the dynamic mathematical model of the low-temperature gas separation, which is a system of differential equations of the first order:

$$\begin{cases} \dfrac{dN_{ij}}{dt} = \sum\limits_{k} {}^{in}F_{ij}^{k} - \sum\limits_{k} {}^{out}F_{ij}^{k} \\ \dfrac{dQ_{ij}}{dt} = \sum\limits_{k} {}^{in}F_{ij}^{k} \cdot H_{ij}^{T^{in}} - \sum\limits_{k} {}^{out}F_{ij}^{k} \cdot H_{ij}^{T^{out}} \pm \Delta Q_{ij}^{*} \\ {}^{in}F_{ij}^{k} = f({}^{out}F_{mj}^{n}) \\ {}^{in}F_{ij}^{k} \cdot H_{ij}^{T^{in}} = f({}^{out}F_{mj}^{n} \cdot H_{mj}^{T^{out}}) \end{cases} \quad (4)$$

Initial conditions: $t = 0, N_{ij} = N_{ij}^{0}, Q_{ij} = Q_{ij}^{0}$.
To solve the system of differential equations we used the Runge-Kutt method of the 4th order.

3 Software implementation and simulator development

Algorithm for two- and three-phase separator calculating the using a simulation of the dynamic model of low-temperature gas separation process includes the following steps:
1. Input data source (flow rates, compositions, temperature, pressure, levels of light and heavy liquid, the geometrical dimensions of devices, etc.). Loading is carried out with the use of databases.
2. Calculation of the devices material balance with determination of debalance value, that allows to calculate the change of device parameters in dynamic conditions.
3. Calculation of the equilibrium constants of the phase considering the temperature, pressure and gas composition. For this, we used the methodology of phase equilibrium constants calculation of Soave-Redlich-Kwong equation for hydrocarbons and Tek-Still for polar substances.
4. Calculation of the output flows composition (the iterative determination of the molar fraction of distillate, calculation of speed of hindered settling of methanol water droplets in the condensate layer and speed of condensate drops surfacing in the layer of methanol water). Definition of residual water content in condensate.
5. Calculation of the process parameters (temperature, pressure, liquid level, considering the geometrical dimensions of devices).
6. The calculation results output (flows rates and compositions, temperatures, pressures, fluid levels). The results are presented in form graphs and tables.

We implemented this algorithm using object-oriented Borland Delphi 7 programming language.

Here is the part of the code implementing the algorithm above.

```
Function TSeparator.Calculate (IDObject:
integer): Boolean;
BEGIN
  GetData(IDObject);
  RecalculationBalance(IDObject);
  RecalculationKfr(IDObject);
  RecalculationParam(IDObject);
END
```

Each included function in the code refers to the shared database, extracts, converts and rewrites the result in appropriate fields according to unique key IDObject, associated with each element of the system.

To move to the implementation phase of the software it is necessary to develop a structure by which the work will be done with a computer model. At this stage, the following provisions are applied:

1) "Client-server" architecture type was chosen for the following reasons:

A) the possibility of access to software from external workstations via network protocols. It also allows you to access an unlimited number of clients;

B) all calculations must be performed on the server, thereby reducing the overall load on your workstation, and the sense of fully bring them closer to the automatic operator workstations that are used in the management of real installations. For this reason, the server provides storage for defined major databases, databases of technological schemes, as well as the ability to connect with (* .dll) libraries, containing the components of the complex simulation models (model specific devices or systems). Building relationships between the components of the structure should be carried out using a special tool to create a specific implementation of a simulation model - the designer, which allows to create any variants of technological schemes of installations);

C) ensuring a high degree of protection, as well as easier organization of access control.

2) The division of management rights. In this case, it is planned to separate functionality to operators and administrators.

The developed computer model should have the following features for the operators of technological installations:

- the manage settings for valves and control equipment;

- setups management;
- monitoring of the system devices parameters.

At the same time, the administrator functional capabilities should be implemented:

- operators monitoring and management during the learning process;
- managing the various emergencies with the ability to create a disturbance in the system at any time;
- changing the raw input parameters (composition, temperature, etc.).

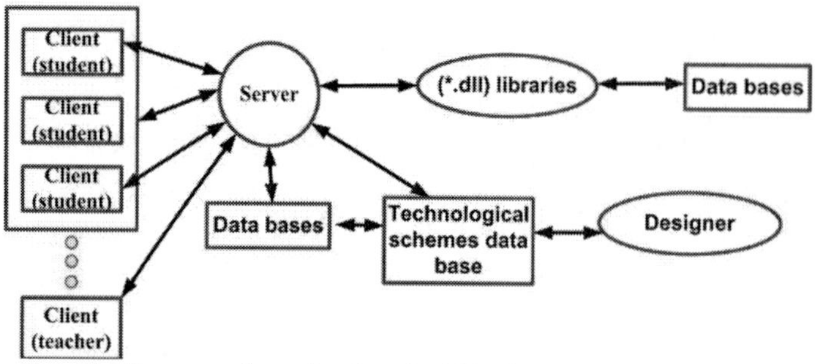

Fig. 2. Scheme of interaction of software products developed within the project

The developed software interface is shown in Fig. 3.

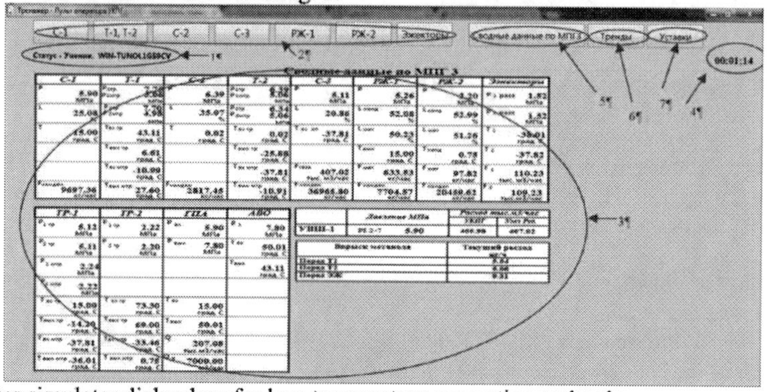

Fig. 3. The main computer simulator dialog box for low-temperature separation technology:
1 - status and connected client's name;
2 - toggle panel between control screens of main devices of low-temperature separation installation;
3 - a summary table of the main devices parameters (temperature, pressure);
4 - hour counter;
5 - switch on the main dialog box;
6 - switch to a graphical representation of the results of calculations;
7 - calling the dialog box with the task of setting control valve devices.

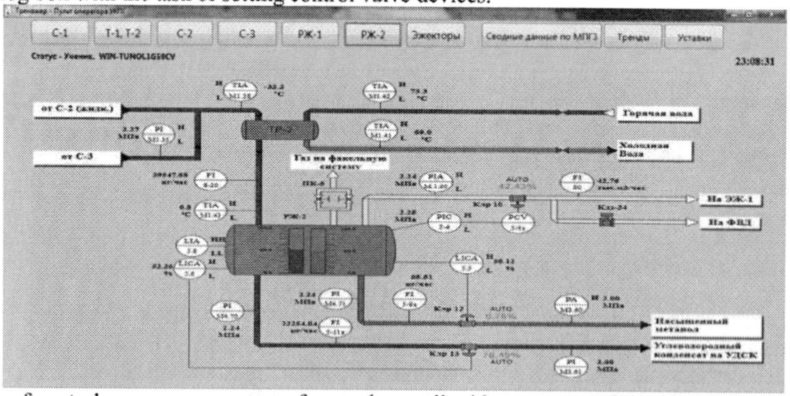

Fig. 4. The screen of control process parameters of second stage liquid separator and liquid heat exchanger

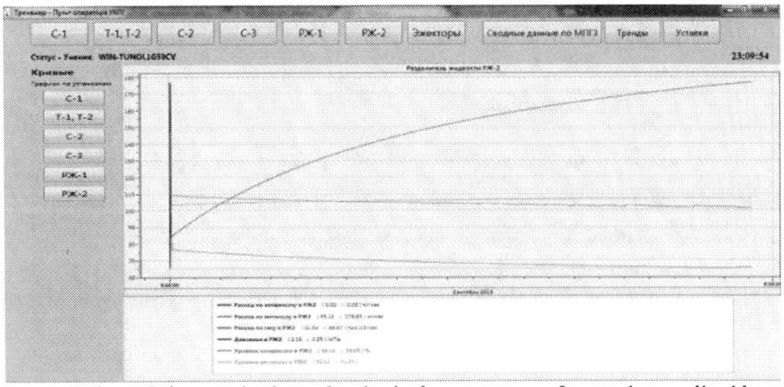

Fig. 5. Graphical representation of changes in the technological parameters of second stage liquid separator

4 Simulation results

The impact on the system was carried out by changing the degree of valves opening of. As constant parameters we adopted: the temperature and composition of the gas inlet; the degree of ejecting device opening; liquid level in the separator, the level of consistency is achieved by changing the degree of control valves opening in an automatic mode.

At that, there were some effects, one of which is shown in Fig. 6.

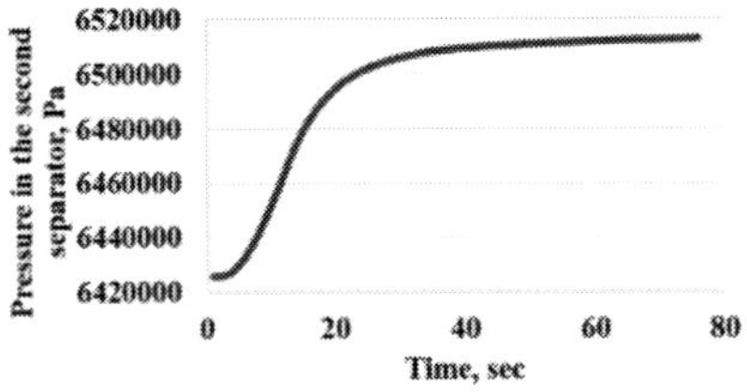

Fig. 6. Dependence of pressure in the second separator on the time

As can be seen from the figure, by increasing the valve opening percent increase in pressure in the separator is not instantaneous and takes some time. The amount of time is related to geometrical dimensions of the separator and the elements included in the system.

4 Conclusions

The considered in the article principle of non-stationary mathematical model development can be used to create similar models for other industries [13-15].

A simulation model for dynamic process of low-temperature gas separation, built on the basis of the laws of phase transformations of the gas-liquid hydrocarbon systems and heat transfer processes in dynamic operating conditions of industrial installations, allows to predict the work of existing units of the technological scheme with regard to their construction.

When you change the control parameters (opening degree of control valves), chemical-technological system retains inertia, and then returns to a stable state. The simulation dynamic model of the low-temperature gas separation process makes it possible to estimate the time to reach a new steady state, to consider all the attendant fluctuations devices operating parameters included in the installation [16, 17].

References

1. S.I. Arbuzov, S.G. Maslov, A.V. Volostnov , S.S. Iljenok, V.S. Arkhipov. Solid Fuel Chem. **46** (2015).
2. N.V. Usheva, O.E. Moyzes, E.A.Kuzmenko , S.F. Kim, E.S. Khlebnikova, S.N. Gizatullina, T.V. Filippova. IOP Conference Series: Earth and Environmental Science. **27** (2015).
3. S.F. Kim, N.V. Usheva, O.E. Moyzes, E.A. Kuzmenko, M.A. Samborskaya, E.A. Novoseltseva. Procedia Chemistry. **10** (2014).
4. A.V. Kravtsov, V.S. Moskvin, O.E. Pleshkova, N.V. Usheva. React Kinet Catal L. **30** (2) (1986).
5. A.I. Levashova, N.V. Usheva, O.E. Moizes, I. Fedjaeva. 8-th Korea-Russia International Symposium on Science and Technology: KORUS. **2** (2004).
6. A.V. Kravtsov, N.G.Sudobin, S.I. Smol'yaninov, S.N. Dneprovskii, A.I. Levashova, Y.V. Maksimov, K.T.

Topliva. Solid Fuel Chem. **15** (1) (1981).
7. Y. Chsherbakova, I. Dolganova, N. Belinskaya. Strategic Technology (IFOST) 7th International Forum. (2012).
8. I. Dolganov, M. Pisarev, E. Ivashkina, I. Dolganova. Petroleum and Coal. **56** (2) (2014).
9. V. Dmitriev, T. Gandzha, I. Dolganov, M. Pisarev, I. Dolganova, E. Sizova, E. Ivashkina. Petroleum and Coal. **57** (6) (2015).
10. I. Dolganov, M. Pisarev, E. Ivashkina, I. Dolganova Petroleum and Coal. **57** (4) (2015).
11. A. Vorobev. Physical Review E. **82** (5) (2010).
12. A. Vorobev, O. Zikanov, P. Mohanty. Journal of Thermal Spray Technology. **17** (5-6) (2008).
13. I.O. Dolganova, I.M. Dolganov, E.N. Ivashkina, E.D. Ivanchina, R.V. Romanovsky. PJCT. **14** (4) (2012).
14. E.D. Ivanchina, E.N. Ivashkina, I.O. Dolganova, V.V. Platonov. Petroleum Chemistry. **54** (6) (2014).
15. I.O. Dolganova, I.M. Dolganov, E.D. Ivanchina, E.N. Ivashkina, N.S. Belinskaya, V.V. Platonov. Procedia Chemistry. **10** (2014).
16. M.A. Samborskaya, O.E. Mityanina, A.V. Kravtsov. International Forum on Strategic Technology. (2010).
17. M.A. Samborskaya, O.E. Mityanina, M.S. Gyngazova. 7th International Forum on Strategic Technology (IFOST). (2012).

MATEC Web of Conferences **85**, 01006 (2016) DOI: 10.1051/matecconf/20168501006
CCT 2016

THE COMPOSITION OF BIOLOGICALLY ACTIVE COMPOUNDS IN PEAT FROM DIFFERENT CLIMATIC ZONES OF WESTERN SIBERIA

Olga Serebrennikova[1,2], Eugenia Strelnikova[2], Yulia Preis[3], Maria Duchko[1,2,a], Irina Rousskikh[2], Petr Kaduchagov[2] and Alexandra Peresedova[1]

[1] National Research Tomsk Polytechnic University, Institute of Natural resources, 634050, Tomsk, Russia
[2] Institute of Petroleum Chemistry SB RAS, 634055, Tomsk, Russia
[3] Institute of Monitoring of Climatic and Ecological System SB RAS, 634055, Tomsk, Russia

Abstract. The paper studies the individual composition of n-alkanes, polycycloaromatic hydrocarbons, steroids, bi-, tri-, and pentacyclic terpenoids of two peat deposits of rich fen Kirek located in Western Siberia. Considering the individual n-alkanes concentrations, some indexes were calculated to estimate the humidity during peat formation. It was shown that the pH of peat medium primarily affects steroids, tri- and pentacyclic terpenoids transformations.

1 Introduction

Peat is an easily available natural material and a source of biologically active substances widely used, not only in agriculture but also in human and animal medicine. [1] The literature refers to a wide spectrum of beneficial effects of peat and various peat preparations on organisms: stimulating effects on digestion, growth, and the immune system of the animals [2].

One of the most important biologically active substances are steroids, terpenoids and tocopherols. Their composition in peat depends on the impact of different peat-forming plant groups to peat formation and gives information and on processes which take place during the peat decomposition.

Steroids are hormones that play a critical role in endocrine systems which control the regulation of metabolic processes. These hormones stimulate biological responses in a wide range of tissues to influence such processes as sexual differentiation, reproductive physiology, and maintenance of salt balance and sugar metabolism. They have also been implicated in responses to stress and behavior [3].

Terpenoids serve as secondary metabolites of mixed biosynthesis, like alkaloids, flavonoids, oligosaccharides etc. There is a growing interest in natural triterpenoids caused as much by the scientific aspects of structural analysis of these compounds, as by the fact of their wide spectrum of biological activities; they are bactericidal, fungicidal, antiviral, cytotoxic, analgetic, anticancer, cardiovascular, antiallergic and so on [4].

Tocopherols (vitamin E) is known for its antioxidant, immune-enhancing, anti-inflammatory, and antiplatelet aggregation effects. Vitamin E is used clinically to prevent cardiovascular disease, cancer, cataracts, complications of diabetes, and to improve immune function. [5]

Optimal choice of peat deposits for the production of biologically active preparations supposes the detailed comparative analysis of peat properties from different deposits.

2 Experimental

The raised bog Sredne-Vasuganskoye is located in the middle taiga subzone of Western Siberia. The peat deposit (4.0 m in depth) is a typical riam with low vegetation and is formed by sphagnum and cotton grass-sphagnum peat. The bog was formed as a result of paludification. Current average annual air temperature is -1.7°C.

The raised bog Temnoye is located in the southern taiga subzone of Western Siberia on the lake shore. The peat deposit (6.5 m in depth) is riam with growthy vegetation. The lower part of the deposit was formed by moss-bog-

[a] Corresponding author: maria.duchko@gmail.com

bean and typha-moss-equisetum peat, in higher layers they are replaced by peat, formed from scheuchzeria, then cotton grass and sphagnum; the top of the reservoir is formed by sphagnum peat. The bog was formed as a result of terrestrialization The average annual temperature of the raised bog Temnoye is +0.6°C.

Peat macrofossil composition was determined by microscopic method, decomposition degree – by centrifugation. All peat samples were dried and powdered before the analysis. Bitumens were concentrated by extraction with 7% methanol in chloroform at 60°C. Group and individual composition of tocopherols, steroids and terpenoids in peat samples were determined by gas chromatography-mass spectrometry (GC-MS). Thermo Scientific gas chromatograph with mass detector fused silica capillary column (30 m length, 0.25 mm bore, 0.25 μm film thickness) was used for biomarker molecule analysis. The operating conditions were: temperature was held at 80°C for 2 min, ramped to 300°C at 4°C/min, with He as carrier gas. The ionization energy of the mass spectrometer was set to 70 eV, with a scanning range from 50 to 550 amu. The compounds were identified by comparison of their mass spectra with those of reference compounds.

3 Results and discussion

Peat of the bog Sredne-Vasuganskoye is characterized by a higher content of studied lipid components in comparison to peat of the bog Temnoye. The average total lipid content in peat of the bog Sredne-Vasuganskoye is 179.3 mcg/g, in peat of the bog Temnoye – 38.6 mcg/g. This may be explained by the differences in ambient temperature during peat formation. Lower temperature along the bog Sredne-Vasuganskoye led to reduced microbiological activity and, as a consequence, inhibited the transformation of organic compounds presented in peat-forming plants.

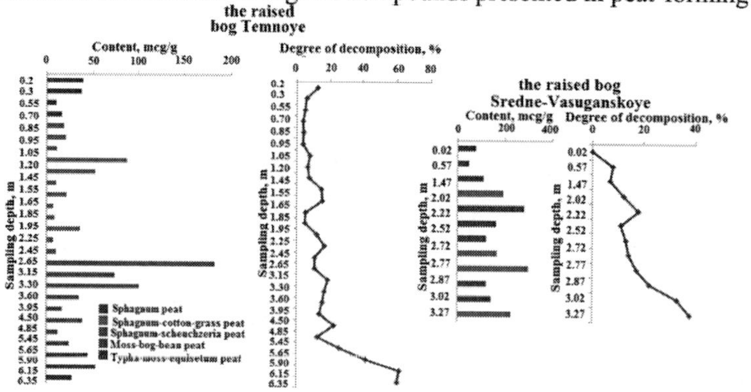

Figure 1. The total content of terpenoids, steroids and tocopherols in peat of bogs Temnoye and Sredne-Vasuganskoye.

The maximum total content of biologically-active compounds in the peat of bog Temnoye was recorded in fibric peat with increased content of herbaceous plants (Scheuchzeria, cotton grass and sedge) and reduced content of sphagnum moss or wood residues in the botanical composition (Figure 1). This may be due to the different total lipid content in peat-forming plants: sphagnum and sedge are characterized by low content of lipids (11.4 and 6.4 mcg/g, respectively), in cotton grass it is slightly higher (13.2 mcg/g). The total lipid content in Scheuchzeria is maximal among the herbaceous plants (41.4 mcg/g).

The highest total lipid concentrations in peat of the bog Sredne-Vasuganskoye was recorded in cotton grass-sphagnum peat and in sphagnum peat (2.22 m) with high proportion of cotton grass in the botanical composition.

The degree of decomposition of bog Sredne-Vasuganskoye peat grows from 0 to 38% alongside with depth increase. The same indicator for the bog Temnoye peat varies slightly with depth increase. The typha-moss-equisetum peat from the lowest part of the deposit is characterized by the maximum degree of decomposition.

The low values of peat decomposition degrees shows that peat was formed in moisture conditions. The raised humidity leads to increased proportion of grass in the bog plant community. It reduces the decomposition of organic matter, which helps to preserve high concentrations of lipids in peat.

The upper layer of sphagnum peat of the bog Temnoye is characterised by lower content of biologically-active substances in comparison to bottom layers and peat of the bog Sredne-Vasuganskoye. This is most likely due to the removal of peat decomposition products, diluvial water coming from the surrounding upland through the bog into a lake and bog water during the periods of climate aridity.

 DOI: 10.1051/matecconf/20168501006

The main sources of geological sesquiterpenoids are terrestrial plants, these compounds were found in resins and essential oils [7]. The isomers of cadinene prevail among sesquiterpenoids in most peat samples of the bog Temnoye, as well as in the major peat-forming plants. The majority of peat samples of the bog Sredne-Vasuganskoye has the maximum content of calacorene isomers, which dominate in Sphagnum fuscum.

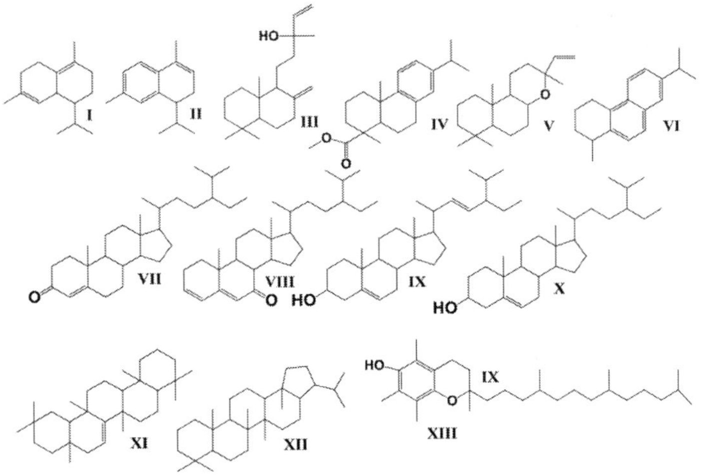

Figure 2. The structures of biologically-active compounds, dominating in peat: I- δ-cadinene, II-calacorene, III-labd-8,14-diene-13-ol, IV-methyldehydroabietate, V-manoyloxide, VI-10,18-bisnorabietapentaene, VII-stigmast-4-en-3-on, VIII- stigmast-3,5-dien-7-on, IX-stigmasterol, X-sitosterol, XI-D-friedoolean-14-ene, XII-homohopane C31, XIII-α-tocopherole.

The origin of diterpenoids may be associated with the transformation of abietic acid, which is widespread in higher plants resin [8]. Manoyloxide prevails among diterpenoids in herbaceous plants as well as in most peat samples of the bog Temnoye. In some sphagnum peat interlayers dominates methyldehydroabietate – the main representative of sphagnum moss diterpenoids. The high content of retene dominating in Scheuchzeria was found in one of the samples of sphagnum-scheuchzeria peat. Manool (labda-8(20),14-dien-3-ol) is present in herbaceous plants and prevails among diterpenoids in the upper part of the bog Sredne-Vasuganskoye deposit. At the depth exceeding 2 m 10,18-bisnorabieta-5,7,9(10),11,13-pentane begins to dominate (Fig. 2). This compound is absent in possible peat-forming plants. At the same time it is the possible product of retene partial hydrogenation.

The bog Temnoye peat differs from the bog Sredne-Vasuganskoye peat in significant prevalence of oxygen-containing compounds among diterpenoids. It can be explained by the intensive microflora development under more genial conditions of a southern taiga. In peat of the bog Sredne-Vasuganskoye such predominance was recorded only at the top of the deposit. At the depth below 2.20 m hydrocarbons begin to dominate. The proportion of diterpenoid hydrocarbons in all peat samples grows alongside with depth increase.

Geosteroids are commonly found in sediment organic matter. They take their origin from sterines, which are included in the structure of the eukaryotes cell walls [9]. Unsaturated stigmastane derivatives mainly with the keto-groups are typical for Sphagnum fuscum and cotton grass. Such structures prevail among steroids in the upper part of the bog Temnoye peat deposit in sphagnum and sphagnum-cotton grass peat (Fig. 2).

Saturated stigmastane derivatives with keto-group (stigmastane-3-one) dominate in the lower part of the bog Temnoye deposit in sphagnum-sheuchzeria peat and in peat with a predominance of green moss, typha, bog-bean and equisetum residues in botanical composition. Such structures can be the products of more deep diagenetic conversion of peat. Sitosterol prevails among steroids in Sredne-Vasuganskoye peat deposit. It can be formed as a result of partial hydrogenation of stigmasterol, which dominates in the near-surface peat layer.

A slight increase in the relative content of alcohol derivatives of steroids was found for sphagnum and sphagnum-cotton grass peat. It may indicate an active biodegradation processes.

The peculiarities in triterpenoid composition are connected, apparently, with the difference in plant associations of bogs located in middle and southern taiga. D-friedoolean-14-ene dominates among triterpenoids in most peat samples of the bog Sredne-Vasuganskoye, as well as in sphagnum and sphagnum-cotton grass peat of the bog Temnoye.

The lower part of the bog Temnoye peat deposit differs from the overlying layers and from peat of the bog Sredne-Vasuganskoye in prevalence of hopane derivatives among triterpenoids. This may indicate high degree of processing peat by aerobic bacterial. It is generally thought that the main source of hopane derivatives is a pentacyclic alcohol bakteriotetrol (C_{35}), which is the part of the lipid cell membrane of aerobic bacteria [10].

Among tocopherols in all peat samples predominates α-form, the content of γ-form is increased, α-tocopherol acetate was not detected. Maximum tocopherols concentrations in the bog Sredne-Vasuganskoye peat was detected in one of the cotton grass peat samples and in complex sphagnum peat, as well as sphagnum and sphagnum-cotton peat of the bog Temnoye.

4 Conclusion

The studied peat samples of the bog Temnoye are characterized by a predominance of cadinene isomers among sesquiterpenoids, manoyloxide and methyldehydroabietate among diterpenoids, stigmastane derivatives with keto-group – among steroids. Peat of the bog Sredne-Vasuganskoye differs in maximum content of calacorene among sesquiterpenoids, 10,18-bisnorabieta-5,7,9(10),11,13-pentane among diterpenoids and sitosterol among steroids. Both deposits are characterized by the dominance of D-friedoolean-14-ene among pentacyclic terpenoids and alpha-form among tocopherols.

The total content of biologically-active compounds is higher in peat, which was formed in the lower-temperature conditions by slowing down the decomposition processes. This peat is also subjected to less intense microbial oxidation. In addition, changes in the content of the studied compounds along the peat deposits are largely due to the diversity of peat botanical composition. The maximum total content of bioactive compounds was observed in peat with a high content of herbaceous plants such as cotton grass and Scheuchzeria and low content of sphagnum moss in the botanical composition. Thus, it is preferable to use for the production of biologically active peat formed in sufficiently cool conditions predominantly by herbaceous plants.

References

1. M. Trckova, L. Matlova, H. Hudcova, M. Faldyna, Z. Zraly, L. Dvorska, V. Beran and I. Pavlik. *Vet. Med.* 50, 361, (2005).
2. O. V. Serebrennikova, E. B. Strelnikova, Y. I. Preis, E. V. Gulaya and M. A. Duchko. *Bul. Of Tom. St. Pol. Univ.* 323, 40, (2013).
3. A. R. McCarthy. *A thesis submitted in partial fulfillment of the requirements for the degree of Doctor of Philosophy University of Canterbury.* p 16, (2006).
4. J. Patočka. *Journal of App. Biomed.* 1, 7, (2003).
5. M. L. Colombo. *Molecules.* 15, 2103, (2010).
6. R. Rainage. *Chemistry of Terpenes and Terpenoids.* p 56, (1993).
7. K. E. Peters, C. C. Walters and J. M. Moldowan. *The Biomarker Guide: Biomarkers in the Environment and Human History.* 1, 492, (2005).
8. J. W. de Leeuw and M. Baas. *Biol. Markers in the Sedimentary Record.* 24, 101–123, (1986).
9. R. A. Andersson and P. A. Meyers. *Org. Geochem.* 53, 63, (2012).
10. X. Huang, S. Xie, C. L. Zhang, D. Jiao, J. Huang, J. Yu, F. Jin and Y. Gu. *Org. Geochem.* 39, 1765, (2008).

Determination of rhodium content by the method of Stripping voltammetry in ores and technogenic raw materials

N.A. Kolpakova[1], E.N. Dyachenko[1]

[1]*Federal State autonomous educational institution of higher education "National Research Tomsk Polytechnic University, Russia*

Abstract. The method of determination of rhodium by the method of Stripping voltammetry in ores and technogenic raw materials has been developed. Determination of rhodium (III) in solution by Stripping voltammetry method was carried out by peak of selective electrooxidation of bismuth from intermetallic compound Bi2Rh. 1 M KSCN was selected as a background electrolyte. The method of melting of the sample with potassium tetrafluoroborate (KBrF4) at a temperature of 500 was used for the conversion of the rhodium solution.The method of deposition of most ions present in solution of alkali metals was used to separate ions of rhodium (III) from the matrix of the samples. Ions of Pt (IV), Rh (III) and Ir (IV) remain in solution after precipitation. Ions of iridium (IV) and 10-fold excess of platinum (IV) do not interfere with the determination of rhodium (III) against the background of 1 M KSCN.

1. Introduction

The biggest obstacle to the research of potential natural sources of platinum group metals is extremely poorly developed base of analytical methods for their quantification. The content of rhodium in the technogenic raw materials ranges from 0.001 mg/kg to 10 mg/kg. Advanced instrumental methods of analysis are necessary to determine such low rhodium content. Papers [1-4] show the capabilities of different analysis methods in determining the rhodium in mineral raw materials. As these papers show all existing methods of analysis require effective methods of sample decomposition and methods for separation of rhodium from the sample matrix.

We propose a method of stripping voltammetry (SV) as a method of analysis for determination of rhodium in mineral raw materials. This method is a highly sensitive instrumental method of analysis, which allows determining the elements after their concentration at the electrode surface, followed by sediment electrooxidation. However, determination of rhodium using this method can cause certain difficulties. We cannot detect the process of electroreduction of rhodium ions (III) in the graphite electrode (GE) on the current-voltage curves. It is connected with hydrogen evolution catalytic process that flows in parallel and results in an overlap of rhodium (III) ions electroreduction process. The process of electrooxidation of rhodium precipitate occurs at potentials greater than 1V and is overlapped by the process of oxygen evolution from water.

Determination of rhodium (III) ions using SV method is carried out by rhodium IV electrodeposition together with a less noble metal (activator), followed by selective electrooxidation of metal activator. The literature describes methods for SV-determining of rhodium (III) by peaks of selective electrooxidation of bismuth or mercury form intermetallic mercury-rhodium [3] and bismuth-mercury [4] compounds (IMC).

[1]Corresponding author: nak@tpu.ru

 DOI: 10.1051/matecconf/20168501007

The aim of this work was to develop a methodology for determination of rhodium in the gold mining mineral raw materials by stripping voltammetry by the peak of the selective electrooxidation of rhodium from intermetallic compound with bismuth and carry out determination of rhodium in a number of non-traditional gold-platinoid minefields.

2. Material and Methods of Research

The research is carried out using voltammetric analyzer TA-4 (LLC Research and Production Enterprise "TomAnalyt", Tomsk), complete with a personal computer. Quartz cups having volume o of 20 cm^3 were used as electrolyzer. All studies were carried out in a three-electrode cell. Graphite electrode (GE) impregnated with polyethylene served as the indicator electrode, silver chloride electrode filled with saturated KCl solution served as the reference electrode and platinum electrode served as the auxiliary one. The studies were conducted on the backgrounds of 1M HCl and 1M KSCN. Electrodeposition of binary electrolytic precipitation was carried out from solutions containing ions of bismuth (III) and rhodium (III). Electrochemical cleaning of the electrode surface was carried out in the background electrolyte within one minute at a potential of 1.2 V or mechanically rubbing the electrode with the filter paper. The surface of the electrode was renewed after each measurement. The solution was stirred automatically during electrolysis by vibrating the working electrode as conditioned by analyzer. All studies were performed using reagents with qualification not lower than "chemically pure" under normal conditions.

3. Results and discussion

Selecting the background electrolyte. Rhodium was introduced into the test solution in the form of chloride complexes. $RhCl_6^{3-}$. In acid background electrolytes bismuth is present in solution in the form of aquacomplexes $[Bi(H_2O)_n]^{3+}$.

Electrodeposition of binary precipitate bismuth-rhodium was carried out from 1M HCl solution containing ions of bismuth (III) and rhodium (III), at a potential of -0.8 V. The characteristic current-voltage curves of selective electrooxidation of bismuth from binary electrolytic precipitate bismuth-rhodium from graphite surface c electrode are shown in Fig. 1.

We observe several peaks on the current-voltage curve of precipitate electrooxidation. The peak at a potential of - 0.05 V corresponds to a process of electrooxidation of bismuth deposited on the surface of GE. When rhodium content in the precipitate increases the peak current of bismuth electrooxidation decreases as bismuth is bound with rhodium in IMC. Selective electrooxidation of bismuth from IMC with rhodium in the solution1M HCl occurs at a potential of 0.1 V and has a wave form. It is difficult to estimate the content of rhodium (III) by current of this wave because it overlaps the process of electrochemical oxidation of oxygen adsorbed on the rhodium. According to the literature review data [5], peak at a potential of + 0.5 V is the peak of oxygen electrooxidation adsorbed on rhodium.

Background 1M KSCN has been studied in order to fix a clearer analytical signal of the process of selective electrooxidation of bismuth from IMS with rhodium.

DOI: 10.1051/matecconf/20168501007

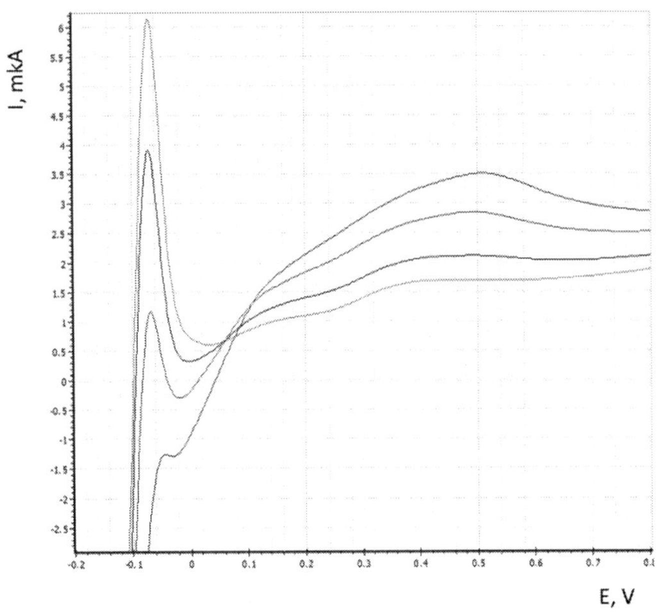

Figure 1. The current-voltage curves of the electrochemical oxidation of precipitate bismuth-rhodium surface from GE. The experimental conditions: background 1M HCl, electrolysis potential is minus 0.8V, electrolysis time is $\tau_э$=120s, scanning speed v = 80 mV/s, $C_{Bi^{3+}}$=2 mg/l; 1) $C_{Rh^{3+}}$=0,02 mg/l; 2) $C_{Rh^{3+}}$=0,04 mg/l; 3) $C_{Rh^{3+}}$= 0,06 mg/l; 4) $C_{Rh^{3+}}$=0,08 mg/l.

Typical current-voltage curves of electrochemical oxidation of binary electrolytic precipitation bismuth-rhodium from the surface of graphite electrode deposited from a solution 1M of KSCN are shown in Fig. 2.

The peak at a potential of - 0.1V on the voltage-current curve corresponds to the process of electrooxidation of hydrogen adsorbed on the rhodium. This peak is well described in the literature [5]. The peak at a potential of - 0.01 V corresponds to the process of electrooxidation of bismuth deposited on the surface of the GE. The peak at a potential of 0.23 V corresponds to the selective electrooxidation of bismuth from IMS with rhodium having composition RhBi2 [4]. This peak current value is used to determine the content of rhodium (III) in solutions using stripping voltammetry method.

DOI: 10.1051/matecconf/20168501007

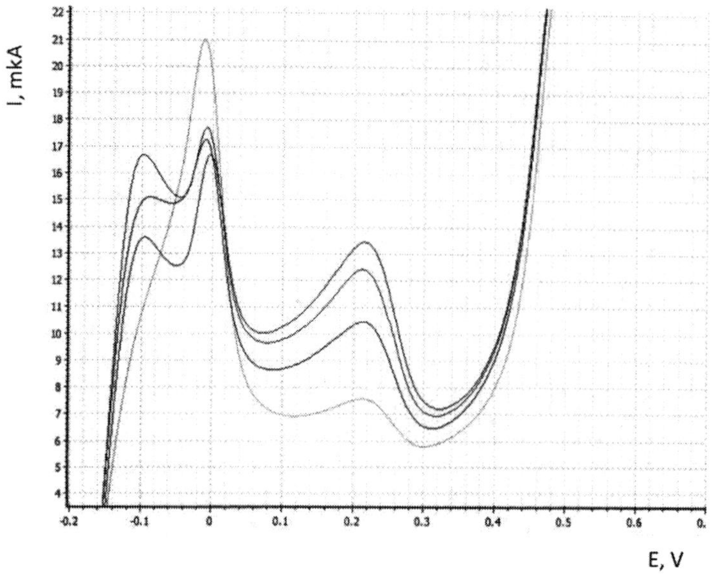

Figure 2. The current-voltage curves of the electrochemical oxidation of precipitate bismuth-rhodium surface from GE. The experimental conditions: background 1 M KSCN, electrolysis potential is minus 0.8V, electrolysis time is $\tau_э$=120s, scanning speed v = 80mV/, $C_{Bi^{3+}}$=2 mg/l; 1) $C_{Rh^{3+}}$=0,02 mg/l; 2) $C_{Rh^{3+}}$=0,04 mg/l; 3) $C_{Rh^{3+}}$= 0,06 mg/l; 4) $C_{Rh^{3+}}$=0,08 mg/l.

Figure 3 shows the current of selective electrooxidation of bismuth Bi_2Rh from IMS depending on the content of rhodium (III) in solution at a constant concentration of bismuth ions (III) in the sample solution.

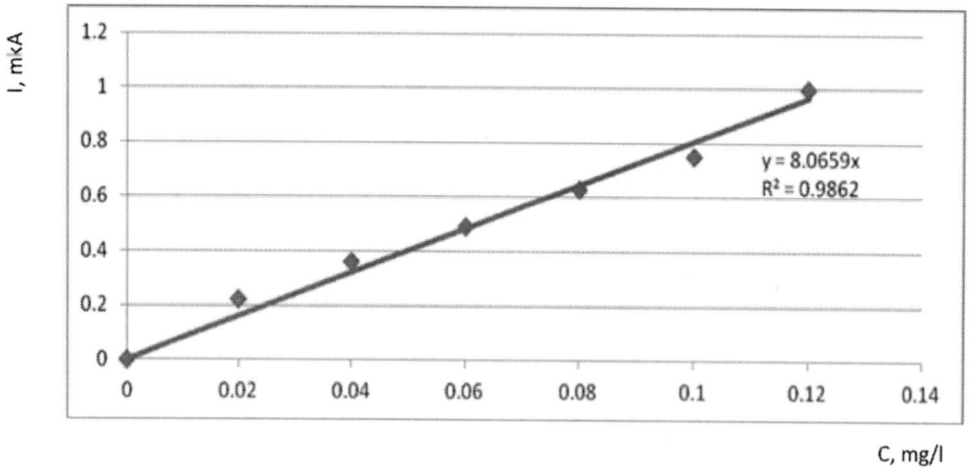

Figure 3. Dependence of peak current of selective electrooxidation of bismuth on rhodium ions (III) concentration in the sample solution. The experimental conditions: background 1 M KSCN, electrolysis potential is minus 0.8V, electrolysis time is $\tau_э$=120s, scanning speed v = 80mV/s., $C_{Bi^{3+}}$=2 mg/l.

Determination of rhodium in the mineral raw materials. A weighed portion of analyzed sample to be analyzed weighing 1 g was placed in a corundum crucible and calcined in a muffle furnace at t = 700 C for one hour. Cooled sample was mixed with five eight-fold amount (by weight) of potassium tetrafluoroborate (KBrF4) directly in a crucible to decompose and coated on top uniformly by two-fold (with respect to the weight of sample) amount of $KBrF_4$. The crucible was capped and placed in muffle electric furnace heated to 500 C. for 1.5-2.0 hours. Potassium tetrafluoroborate was taken with excess against stoichiometric reactions not only to ensure complete oxidation of the sample, but also to form a flux in which fluorocomplexes of noble metals "dissolve". After completion of decomposition process, the product was mostly in a form of pale yellow melt. The solution was evaporated on electric heater to wet salts, then dissolved in 1 M HCl and heated again for 15 minutes. Here all platinum metals pass from sample into solution in the form of chloride complexes. The resulting solution was filtered through a "blue tape" filter. After filtration, the filter was washed with bidistilled water. The resulting solution was treated with 3% NaOH solution to form a precipitate of metal hydroxides. The resulting metal hydroxides were separated from the solution by filtration through a "blue tape" filter. Ions of Pt (IV), Rh (III) and Ir (IV) stay in the solution after precipitation of the hydroxides. The solution was evaporated to wet salts, dissolved in 10 ml of 1 M KSCN and polarographised. Iridium in any ratio with rhodium does not interfere with determination of rhodium (III) on the background of 1M KSCN, because it is not able to settle on the electrode surface. According to literature review data, platinum and rhodium precipitate is a solid solution of these metals. By increasing the platinum content in electrolytic precipitate, the current of selective electrooxidation of bismuth increases, but the proportionality of current to rhodium (III) ion content remains to a tenfold excess of platinum (IV) ions in the solution.

Performance accuracy of methods of determination of rhodium was evaluated by comparing the results of IV-determination of rhodium in the standard samples of composition (SS).

Table 1. Average values of rhodium content in standard samples obtained by stripping voltammetry method (g/t, ppm, 10^{-4}%)

Sample	Standard sample characteristics	Rhodium content, g/t	
		IV method	In SS
G-3	Copper-nickel sulfide ore	0,64±0,03	0,66
ST-1	Concentrations tailings of copper nickel sulphide ore	0,095±0,024	0,094
ShT-1	Matte of ore-heating smelting	3,68±0,024	3,72

Table 2 shows data on the content of rhodium in the quartz-gold-sulfide ore and flotation concentrate of this ore obtained by stripping voltammetry method.

Table 2. Average values of the content of precious metals in standard samples (g/t, ppm, 10^{-4}%)

Sample characteristic	Rhodium content, g/t	
	IV-method	AAS method
Quartz-gold-sulfide ore	0,007±0,002	0,009±0,003
Flotation concentrate of quartz-gold-sulfide ore	0,065±0,008	0,072±0,006

4. Conclusions

1. The method of determination of rhodium by stripping voltammetry in ores and technogenic raw materials was developed. Determination of rhodium (III) in solution by stripping voltammetry was carried out by the peak of selective electrooxidation of bismuth from intermetallic compound Bi2Rh. Electrolyte 1 M KSCN was selected as background.
2. The method of melting of the sample under analysis with potassium tetrafluoroborate (KBrF4) at a temperature 500 $^{\circ}C$ was used to convert rhodium into solution.
3. Method of precipitation of ions of majority metals in the solution by alkali was used for the separation of rhodium (III) ions from the sample matrix. Ions Pt (IV), Rh (III) and Ir (IV) stay in the solution after precipitation.
4. Ions of iridium (IV) and 10-fold exceed of platinum (IV) do not interfere with rhodium (III) determination against 1M KSCN background.

References

1. Torgov V.G., Korda T.M., Demidova M.G., Gus'kova E.A., Bukhbinder G.L. ICP AES determination of platinum group elements and gold in collective extract and 21 strip product solution in analysis of geological samples // J. Anal. At. Spectrom. – 2009. – V. 24. – P. 1551–1557.
2. Arabova Z. M., Dedkov Yu. M., Arabs M. sh., Krigman L. V. Methods for the determination of rhodium (review) //Zavodskaya laboratoriya. Diagnostics of materials. - 2014. T. **80**, № 8. - C. 5–16.
3. Kolpakova N. A. Determination of the platinum metals in minerals by method of Stripping voltammetry (review). - 2014 - T. 80 - №. 9. - P. 5-13.

4. Dyachenko E. N., Kolpakova N. And. Method of determination of rhodium (III) in water solutions by method of Stripping voltammetry by peak of selective electrooxidation of bismuth from intermetallic compounds with the rhodium //Abstracts of IX all-Russian conference on electrochemical methods of analysis with international participation and youth scientific school. Ekaterinburg: 29 may-3 June 2016. P. 57.
5. Mansurov, G. N., Petrov O. A. the electrochemistry of thin metal films. Monograph. -M.: MGOU, 2011. -351 p.
6. Beamish, F. Analytical chemistry of noble metals. M: Peace. 1969,part 1. 297 p.

MATEC Web of Conferences 85, 01008 (2016) DOI: 10.1051/matecconf/20168501008
CCT 2016

The Resource-Saving Technology of Aluminum Nitride Obtaining During Combustion of Aluminum Nanopowder in Air

Alexander Ilyin[1,*], *Andrei* Mostovshchikov[1] , and *Liudmila* Root[1]

[1] Tomsk Polytechnic University, 634050 Tomsk, Russia

Abstract. The resource-saving technology of aluminum nitride obtaining during the combustion of aluminum nanopowder in air has been analyzed in the article. The investigation of the crystal phases of aluminum nanopowder combustion products obtained under the magnetic field exposure has been made. The experimental results showed the increase of aluminum nitride content up to 86 wt. % in comparison with the aluminum nitride content in combustion products without any exposure. The mechanism of aluminum nitride formation and stabilization in air was due to the oxygen molecules deactivation by light emission during combustion.

1 Introduction

It is known that aluminium nitride has the following unique properties: a high thermal conductivity ($\lambda > 250$ W/m^2·s) and low electrical conductivity ($\rho > 10^{13}$ Ohm·m) [1]. At the same time, the technology of aluminum nitride synthesis is complicated by the absence of the liquid phase during heating (i.e. above 2400 °C aluminum nitride evaporates). Therefore, the technology of aluminum nitride production includes the stages of γ-Al_2O_3 with the coke prolonged heating at 1600–1700 °C, grinding the product, and its re-nitriding. The sintering of items made from aluminum nitride is realized using a hot pressing method in the high-purity nitrogen medium [2, 3]. The synthesis of aluminum nitride is also possible by the combustion of aluminum powder in pure nitrogen medium [4–6].

It was experimentally established that as the result of combustion of the powdered aluminum in air more than 60 wt. % of aluminium nitride is stabilized in the final combustion products [7–10]. Even the first investigations of aluminum nanopowder combustion [11] showed, that the combustion of relatively small samples (less than 0.5 g) took place in a single step (600–800 °C), during which aluminum nitride did not form. If the sample mass of nanopowder exceeded 1 g, the low-temperature combustion stage exceeded to the high-temperature stage (2200–2400 °C), accompanied by a bright irradiation in the visible and infrared spectral ranges. This combustion stage passed in the thermal explosion mode. The combustion rate increased with the sample mass raise. The combustion conditions approximated adiabatic as a result of temperature and luminosity increase of the Al–O_2–N_2 burning system. When the sample weight reached 15 g, the

* Corresponding author: ilyin@tpu.ru

aluminum nitride yield increased up to 66 %, taking into account that even at 10–12 g the growth of aluminum nitride yield in aluminum nanopowder combustion products in air slowed down. Furthermore, it was established in early studies that high-power exposure on the aluminium nanopowder during the combustion increases the thermal effect of its oxidation [12–14]. This is a promising trend for increasing the burning power of the aluminum nanopowder that can be also used to raise the aluminum nitride yield in the combustion synthesis [15].

The aim of this work was to improve the resource efficiency of aluminum nitride synthesis by increasing the yield of aluminum nitride in aluminum nanopowder combustion products under the magnetic field exposure.

2 Experimental procedure

The aluminum nanopowder obtained by the wire electrical explosion in argon, has been studied. The aluminum nanopowder was obtained with the help of UDP-4G machine, developed in National Research Tomsk Polytechnic University [16]. The concept scheme of the machine for nanopowders production is shown in Fig. 1.

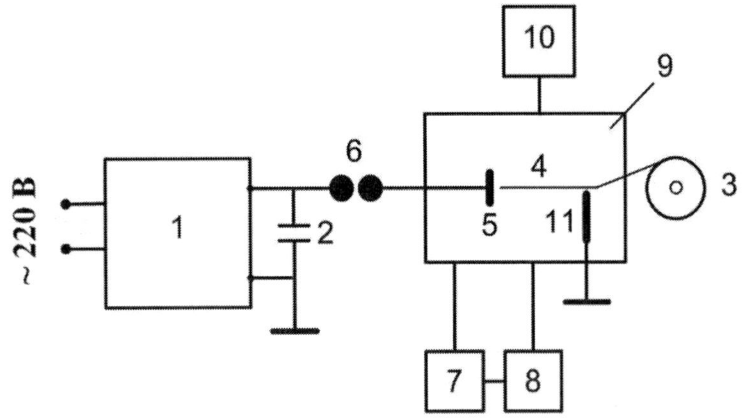

Fig. 1. The concept scheme of the wire electrical explosion machine for nanopowders production.

The machine operates in the following way. The capacitive energy storage (2) is charged from a high voltage power supply (1). The wire feeder (3) automatically moves the wire segment being exploded (4) to the inter-electrode gap. When the wire reaches the high-voltage electrode (5), the commutator (6) operates. The capacitive energy storage discharges on the wire segment between the high-voltage electrode and the ground electrode (11), as the result, the wire segment explodes. The obtained powder is collected in the powder collector (7). The gas purified from the powder is fed back into the explosion chamber (9) by means of the gas circulation system (8). The volume of the explosion chamber (9) before exploitation is evacuated and then filled with the required gas. These functions are performed by the gas supply system (10).

Aluminum nanopowder obtained in argon is pyrophoric, i.e. it ignites when exposed to air. It was characterized by the following activity parameters: oxidation start temperature – 450 °C; oxidation degree – 63.8 %; maximal oxidation rate – 0.13 wt. %/deg, specific heat effect – 4995 J/g. The particle size distribution was close to the regular-logarithmic with a maximum of 120 nm. The content of the metallic impurities in nanopowder did not exceed 1.0 wt. % [17].

Alundum pot was used for the experiments on aluminum nanopowder combustion. The combustion of some nanopowder samples was performed in the conical shaped form on polycor or duralumin base in air, as well. To conduct experiments on combustion the optimal sample weight of 3 g was chosen, since at the weight less than 0.5 g the combustion process took place in a single step, and at the weight of 4 g the thermal destruction of alundum pots occurred. The process of aluminum nanopowder combustion in air occurred in two stages. On the first stage, after ignition, the heat waves spread over the sample, the temperature maximum reached 600–800 °C. After that the burning process accelerated and passed in the thermal explosion mode, accompanied by the temperature rise up to 2000–2100 °C (on the base and in the pot, correspondingly). The temperature pyrometric measurements were realized at the Laboratory of thermal control methods (Laboratory no. 34) of National Research Tomsk Polytechnic University by the thermal imager ThermaCam P65 HS.

To determine the influence of a homogeneous magnetic field on the morphology of combustion products the samples were ignited in the pot between the poles of a magnet (0.34 T). And for determination the influence of an inhomogeneous magnetic field the samples were ignited between two magnets each of the same induction (0.37 T).

To determine the influence of a constant electric field on the morphology of combustion products the samples were ignited between the plates of air-dielectric capacitor; the field potential gradient in the capacitor was 10 kV/m.

In order to establish the sequence of crystalline phases formation during the combustion of aluminum nanopowder in air, the equipment of the precision diffraction analysis center with the electron storage ring VEPP-3 of NPI n. a. G. I. Budker SB RAS (Novosibirsk, Russia) was used. The synchrotron radiation with a wavelength of 0.10173 nm was focused on the compacted sample surface. After initiation of the combustion process, the diffraction patterns were recorded by a single-axis detector OD-3M, which simultaneously detects radiation in the range of angles about 60 grad. The detector was set to the angular range from 30 to 90 grad. As only in this range the reflections of maximum intensity from the crystalline phases, formation of which is possible by aluminum nanopowder combustion, are located. The time of one frame making was 0.3 s that allowed getting 200 shots in a burning time of 60 s [10].

3 Experimental results and discussion

As a result of aluminum nanopowder combustion in a constant magnetic field (0.34 T) in alundum pot the output of aluminum nitride was 83 wt. %. Due to the additional thermal insulation during the combustion in the pot a higher temperature and luminosity were observed.

DOI: 10.1051/matecconf/20168501008

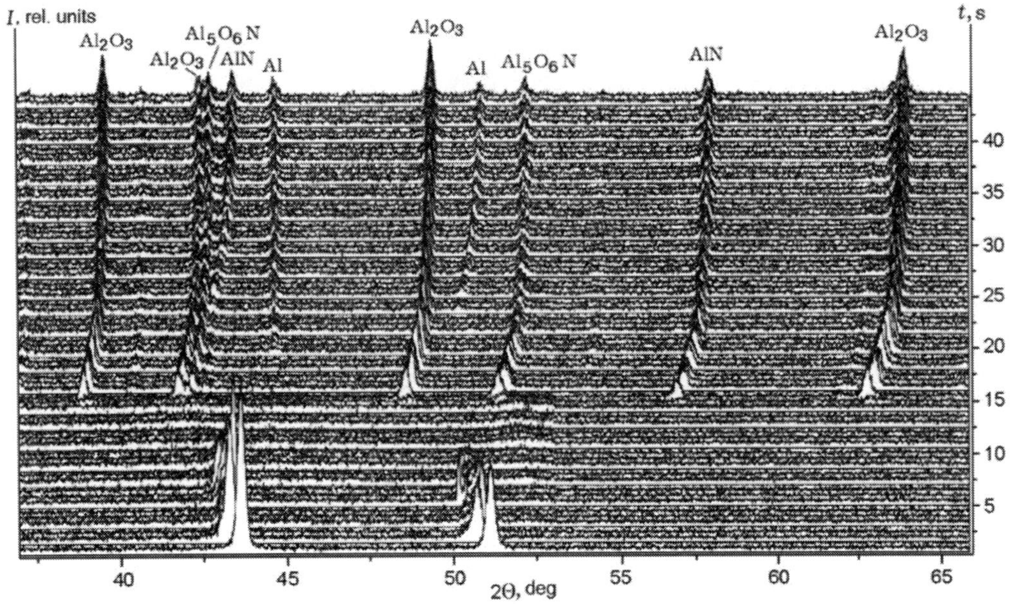

Fig. 2. Time scan of diffractograms of the compacted aluminum nanopowder surface while combustion.

When the compacted aluminum nanopowder (porosity of 66 vol. %) was combusted in the field of synchrotron radiation (λ = 0,10173 nm), an average of 86 wt. % of aluminum nitride was stabilized in the combustion products. According to the diffraction patterns, one of the mechanisms of the crystalline phase of aluminum nitride formation (Fig.2) is the consistent replacement of oxygen with nitrogen in aluminium oxide ($Al_2O_3 + N_2 \rightarrow Al_5O_6N + N_2 \rightarrow AlN$).

The formation of the independent crystalline phase of aluminum nitride involves the reaction of aluminum with nitrogen only at a certain stage of the combustion process. Non-participation of oxygen in aluminum oxidation can be explained by the deactivation of the oxygen molecules under the electromagnetic radiation exposure, namely, due to the photochemical transition of paramagnetic biradical $O^{\cdot\cdot}$ into a stable inactive molecule $O \equiv O$ [18].

Thus, the realization of the combustion process in a limited volume (pot) and in the magnetic field increased the yield of aluminum nitride in the combustion products 1.5–2 times, that results in the increase of the resource efficiency of the developed technology.

The research is supported by the Russian Foundation for Basic Research, project No 15-03-05385.

References

1. R. Bachelard, P. Joubert, *Mater. Sci. Eng.* **109A** (1989)
2. G. Selvaduray, L. Sheet, *Mater. Sci. Technol.* **9** (1993)
3. M. Boćkowski, B. Łucznik, I. Grzegory, S. Krukowski, M. Wróblewski, S. Porowski, *J. Phys.: Condens. Matter,* **14** (2002)
4. C. Lin, S. Chung, *J. Mater. Res.* **16** (2001)
5. A. Wilmański, M. Bućko, Z. Pędzich, J. Szczerba, *J. Mater. Sci. Chem. Eng.* **2** (2014)
6. T. Khabas, *Glass and Ceramics,* **59** (2002)

7. A. Ilyin, L. Root, A. Mostovshchikov, *Key Eng. Mat.* **685** (2016)
8. A. Mostovshchikov, A. Ilyin, I. Zabrodina, *Key Eng. Mat.* **685** (2016)
9. A. Il'in, A. Mostovshchikov, L. Root, *Tech. Phys. Lett.* **37** (2011)
10. A. Il'in, A. Mostovshchikov, N. Timchenko, *Combust. Explos. Shock,* **49** (2013)
11. A. Il'in, L. Proskurovskaya, *Combust. Explos. Shock,* **26** (1990)
12. A. Il'in, L. Root, A. Mostovshchikov, *Techn. Phys. Lett.* **57** (2012)
13. A. Mostovshchikov, A. Ilyin, A. Azanov, I. Egorov *Key Eng. Mat.* **685** (2016)
14. A. Mostovshchikov, A. Il'in, P. Chumerin, Yu. Yushkov, V. Vaulin, B. Alekseev, *Tech. Phys. Lett.* **42** (2016)
15. A. Gromov, L. Chukhlomina, *Nitride Ceramics: Combustion Synthesis, Properties and Applications,* Weinheim: Wiley-VCH (2014)
16. A. Gromov, U. Teipel, *Metal Nanopowders: Production, Characterization, and Energetic Applications*, Weinheim: Wiley-VCH, (2014)
17. A. Ilyin, A. Gromov, V. Merkulov, A. Mostovshchikov, *Russ. Phys. J.* **58, 6/2** (2015)
18. A. Il'in, L. Root, *Russ. Phys. J.* **54, 11/3** (2011)

The study of the process of alkaline precipitation purification of solutions from silicon macroscales in the ammonium-fluoride processing of beryllium-containing materials

Alexander Dyachenko[1], *Roman* Kraydenko[1], *Margarita* Lesnikova[1], *Lev* Malyutin[1] and *Ilya* Petlin[1]

[1]Tomsk Polytechnic University, Lenin av. 30, Tomsk, 634050, Russia

Abstract. The ammonium-fluoride method of beryllium materials processing is examined. An analysis of the existing sulfuric-acid and fluoride processing scheme of beryllium-containing concentrates processing is described; advantages of the proposed ammonium-fluoride scheme and possible problems that may occur when testing a new technique are discussed. Studies on determining the effect of silicon macroscales on the behavior of beryllium in fluoride solutions at an increased pH, as well as on establishing the distribution of beryllium and silicon by phases in laboratory testing of the technology on model mixtures, are described. As a result of the studies, it is found that precipitation purification of solutions from silicon with a solution of ammonia is possible at a temperature of 20-25 °C. The paper presents data on the effect of an excessive concentration of ion-fluoride in a solution on the pH level of silicon precipitation; and on the hydrolysis degree of beryllium and silicon during the precipitation.

1 Introduction

Beryllium in nature mainly represented about 30 minerals, the most important of them are: beryl, chrysoberyl, behoit, bertrandite, phenacite [1-3]. The volume of metallic beryllium production amounts to nearly 300 tons/year. The main source of beryllium is a bertrandite-phenacite concentrate which is produced using methods of flotation from the beryllium-fluorite-containing ore; the concentrate from the Ermakovskiy deposit, which has been used for following researches, contains up to 4 % of beryllium (Table 1) [4].

DOI: 10.1051/matecconf/20168501009

Table 1. Chemical composition of phenacite concentrate

Substance	Be	SiO_2	CaF_2	F	Al_2O_3; Fe_2O_3; CaO
Content, %	4	29	20	5	40

To date, two ways of hydrometallurgical processing of beryllium concentrates have been designed and developed: sulfuric-acid and fluoride (Table 2) [2].

Table 2. Description of existing methods of hydrometallurgical processing of beryllium concentrates

Parameters	Processing method	
	Sulfuric-acid	Fluoride
1	2	3
Basic reagents	93% sulfuric acid, water, ammonium sulphate, caustic soda, EDTA, ammonium bifluoride solution	Sodium hexafluorosilicate, soda ash, caustic soda, iron sulfate (III), water, ammonium bifluoride
Stages	Melting in an arc furnace, thermal activation of the concentrate, sulfuric-acid interaction, water leaching of beryllium, purification from Al and Fe, formation and hydrolysis of sodium beryllate, production of beryllium hydroxide	Mixing of reagents, briquetting, sintering, water leaching, alkaline precipitation, production of beryllium hydroxide
Process conditions	Melting point of the concentrate in an arc furnace – 1700 °C, heat treatment – 900-950 °C, sulfation – 200 °C; Process duration – 20 hours.	Sintering temperature – 750 °C; Process duration – 32 hours.
Energy intensity	High-energy	High-energy
Environmenta l risks	High Utilization of volatile fluorides (SiF_4, HF), utilization of ferrous sulphate solutions	High Potential emissions of beryllium compounds during crushing and grinding
Possibility of reagent regeneration	Absent	Partially in the form of $(NH_4)_2FeF_6$
The volume of the liquid phase in the process	High	Average
Beryllium recovery, %	80	70
Commercial products	Beryllium metal, plaster	Beryllium metal

Disadvantages of the sulfuric-acid method: huge water consumption; multistage processing; necessity in high-temperature two-stage activation of concentrates; large amount of impurities passes into the target solution in the form of sulphates; necessity to produce intermediate highly-toxic beryllium compounds ($Be(OH)_2$).

As compared to the sulfuric-acid scheme, the fluoride processing does not include high-temperature, energy-intensive stages of concentrate activation, and there is a possibility to regenerate the fluorinating reagent in the form of ferrous cryolite (Na_3FeF_6) evolved from mother solutions of sodium fluoride which remained after separation of the amorphous $Be(OH)_2$. Disadvantages of the fluoride method: multistaging of the hydrometallurgical process, costly leaching reagent – Na_2SiF_6, high consumption of sodium hexafluorosilicate as a result of its intensive decomposition and volatilization of the produced SiF_4 from the reaction zone, necessity in a high-temperature stage of concentrate fluorination, necessity to produce highly toxic beryllium compounds – $Be(OH)_2$.

Today, only the sulfuric-acid method is commonly used in processing of concentrates, because it is more economically viable, environmentally sound, and allows producing a cleaner beryllium hydroxide.

2 Theory section

In order to improve the profitability and to reduce the risks at production of beryllium metal, a method of ammonium-fluoride processing of beryllium concentrates is being developed. The method is based on the interaction of the concentrate components with ammonium bifluoride (NH_4HF_2) at 210 °C, omitting stages of concentrate activation. The main component of the ore concentrate from the Ermakovskiy deposit is fluorite – CaF_2, which does not react with ammonium bifluoride and is separated at the filtration stage after water leaching of the sinter. The remaining components of the concentrate come into the following interactions [5-6]:

$$Be_2SiO_4 + 7NH_4HF_2 \rightarrow 2(NH_4)_2BeF_4 + (NH_4)_2SiF_6 + 4H_2O\uparrow + NH_3\uparrow, \quad (1)$$

$$Al_2O_3 + 6NH_4HF_2 \rightarrow 2(NH_4)_3AlF_6 + 3H_2O\uparrow, \quad (2)$$

$$Fe_2O_3 + 5NH_4HF_2 \rightarrow 2(NH_4)_2FeF_5 + 3H_2O\uparrow + NH_3\uparrow, \quad (3)$$

$$CaO + NH_4HF_2 \rightarrow CaF_2 + H_2O\uparrow + NH_3\uparrow, \quad (4)$$

$$SiO_2 + 3NH_4HF_2 \rightarrow (NH_4)_2SiF_6 + 2H_2O\uparrow + NH_3\uparrow, \quad (5)$$

With further aqueous leaching of the sinter the following compounds go into the solution: beryllium, silicon, partially iron and aluminum, excessive ammonium bifluoride. Fluorides of calcium, partially of aluminum and iron, remain in the insoluble precipitate.

The next stage of the process is the alkaline precipitation purification of the beryllium solution from impurities. Beryllium remains in the solution in the form of BeF_4^{2-} - ion, and impurities (the main impurity is silicon) precipitate as a result of hydrolysis under the following reactions:

$$(NH_4)_2FeF_5 + 3NH_4OH = Fe(OH)_3\downarrow + 5NH_4F \quad (6)$$

$$(NH_4)_3AlF_6 + 3NH_4OH = Al(OH)_3\downarrow + 6NH_4F \quad (7)$$

$$(NH_4)_2SiF_6 + 4NH_4OH = Si(OH)_4\downarrow + 6NH_4F \quad (8)$$

The purified solution $(NH_4)_2BeF_4$, omitting the stage of production of highly toxic $Be(OH)_2$, goes to crystallization. The formed crystals of ammonium fluoroberyllate are subjected to thermal decomposition to produce BeF_2 [5].

The aim of this research was the study the influence of silicon macroscales on the behavior of beryllium in fluoride solutions at an increased pH, as well as the establishment of the distribution of beryllium and silicon by phases in laboratory testing of the described ammonium-fluoride technology on model mixtures.

 DOI: 10.1051/matecconf/20168501009

3 Material and methods

The prepared model mixture, consisting of BeO and SiO_2, was combined with ammonium bifluoride taken with a 50 % excess relative to the stoichiometric amount. The load was placed into a laboratory shaft furnace, where the fluorination process of the model mixture took place during 1 hour at a temperature of 210 °C. The produced sinter was crushed and dissolved in 200 ml of water. The sinter dissolved completely. The pH level of the solution was equal to 3. A complete dissolution of the fluorinated product demonstrates the importance of the fluorination degree approximately equal to 100%.

Precipitation was carried out with an ammonia solution at room temperature; the pH value was gradually increased until a white precipitate. The pulp was settled at each pH unit for 1 hour. Precipitation was observed during the transition of pH level from 7 to 8. The produced pulp was filtered, and the clarified solution was brought up to pH = 9, the precipitate no longer formed. The precipitate after the filtration was pulped in 100 ml of water; the formed on the filter cake was washed with another 100 ml of water. The washed precipitate was removed together with the filter and transferred to a platinum cup. The precipitate was dried at 150 °C for 2 hours, then calcined at 700 °C.

Crystallization of $(NH_4)_2BeF_4$ from the filtrate was carried out until a gelatinous mass was obtained. Then, amorphous crystals were dried for 3 hours without heating using an air extraction system. Thermal decomposition of $(NH_4)_2BeF_4$ was carried out at 500 °C in a platinum cup.

The elemental analysis of liquid and solid samples was carried out using an atomic-emission spectrometer with an inductive coupled plasma iCAP 6300 Duo. To dissolve solid samples, the microwave system MARS 6 was used. A scandium solution with a concentration of 100 ppm was used as an internal standard. To calibrate the method, state standard reference samples (GSO) were used: for silicon – GSO 8934-2008, beryllium – GSO 7759-2000. In sample preparation the following acids were used: hydrochloric (GOST 14261-77), nitric (GOST 4461-77), and hydrofluoric (TU 2612-007-56853252-2010). The analysis was carried out under the following lines: for beryllium – 234.8; 249.4; 265.0 nm; silicon – 198.8; 251.6; 288.1 nm.

4 Results and discussion

As a result of the precipitation purification of beryllium fluoride solutions from silicon macrocsales, more than 21 % of beryllium removes together with a gelatinous precipitate of polysilicon acids. But, during washing of the produced cake with ordinary distilled water it is possible to recover 18.5% of beryllium into the solution, which indicates the absence of a large-scale hydrolysis of BeF_4^{2-} - ion at pH equal to 9 and a ratio of a free fluoride-ion to beryllium equal to Be:F = 1:6. Fluoroberyllate ion in the basic amount is drawn into a jelly-like precipitate together with moisture; as a result of alkali hydrolysis 2.7 % of beryllium is precipitated.

Approximately 2 % of silicon remains in the clarified solution at a pH level equal to 9.

An incomplete removal of silicon, as well as a very high pH of silicon hydrolysis (pH = 7), is associated with a large excess of fluoride-ion present in the solution. Part of silicon can be removed in the form of $(NH_4)_2SiF_6$ at the stage of thermal decomposition of $(NH_4)_2BeF_4$ (Fig. 1).

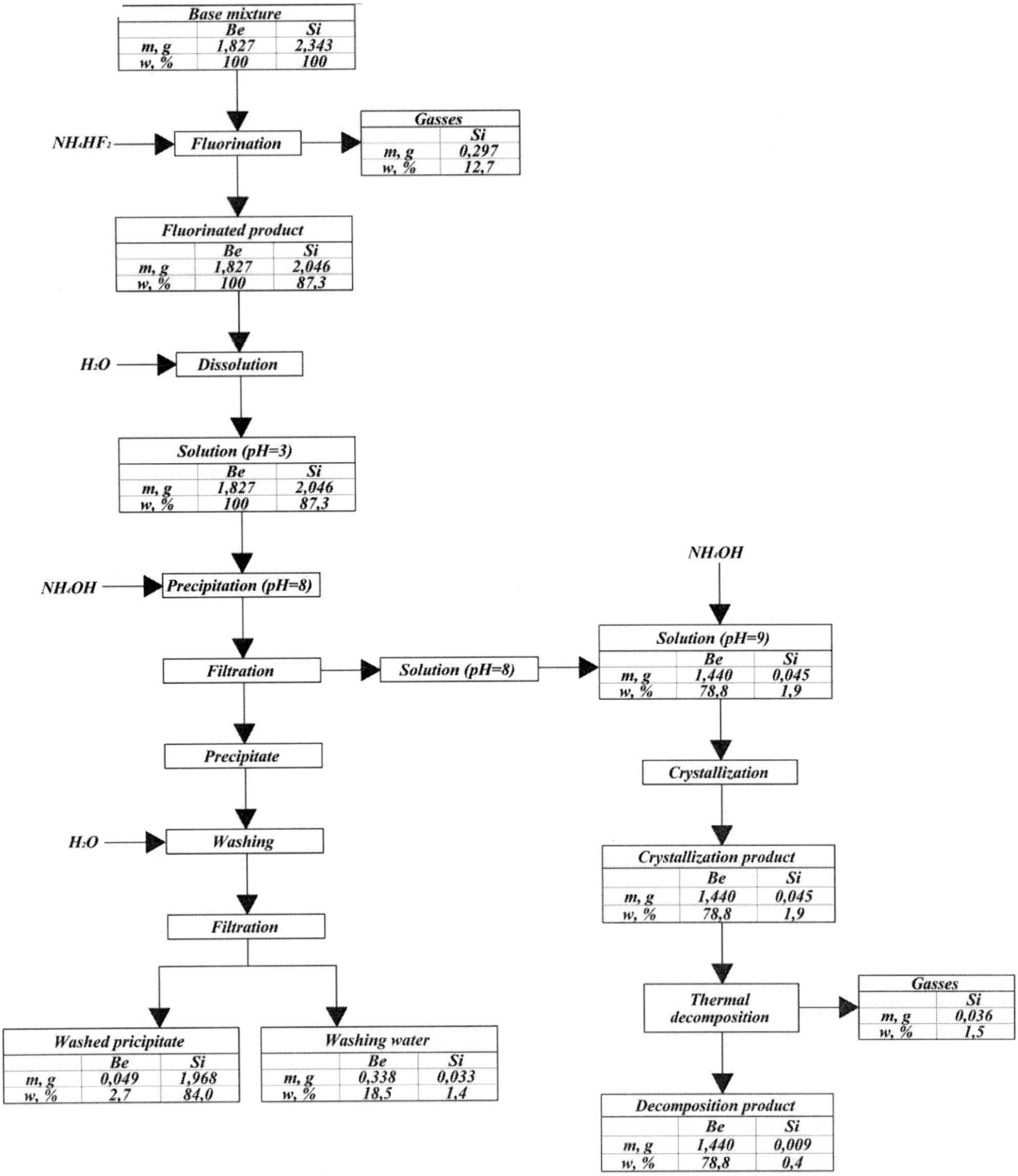

Figure 1. Distribution of beryllium and silicon by phases

5 Conclusion

1. Precipitation purification of beryllium fluoride solutions from silicon macroscales is feasible at room temperature (20-25 °C) and at a ratio Be:F = 1:6.
2. Co-precipitation of beryllium and silicon was not established. Approximately 18.5 % of beryllium removes together with the precipitate of polysilicic acids in the composition of the adsorbed by the cake moisture. During the purification 2.7 % of beryllium hydrolyzes and precipitates in the form of $Be(OH)_2$.
3. Due to a significant amount of fluoride-ion present in the productive solutions, the pH of hydrolysis SiF_6^{2-} - ion increases up to 7.
4. It was impossible to completely remove silicon at the stage of precipitation purification: 1.9 % of silicon remains in the solution of ammonium fluoroberyllate. It is possible to remove the remaining portion of silicon at the stage of thermal decomposition in the form of $(NH_4)_2SiF_6$.
5. The purity of beryllium fluoride produced during the study was 99.7 %. The amount of SiO_2 in the product was 0.3 %.

References

1. D.D. Thorat, B.M. Tripathi, D. Sathiyamoorthy. Extraction of beryllium from Indian beryl by ammonium hydrofluoride // Hydrometallurgy, 109 (2011). – P. 18-22.
2. E.E. Zaki, Z.H. Ismail, J.A. Daoud, H.F. Aly. Extraction equilibrium of beryllium and aluminium and recovery of beryllium from Egyptian beryl solution using CYANEX 921 // Hydrometallurgy, 80 (2005). – P. 221-231.
3. D.W. White, J. E. Burke. The Metall Beryllium. American Society for Metals, 1955 – 703 p.
4. Samoylov V.I, Borsuk A.N. Methods for co-processing of phenacite, bertrandite, and beryl in beryllium hydrometallurgy, Media Alliance, Ust-Kamenogorsk, 2006.
5. V.S. Rimkevich, A.A. Pushkin, A.A. Malovitskii, Yu.N. Dem'yanova, I.V. Girenko. Fluoride processing of non-bauxite ores // Russian Journal of Applied Chemistry. 2009. V. 82. № 1 – P. 6-11.
6. Andreev A.A., Dyachenko A.N., Kraydenko R.I., *Russian Journal of Applied Chemistry*, **81** (2), (2008), 178-182.
7. Silina G.F., Zarembo Yu.I., Bertina L.E. Beryllium. Chemical technology and metallurgy, Atomizdat, Moscow, 1960.

MATEC Web of Conferences **85**, 01010 (2016) DOI: 10.1051/matecconf/20168501010
CCT 2016

CALCULATION METHOD TO DETERMINE THE GROUP COMPOSITION OF VACUUM DISTILLATE WITH HIGH CONTENT OF SATURATED HYDROCARBONS

Galina Nazarova[1,*], *Elena* Ivashkina[1], *Tatyana* Shafran[1], and *Nataliya* Belinskaya[1]

[1]Tomsk Polytechnic University, Lenin av. 30, Tomsk, 634050, Russia

Abstract. Calculation method to determine the group composition of the heavy fraction of vacuum distillate with high content of saturated hydrocarbons, obtained by vacuum distillation of the residue from the West Siberian oil with subsequent hydrotreating, are given in this research. The method is built on the basis of calculation the physico-chemical characteristics and the group composition of vacuum distillate according to the fractional composition and density considering with high content of saturated hydrocarbons in the fraction. Calculation method allows to determine the content of paraffinic, naphthenic, aromatic hydrocarbons and the resins in vacuum distillate with high accuracy and can be used in refineries for rapid determination of the group composition of vacuum distillate.

1 Introduction

Heavy petroleum fractions are the multicomponent mixture with a predominance of high molecular weight hydrocarbons, resins and asphaltenes, carbenes, carboids, organic compounds with heavy metals (V, Ni, Fe, Mn, etc.) and heteroatoms (S, O, N, F) [1,2]. Hydrocarbon components of heavy petroleum fractions are divided into the three main types: paraffinic, aromatic and naphthenic structures.

A group composition of petroleum fractions during an oil refining processes is the most important factor influencing in the yield and composition of products, and an efficiency of the catalysts. Information about the group composition of petroleum fractions is required during a design of industrial installations, optimization of oil refining processes, prediction of the composition, yield and quality of products from industrial units and the development and improvement of highly efficient catalyst compositions, evaluating the deactivationetc.

In the condition of intensive development of advanced petroleum refining processes and catalysts, there is much tension around the issue of optimization of industrial unit by experimental ways and mathematical modeling [3-6]. Mathematical models should be based on the physical-chemical patterns and taken into account the constantly changing group composition of heavy petroleum fractions to adequately describing the processes. It is known that the regular analysis of the fractional composition of heavy fractions do not reliably reflect the group composition.

There are different calculation methods to determine the physico-chemical characteristics, structural-group and group composition of petroleum fractions [7-10]. Such methods are based, for the most part, on the correlation dependencies depending on the boiling point and molecular weight of the fractions.

N-d-M method was proposed in 1951 for the calculation of structural-group composition of petroleum fractions according to the refractive index, density and molecular weight. The method allows to determine the carbon content in naphthenic, aromatic and paraffinic structures and average number of naphthenic and aromatic rings [11]. TOTAL correlation allows to determine the weight content of aromatic hydrocarbons in fraction on the basis of the fractional composition, density, and aniline point [12].

Calculation method to determine the group composition of petroleum fractions based on the viscosity gravity constant and refractivity intercept was proposed in 1980. Laboratory results on the fractional composition, density and kinematic viscosity are required as input data for the calculation of the group composition according to the API method. It allows calculating the group composition for light and heavy petroleum fractions with molecular weight in the range of 70–200 and 200–600 respectively [13].

Calculation method to determination of the group composition of petroleum fractions on the basis of laboratory results on the fractional composition, density, refractivity intercept and ratio C/H was proposed in 1986 [14]. In [15] authors offered to calculate the group composition of petroleum fractions on the basis of density and refractive index. In [16], a new method was proposed for correlations elaboration based on the use of pure hydrocarbons specific data using normal boiling point, density, refractive index and carbon to hydrogen weight ratio.

According to the high content of saturated hydrocarbons in vacuum distillate from West Siberian oil, the molecular weight and refractive index should be calculated considering with characteristic factors of vacuum distillate.

Therefore, the aim of the research is the development of calculation method to determine the group composition of vacuum distillate based on the data of the fractional composition and density considering with high content of saturated hydrocarbons.

2 Research methods

Fractional composition and density of vacuum distillate was determined according to ASTM D 2887 and GOST 3900.

Liquid adsorption chromatographic separation on silica gel ASA with a grain size of 0.2 - 0.3 mm was performed according to VNII NP algorithm with the aim to determine the group composition of vacuum distillate. Structural-group composition of paraffin-naphthenic fractions was determined by the n-d-m-method, for aromatic hydrocarbon fraction by the Hazelvude-method using refractometer IRF-22.

The research in the determination of molecular weight vacuum distillate was performed by device CRYETTEWR, determining the freezing point of the samples (accuracy of the cryoscope temperature measurement is ± (2-5) millidegree Celsius).

3 Results and discussion

3.1 Development of calculation method for determining the group composition of heavy fraction of vacuum distillate

Initial data for calculation of the group composition of vacuum distillate with high content of saturated hydrocarbons are fractional composition and density. According to experimental data the density of vacuum distillate is changed in the range of 0.8990–0.9055 g/cm3, Boiling temperature is changed in the range of 282–306 for initial boiling point, of 360–373 for 10 % by volume, of 435–445 for 50 % by volume, and of 515-540 for 90 % by volume.

Laboratory results for determining the group and structural-group composition and molecular weight of vacuum distillate from West Siberian oil is presented in Table 1.

Table 1. The group and structural-group composition and molecular weight of vacuum distillate

Characteristics	**Values**		
Molecular weight, g/mol	382.43		
Group composition, wt.%	Saturated hydrocarbons	Aromatic hydrocarbons	Resins
	61.20	35.57	3.23
Content of carbon in:	n-d-m-method	Hazelvude-method	-
- paraffinic structures	66.14	16.76	-
- naphthenic structures	31.27	55.73	-
- aromatic structures	2.59	27.52	-
Number of aromatic rings	0.14	1.40	-
Number of naphthenic rings	2.05	3.90	-
Total number of rings	2.19	5.30	-

According to Table 1, vacuum distillate are characterized by high content of saturated hydrocarbons (61.2 wt. %), molecular weight is 382.43 g/mol. The content of carbon in paraffinic and naphthenic structures is 66.14 % and 31.27 % respectively, the average number of rings is 2.05 % in paraffin-naphthenic fraction

According to the analysis of aromatic concentrates by the Hazelvude-method the aromatic structure is presented by hybrid structures. In aromatic fractions isolated from vacuum distillate the average number of naphthenic rings (3.9 units) is more than the number of aromatic rings (1.4 units), the total content of rings is 5.3 units. Carbon content in paraffinic fragments of aromatic concentrate indicates the presence of alkyl substituted of aromatic hydrocarbons and naphthenes; the content of paraffins fragments – 16.76 %.

Method API [13] is accepted as the basis for the calculations. The following are the main stages of the calculation to determine the group composition of vacuum distillate according to the developed method.

1. Definition of average volume boiling point (T_{vol}, °F) and molecular boiling point (T_m, °F) according to the the results of fractional composition of vacuum distillate [10]:

$$T_{vol} = (T_{10\%} + T_{50\%} + T_{90\%}) / 3 \quad (1)$$

$$T_m = T_{vol} - [(T_{90\%} - T_{10\%}) / (170+0.075 \cdot T_{vol}) + 1.5)] + 2 \quad (2)$$

here $T_{10\%}$; $T_{50\%}$; $T_{90\%}$, °C – boiling point of 10%, 50% and 90% volume.

3. Calculation of characteristic factor by the UOP method (K_{UOP}):

$$K_{UOP} = [(T_{vol} + 460) / p]^{1/3} \quad (3)$$

here p – density of vacuum distillate at 15^0C, g/cm^3.

4. Calculation of molecular weight (MW) by refined Voinov equation depending on the characteristic factor and molecular boiling point of fraction (°C) [17]:

$$MW = (7 \cdot K_{UOP} - 21.5) + (0.76 - 0.04 \cdot K_{UOP}) \cdot T_m + (0.0003 \cdot K_{UOP} - 0.00245) \cdot T_m^2 \quad (4)$$

5. Defined the refractive index at 20°C (n^{20}) according to R. Hersch and M. Fenske equation connecting the molecular weight and molecular boiling point of fraction (°C):

$$n^{20} = 2.1500 - 10^{(\lg MW - 1.9939436 - 0.0019764 \cdot Tm)} \quad (5)$$

6. Calculation of refractivity intercept (Ri) depending on refractive index and density and kinematic viscosity at 100°C (υ_{100}, St) from the experimental dependence:

$$Ri = n^{20} - p / 2 \quad (6)$$

$$\upsilon_{100} = 0.1488 \cdot exp^{0.0087 \cdot Tvol} \quad (7)$$

8. Determination of viscosity gravity constant (VGC) depending on the density and kinematic viscosity of fraction:

$$VGC = [(p - 0.24 - 0.022 \cdot lg(\upsilon_{100} - 35.5)] / 0.755 \quad (8)$$

υ_{100} – kinematic viscosity fraction at 100^{0}C, SSU.

9. Compilation of the three equations system for calculation of the group composition of of the vacuum distillate from the West Siberian oil is the final stage. The first equation is the molar balance of the group composition of vacuum distillate, the second and third equations are additive law for refractivity intercept and viscosity gravity constant.

$$\begin{cases} x_p + x_N + x_{A+R} = 1 \\ 1.0482 \cdot x_p + 1.0359 \cdot x_N + 1.0893 \cdot x_{A+R} = Ri \\ 0.7406 \cdot x_p + 0.9289 \cdot x_N + 0.9505 \cdot x_{A+R} = VGC \end{cases} \quad (10)$$

Here x_P, x_N, x_{A+R} are the molar concentration of paraffinic, naphthenic and aromatic hydrocarbons with resins; 1.0822; 1.0359; 1.0893 and 0.7406, 0.9289, 0.9505 are the average refractivity intercept and viscosity gravity constant for paraffinic, naphthenic and aromatic hydrocarbons respectively.

Equation system is solved by Cramer method relatively x_P, x_N and x_{A+R}. and the expressions for the calculation of the molar concentration of the hydrocarbon groups is obtained:

$$\begin{cases} x_p = 2.6386 + 2.09302 \cdot Ri - 5.1744 \cdot VGC, \\ x_N = 18.3721 - 20.3391 \cdot Ri - 3.9826 \cdot VGC, \\ x_{A+R} = -20.009690 + 18.2461 \cdot Ri - 1.1919 \cdot VGC. \end{cases} \quad (10)$$

According to experimental data, content of aromatic hydrocarbons is 90.25% from the total content of aromatic hydrocarbons with resins.

3.2 Adequacy of calculation of the group composition of vacuum distillate with high content of saturated hydrocarbons

Adequacy evaluation of the calculation method is performed by comparison between calculated and experimental data of the group composition of vacuum distillate (wt.%) obtained in the laboratory on the device "GRADIENT-M". The content of resins, saturated and aromatic hydrocarbons in vacuum distillate depending on the average volume boiling point are presented in Fig. 1-3.

According to Fig. 1-3, the content of saturated hydrocarbons decreases and content of aromatic hydrocarbons and resins rises with increasing the average volume boiling point of vacuum distillate. Error between calculation and experimental data by the group composition in vacuum distillate is not more than 5.0 wt.%.

DOI: 10.1051/matecconf/20168501010

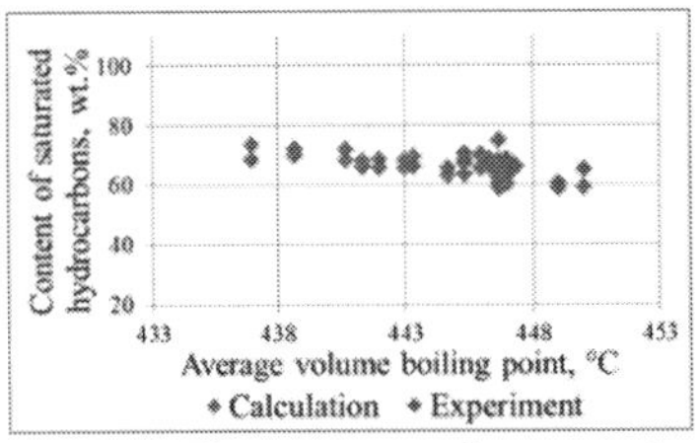

Fig. 1. The content of saturated hydrocarbons depending on the average volume boiling point.

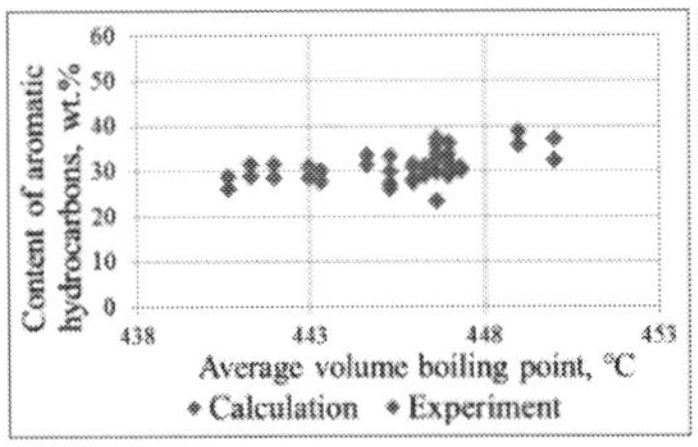

Fig. 2. The content of aromatic hydrocarbons depending on the average volume boiling point.

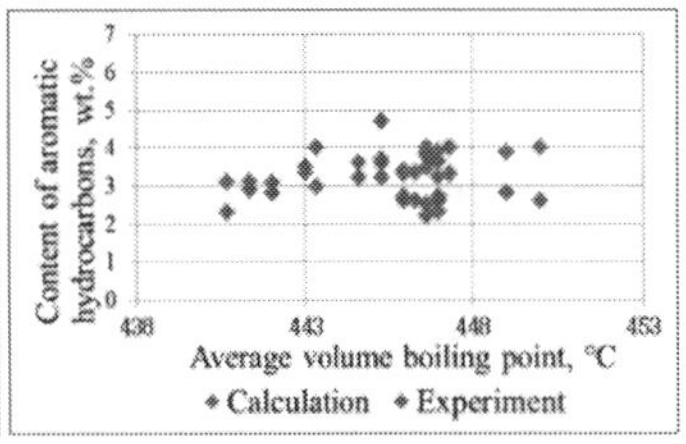

Fig. 3. The content of resins depending on the average volume boiling point.

4 Conclusion

The presented method allows to calculate with high accuracy (calculation error is not more than 5%) the group composition of vacuum distillate with high content of saturated hydrocarbons (280-575°S fraction) on the basis of the fractional composition and density. The developed method takes into account the high content of saturated hydrocarbons and allows to calculate the content of saturated (paraffinic and naphthenic hydrocarbons), aromatic hydrocarbons and resins in vacuum distillate. Calculation method can be used in oil refineries for rapid determination of the group composition of vacuum distillate with high content of saturated hydrocarbons and in the complex with mathematical methods to improve the operating efficiency of industrial units of advanced petroleum processes and catalysts. Program realization of calculation methods is made using Delphi 7.0 software.

References

1. C. A. Hughey, R. P. Rodgers, A. G. Marshall, Anal. Chem. **74**, 4145-4149 (2002)
2. N. Aske, Harald Kallevik, J. Sjoblom **15**, 1304-1312 (2001)
3. V. A. Chuzlov, K. V. Molotov, Petroleum and Coal. **58(1)**, 47-55. (2016)
4. V. A. Chuzlov, E. D.Ivanchina, N. V. Chekantsev, K. V. Molotov, Procedia Engineering. **113**, 131-137 (2015)
5. N. S. Belinskaya, E. D. Ivanchina, E. N. Ivashkina, V. A. Chuzlov, S. A. Faleev Procedia Engineering. **113**, 68-72 (2015)
6. N.S. Belinskaya, E.V. Frantsina, E.D. Ivanchina, N.V. Popova, I.V., Zyryanova, E.V. Averyanova, IOP Conf. Series: Earth and Environmental Science **43**, 1-6, (2016)
7. A. I. Levashova, N. V. Usheva, O. E. Moizes, I. Fedjaeva, Proceedings: KORUS 2004. **2**, 63-64 (2010)
8. T. A. Albahri, Fluid Phase Equilibria, 315, 113–125 (2012)
9. M. R. Riazi, T. E. Daubert, Ind. Eng. Chem. Res. **26**, 755-759 (1987)
10. L. P. Guilyazetdinov, Ind. Eng. Chem. Res. **34,** 1352-1363 (1995)
11. R. Sadeghbeigi,Elsevier Inc. 361 (2012)
12. ASTM, Standard Test Method for Calculation of Carbon Distribution and Structural Group Analysis of Petroleum Oils by the n-d-M Method, ASTM Standart D 3238-5, ASTM, West Conshohocken (1985)
13. H.Dhulesia, Oil Gas J. **84,** 51-54 (1986)
14. M. R. Riazi, T. E. Daubert, Ind. Eng. Chem. Process Des. Dev. **19**, 289-294 (1980)
15. M. R. Riazi, T. E. Daubert, Ind. Eng. Chem. Process Des. Dev. **25**, 1109-1115 (1986)
16. C. A. Nwadinigwe, K. A. Okoroji, Fuel **69**, 340-343 (1990)
17. D. El-Hadi, M. Bezzina, Fuel **84**, 611–617 (2005)
18. Yu.G. Kirsanov, Ural. University – Ekaterinburg, 136 (2014)

Engineering-geological conditions of surface mining in Ainskoe gold field and pit wall stability

V.E. Ol'khovatenko[1] and E.G. Pryanishnikov[1]

[1]*Tomsk State University of Architecture and Building, Tomsk, Russian Federarion*

Abstract. The paper presents the estimation of the engineering-geological conditions of the surface mining in Ainskoe gold field. The mechanical-and-physical properties of the rock are described in this paper, and the stability calculations are provided for the pit walls. Recommendations are given for the geoecological safety measures during the surface mining.

1. Introduction

The exploration of mineral deposits using the surface mining poses a lot of problems related to the geodynamical disbalance of the environment, contamination of the atmosphere, surface and ground waters. The surface mining has an effect on the geological environment resulting in the modification of the stress-strain state of subsurface rocks, instability of pit walls and the development of geological hazards. In this connection, a predictive estimate of the engineering-geological conditions is very important for the surface mining and the stability of pit walls.

The Ainskoe gold field is located in the South part of Urup Island which is the integral part of the Kuril Islands chain. Its natural and climatic factors affecting the surface mining include the climate, geological structure, composition and mechanical-and-physical properties of rocks.

2. Climate

The climate on the Kuril Islands is influenced by the Pacific Ocean and the Sea of Okhotsk. The climate on the Kuril Islands is maritime, generally cold with stormy but rather soft winters and short and cold summers. It is also characterized by strong winds, precipitations and fogs observed all the year round.

Urup Island locates in the climatic region of the Kuril Islands with the average annual precipitation of 1418 mm. Winter is rather warm and windy. The transition of the average daily temperature across 0 °C towards negative values is observed at the end of November and the beginning of December. February is the coldest month in the year with the average temperature range of –5–7 °C. Thaw periods and cloudiness are typical for the Urup climate. The spring on Urup is cold, rainy and windy. The reverse transition towards the positive temperatures starts in April. A strong snow blanket prevents the soil from freezing.

The average temperature in January is –4,8 °C, while in August it is 8,4 °C. The minimum and maximum of the annual absolute temperature come to –19 °C and 32 °C, respectively. The average annual temperature on Urup is 2,2°C. According to the long-term observations of the hydrometeorological services, there are 20–29 foggy days per month within the period from May to September, and only in October this number decreases down to 10 days; the number of rainy days per month comes to 11–17. The average monthly temperature in summer does not exceed 10^{0}C. Autumn begins mildly and is the best season characterized by many clear days. The amount of precipitations is reduced. The average monthly temperature is 5^{0}C. The autumn period lasts for 2–2,5 months and covers late September, October and November.

The amount of precipitations during the warm period (April–October) achieves 866 mm, while in the cold period (November–March) it comes to 552 mm. The maximum daily rain precipitation amounts to 221 mm. The number of snow-stormy days is 67. Northwestern and south-easter winds prevail on Urup having the average annual velocity of 7,8 m/s. Storm and hurricane winds can achieve 58 m/s leading to the building destruction, while narrow beaches and benches are exposed to hurricane tides. In southern Urup, the average water temperature in August is not over 9^{0}C, while in February it is 0,5^{0}C.

3. Geological structure of Ainskoe gold field

The following genetic stratigraphic sequence of the rock is involved in the geological structure of Ainskoe gold field:

- Modern quaternary deposits (Q_{III}).
- Volcanic-sedimentary and sedimentary deposits of Rybakov formation (N_3rb).
- Subvolcanic deposits of Rybakov formation of volcanic andesitic complex (eN_3rb).

[1] Valentin Ol'khovatenko: igg@tsuab.ru

- Hydrothermally altered rocks.

Quaternary deposits are represented by clayey soils and sandy cobble. They are weak and have no an effect on the pit wall stability. *Deposits of Rybakov formation* are represented by tuffites characterized by horizontal and oblique (7-10%) deposits. *Subvolcanic deposits of Rybakov formation of volcanic andesitic complex* contain andesites, basalts, diorites, shastalites. *Hydrothermally altered rocks* include quarzites, quartz metasomatites, argillizated rocks and fault breccia. In the tectonic structure of Cainozoic formations, the faulting plays the important part, while folding is present insignificantly. Volcanic-sedimentary and sedimentary deposits of Rybakov formation have a flat monoclinal folding with the north-eastern course that is a favorable factor for the surface mining. The predictive estimate of the pit wall stability of all engineering-geological types of rock determined in the open-pit was accompanied by the detailed investigations of their mechanical-and-physical properties, construction of strength certificates, determination of standard and design values of strength properties.

4. Determination of strength properties

The ultimate uniaxial compression strength of rocks is detected by GOST 21153.2-84 using the hydraulic press IP-1000 with the automatic setting of the loading rate and fracture load recording. Compressive tests were carried out for the cylindrical rock samples having 47 mm diameter. Their end surfaces were polished on the surface-grinding machine. At the same time, the size, cross-sectional area, and the weight of the samples were identified. Compressive tests included the automatic recording of the facture force, kN. The ultimate uniaxial compression strength can be obtained from $\sigma_c = K_h \frac{P}{S} \cdot 10$, where σ_c is the ultimate uniaxial compression strength, MPa; P is the fracture force, kN; S is the area of fracture, cm^2; K_h is the dimensionless factor of the sample height equaling to 1,0 at the diameter ratio m = 2+0,5. In this work, the dimensionless coefficient is taken from GOST 21153. 2-84.

The data obtained at the ultimate uniaxial compression strength of rocks in the Ainskoe gold field are presented in Table 1.

According to GOST 21153.3-85, the uniaxial tensile strength can be obtained from $\sigma_t = \frac{P}{S} \cdot 10\ \text{MPa}$, where P is the fracture force, kN; S is the area of fracture equaling to the product of the sample diameter by length, cm^2.

The uniaxial tensile strength for the arbitrary-shaped samples is obtained from $\sigma_t = 7{,}5 \frac{P}{S} \cdot K$, where P is the fracture force, kN; S is the cross-sectional area of the sample, cm^2; K is the dimensionless scaling factor equaling to 1,00 at $S = (15\pm)$ cm^2. For other S values, the scaling factor K is taken from GOST 21153.3-85.

The strength certificates are constructed in accordance with the results of compressive strength testing as well as the detection of the rock strength properties.

5. Mechanical-and-physical properties

The investigation of such strength properties as the angle of internal friction and specific cohesion of soil is of paramount importance in evaluating the engineering-geological conditions of the mineral deposits. These properties are required to calculate the pit wall stability.

The research investigations allow classifying the engineering-geological conditions of rocks in Ainskoe gold field that includes 13 rock types the mechanical-and-physical properties of which are given in Table 1.

Table 1. Mechanical-and-physical properties of rocks in Ainskoe gold field

Types of rock	Mechanical-and-physical properties				
	ρ, g/cm^3	σ_c, MPa	σ_t, MPa	φ, degree	Cohesion, MPa
Psephitic tuffites	2,00-2,50 2,21	1,38-26,80 8,23	0,23-4,50 1,39	22-35 30	0,37-8,00 2,28
Psammitic tuffites	2,05-2,27 2,17	1,80-36,0 11,38	0,30-6,00 1,89	27-42 33	0,50-2,25 1,58
Silt-rich psammitic tuffites	2,37-2,68 2,52	1,79-36,4 19,09	0,29-6,07 3,18	34-36 35	0,42-8,70 4,56
Argillaceous tuffites	2,30-2,36 2,34	3,96-16,88 11,33	0,66-2,81 1,88	30-38 32	0,75-5,00 3,18
Argillizated rocks in tuffites	2,07-2,59 2,24	1,48-21,0 6,31	0,24-5,32 1,32	25-45 33	0,60-5,80 1,72
Argillizated rocks in psephitic tuffites	2,18-2,42 2,32	0,72-6,19 3,25	0,12-1,26 0,59	29-36 32	0,18-1,75 0,86

DOI: 10.1051/matecconf/20168501011

Argillizated rocks in psammitic tuffites	2,36	0,72-5,33 3,02	0,12-0,75 0,43	32-34 33	0,18-1,50 0,84
Advanced argillizated rocks	1,94-2,53 2,25	2,60-21,50 8,76	0,43-3,50 1,47	30-43 35	0,60-4,30 2,00
Andesites	2,39-2,70 2,26	8,13-27,81 18,68	2,32-8,61 4,19	34-40 35	2,00-7,00 4,85
Shastalites	2,31-2,40 2,36	2,10-6,72 3,99	0,35-1,12 0,79	27-33 31	0,62-1,75 1,04
Quartz metasomatites	2,18-2,32 2,24	2,56-22,37 12,69	0,43-3,71 2,20	38-40 39	0,55-6,00 2,68
Fault breccia	2,47-2,63 2,55	2,30-6,60 4,45	0,60-1,10 0,85	39	0,50-1,50 1,0
Secondary quartzites	2,12-2,25 2,18	14,7-24,10 19,40	2,40-4,00 3,20	35	4,00-7,00 5,50

According to Table 1, secondary quartzites are the strongest, i.e. their uniaxial compression strength ranges from 14,7 to 24,10 MPa at its average value of 19,40 MPa. The strength properties of andesites are close to that of secondary quartzites, namely: 18,68 MPa uniaxial compression strength; 35^0 angle of internal friction; and 4,85 MPa specific cohesion. Among other tuffites, silt-rich psammitic tuffites possess the highest strength properties, the uniaxial compression strength of which is 19,09 MPa; 35^0 angle of internal friction; and 4,56 MPa specific cohesion. Argillizated rocks in psammitic tuffites are characterized by the lowest strength properties, uniaxial compression strength of which does not exceed 5,33 MPa, while its average value of 3,02 MPa. Also, the low uniaxial compression strength of 3,25 MPa is characteristic to psephitic tuffites.

Thus, Table 1 shows that silt-rich psammitic tuffites, andesites and secondary quartzites are the most stable in the pit walls since they possess the highest strength. Less stable are argillizated rocks in psephitic and psammitic tuffites, while psammitic and argillaceous tuffites and advanced argillizated rocks take the intermediate position. It is worth noting that the strength properties of the rock mass differ from that in a rock. Therefore, in calculating the pit wall stability, the reduction factor of the structural rock cohesion is introduced in the design value of specific cohesion. In this work, the reduction factor is calculated using the Fisenko method [3].

6. Pit wall stability analysis

The stability analysis of the pit walls is a complex engineering problem connected with the safety of production mining.

In addition, the rationale for the optimum angle of the pit wall defines not only the safety of mining but also its efficiency and the reduction of areas intended for the production mining. The latter defines one of the most important problem such as the sustainable land use.

The computational methods of evaluation of the pit wall stability are based on the limit equilibrium method illustrated in the works of Demin, Ol'khovatenko, Fisenko [1–3] and others. The main quantitative parameter used for the local evaluation and prediction of the slope stability is the stability factor represented by the ratio between the sum of confining and shearing forces acting on the surface of the expected landslide. Taking the engineering-geological conditions of the gold field, a logarithmic spiral is used as a computational method. This method is based on the limit equilibrium method, while the radii with the minimum stability factor are computed for constructing the sliding surface. The design parameters of the rock shear resistance are accepted for each estimated section as weighted average values of the mass computed by the exploratory hole columns typical for each pit wall.

The investigations of mechanical-and-physical properties of rocks allow the determination of their estimated properties that, in turn, are used to compute the pit wall stability. Since the Ainskoe gold field is located in seismic area with the earthquake intensity of 9, the pit wall analysis is carried out accounting for seismic forces. Seismic forces induce the additional horizontal and vertical loads in the rock mass. The contribution of the vertical components to the seismic force is rather small and can be neglected in the stability analysis.

The direction of the horizontal load corresponds to that of the expected landslide, while its value for each slip block equals to the product of the slip block weight by the seismicity coefficient that depends on the estimated seismicity of the Ainskoe gold field.

The earthquake intensity of the Ainskoe gold field is 8 that is taken into account during the stability analysis of the pit walls (Table 2).

Table 2. Results of pit wall stability analysis

E.L.*	Well	Pit walls	Weighted design parameters			Stability factor	
			υ; kN/m^3	φ, degree	Cohesion, kPa	Without seismicity	With seismicity

DOI: 10.1051/matecconf/20168501011

82	281	Right	23,8	29,7	20,6	1,59	1,23
	282	Left	23,2	31,8	25,5	1,62	1,25
90	215	Right	22,1	29,7	19,6	1,92	1,41
	225	Left	28,6	28,6	18,3	1,60	1,21

* Exploration line

According to this Table, the stability factor for the right wall (exploration line 82, well 281) is 1,59 and for the left wall it is 1,62 without accounting for the seismicity coefficient. The stability factors for the right and left walls (exploration line 90) are respectively 1,92 and 1,60. With the account for the seismicity coefficient, the stability factor decreases down to 1,21 for the left wall in the exploration line 90 and 1,23 for the right wall in the exploration line 82. The stability factor for the right wall in the exploration line 90 is 1,41. The obtained stability factors are higher than their standard values. Thus, the Ainskoe gold field will be provided with the pit wall stability for the whole period of its mining.

7. Conclusions

Geological investigations carried out for the Ainskoe gold field showed that the surface mining depends on the geological structure, composition, bedding and mechanical-and-physical properties of rocks.

The Ainskoe gold field is rich in volcanic-sedimentary and sedimentary deposits of Rybakov formation, subvolcanic deposits of Rybakov formation of volcanic andesitic complex, and hydrothermally altered rocks. These rocks differ by their litho-mineralogical composition and strength properties as well. Test results showed that silt-rich psammitic tuffites, andesites and quartz metasomatites are the strongest rocks having 18–19 MPa uniaxial compression strength. Argillizated rocks in psammitic and psephitic tuffites possess the lowest strength properties, the uniaxial compression strength of which does not exceed 6,19 MPa.

The stability of pit walls mostly depends on the natural and climatic factors, geological structure of the gold field and mechanical-and-physical properties of rocks. The seismicity of the area under research also has an effect on it that should be accounted during the surface mining.

It is advisable to monitor the geological conditions of the Ainskoe gold field so as to detect the changes in the stress-strain state of rocks in the pit walls and prevent the development of geological hazards. This will assist in the elimination of the non-equilibrium conditions of the natural and engineering system during the surface mining.

References

1. A.M. Demin. *landslide in pits: analysis and prediction*, (Moscow: GEOS) 79, (2009) (in Russian)
2. V.E. Ol'khovatenko. *Engineering geology of coal deposits of Kuznetsk Basin,* (Tomsk: TSUAB Publ.) 150, (2014) (in Russian)
3. G.L. Fisenko. *Stability of pit walls,* 2nd edition (Moscow: Nedra) 376, (1965) (in Russian)

MATEC Web of Conferences 85, 01012 (2016) DOI: 10.1051/matecconf/20168501012
CCT 2016

Determination of Au, Pt, Pd in gold ore mineral raw materials by stripping voltammetry

N.A. Kolpakova[1], Y. A. Oskina[1a], S.M. Panova[1], A. Ya. Pshenichkin[1]
[1]Tomsk Polytechnic University, 634050 Tomsk, Russia

Abstract. The paper considers the possibilities of use of the method of stripping voltammetry for finding platinum metals in mineral gold and ore raw material. A review of new options of electro-concentration of platinum metals on the surface of graphite electrode with the following sediment electro-oxidation and receipt of an analytical signal is presented: platinum finding was carried out by picks of selective electro-oxidation of iridium from intermetallic compound with platinum; gold finding was carried out by picks of gold electro-oxidation on the surface of graphite electrode modified by bismuth; palladium finding was performed by picks of palladium electro-oxidation on the surface of graphite electrode. 1M HCL solution was selected as a supporting electrolyte. Gold and hydrogen elimination on the process of palladium electro-oxidation was performed by means of UV irradiation of solution in the process of electro-concentration of palladium sediment. Gold, platinum and palladium determination was carried out in mineral gold and ore raw material of Verkhneamylskiy gold and ore district.

1 Introduction

Despite extremely low content of precious metals in natural site, big investments are made in exploration of new gold and ore fields. These costs are covered not only by a high price of precious metals but by their priority use in the number of engineering branches, which are of defense importance in particular. Identification of new raw material sources and reliability of estimation of gold, platinum and palladium reserves are closely related to degree of sensitiveness and accuracy of analytical methods as well as to economic indicators of their implementation. State of analytical chemistry of precious metals is reflected in reviews [1-6]. It is noted in these papers that extremely underdeveloped basis of analytical methods of quantitative evaluation of potential platinoid natural sources is the greatest obstacle of their research. Content of platinum metals in industrial raw material fluctuates from 0.1 mg/t up to 50.000 mg/t. To determine such low contents of platinum methods, highly instrumental procedures of analysis are required. As is clear from these papers, all the existing methods of analysis demand effective methods of specimen deterioration and techniques for separation of platinum metals from sample matrix.

The method of stripping voltammetry (SV) is proposed hereby as a method of analysis which can be used to determine platinum, palladium and gold in mineral raw material. This method refers to sensitive instrumental procedures of analysis which allows to determine elements after their concentration on the surface of an electrode with the following electro-oxidation of sediment. To determine gold and palladium, it is enough to carry out electrolytic reduction of their ions to the surface of a graphite electrode. The following process of electro-oxidation of sediment allows to receive analytical signal and to determine content of the element in a specimen. There are certain difficulties during the process of determination of platinum by means of the method of stripping voltammetry. The process of electro-oxidation of platinum sediment proceeds at potentials bigger than 1 *B* and is overlapped by the process of oxygen separation from water. Determination of platinum ions (IV) by the SV method is carried out by means of electro-deposition of these metals into an alloy with a less precious metal (activator metal) with the following selective electro-oxidation of the activator metal. There have been recorded instances of SV determination of platinum ions (IV) by picks of selective electro-oxidation of mercury [7] or indium [8, 9] from intermetallic compounds with platinum.

The purpose of this work is to estimate the content of gold, platinum and palladium in gold and ore raw material using the method of stripping voltammetry.

2 Experimental

All research was conducted with use of voltammetry analyzers TA–4 («Research and Production Enterprise "Tom'Analit"» LLC, Tomsk) as well as with a personal computer. Quartz glasses of 20 sm^3 volume were used as an

[a] Corresponding author: iuoskina@mail.ru

electrolyzer. All the research was conducted in three-electrode cell. Graphite electrode (GE) impregnated with polyethylene served as an indicator electrode; chloride and argental electrode (ch.a.e.) filled with saturated KCl solution served as a reference electrode; and platinum electrode was an auxiliary one. Electro-deposition of platinum was conducted from solutions containing ions of indium (III). Determination of platinum was performed by a pick of selective electro-oxidation of indium from intermetallic compound with platinum [8, 9]. To determine gold, a graphite electrode modified by bismuth was used [10]. Electrochemical electrode surface cleaning was carried out either in a supporting electrolyte during a minute with potential +1.2 *V* or mechanically by rubbing an electrode against filter paper. Electrode surface was renewed after each measurement. Solution mixing during electrolysis was carried out automatically by vibration of a working electrode, which is provided by the used analyzers. All the research was conducted with use of reagents which qualification is not lower than «c.p.».

Extraction of gold ions (III) from the sample matrix was carried out by means of extraction of diethyl ether [11]. Palladium was extracted from a specimen by chloroform by use of extraction of dythemil-glyoxysome complex [3, 12]. The separation of the sample matrix from platinum ions was performed by means of hydrolysis method by treating the sample with 3% NaOH solution [13]. 1 M HCl solutions were used as a supporting electrolyte. Gold and hydrogen elimination on the process of palladium electro-oxidation was performed by means of UV irradiation of solution in the process of electro-concentration of palladium sediment [14, 15]. Analysis procedures are developed and certified in Innovative Research and Educational centre «Gold-platinum» in TPU [16].

3 Results and discussion

3.1 Brief description of objects of analysis.

25 specimens of rocks from the right side of the river Izinzyul of Verkhneamylskiy gold and ore district were selected as objects of testing for Pd, Pt, and Au. Verkhneamylskii gold and ore district is situated in the north-eastern part of the Western Sayans within the boundaries of Kurtushibinskiy volcanogenic and ophiolitic belt. Ore-grade gold in the region is represented by silica gold and gold and sulphide vein-disseminated rock formation, and it is paragenetically related to small subvolcanic intrusions of -gabbro diorite-diabase and gabbro and plagiogranite formations of basic and acid composition of lower Cambrian and lower Devonian.

Table 1. The content of Au, Pt, Pd in rock Izinzyul complex according to the method of stripping voltammetry. The n symbol represents the number of laboratory tests; The X symbol represents average values; laboratory test results are carried out in Innovative Research and Educational centre «Gold-platinum» in TPU.

Rock	n	Content, mg/t								
		Au			Pt			Pd		
		from	to	X	from	to	X	from	to	X
Gabbrodiorite	3.0	3.3	3.8	3.6	1.0	2.0	2.0	8.0	9.3	8.7
Gabbrodiorite albitized	12.0	12.5	69.8	34.0	3.0	40.0	21.4	6.3	10.0	7.6
Albitite of gabbrodiorite	19.0	0.3	3.3	2.2	7.0	8.8	8.0	4.5	8.8	6.8
Quartz-albite metasomatic	4.0	4.9	26.7	21.0	2.0	10.0	5.3	3.4	5.9	4.4
Gabbrodiorite serpentized	10	-	-	5.2	-	-	8.0	-	-	2.5
Diorite	8.0	23.6	122.1	63.0	9.0	30.0	16.3	3.7	10.9	6.6
Metasomatite of diorite	4.0	-	-	29.0	-	-	50.0	-	-	3.3
Diabase porphyrite	13.0	-	-	23.0	-	-	11.0	-	-	4.9
Metasomatite of diabase porphyrite	6.0	-	-	2.7	-	-	8.0	-	-	5.7
Quartziteof diabase porphyrite	3	-	-	3.8	-	-	40.0	-	-	11.5
Talcose metaschists	3.0	-	-	0.1	-	-	10.0	-	-	4.9
Metaschists hematitized	12.0	12.4	27.0	20.0	-	-	90.0	-	-	7.2
Quartz vein	19.0	2.0	16.4	9.0	7.0	60.0	33.5	2.5	6.7	4.6

3.2 Results of determination of gold by SV method.

Gold is noticed in all the analyzed specimens with contents from 0.1 to 122.1 mg/t (Table 1). At the same time the largest concentrations of 63.0 mg/t (ranging from 23.6 to 122.1 mg/t) are observed in diorits, and albitized gabbro-diorits — 34.0 mg/t. Close concentrations are observed in metasomatites by diorites (29.0 mg/t). In diobase porphyrites

(23.0 mg/t), albite metasomatites and hematitized metashales (21.0 and 20.0 mg/t). Content of gold less than bulk earth values is noted in gabbro-diorits (3.6 mg/t), quartzites (3.8 mg/t) and metasomatites by diabase porphyrites (2.7 mg/t), albitites by gabbro-diorites (2.2 mg/t) gabbros, and talcose metashales (0.1 mg/t).

3.3 Results of determination of platinum by SV method.

Platinum is found in almost all analyzed samples in concentrations of from 1.0 mg/t in gabbro- diorites, and to 90.0 mg/t in hematitized metashales (Table 1). The increased contents of platinum are noted in unaltered diorites (16.3 mg/t) and diabase porphyrites (11.0 mg/t). While the lowest contents of platinum (2.0 mg/t) are found in unaltered gabbro-diorites. During metasomatic engineering of rocks, accumulation of platinum occurs in albitited gabbro-diorites (21.4 mg/t), metasomatites by diorites (50.0 mg/t), quartzites by diabase porphyrites (40.0 mg/t). High contents of platinum are found in specimens of gangue quartz (33.5 mg/t) and, especially, in hematitized metashales (90.0 mg/ton).

3.4 Results of determination of palladium by SV method.

The levels of content of palladium in rocks are not high – they are from 2.5 to 11.5 mg/t at close average contents for different types of rocks and metasomatites, that is, 4.6-8.7 mg/t (Table 1). Some decrease of concentrations of palladium is observed during metasomatic engineering of gabbro-diorites (from 8.7 for unaltered rocks to 2.5 mg/t for serpentized ones) and diorites (from 6.6 mg/t for unaltered ones to 3.3 for altered rocks). However, palladium concentration during carbonization occurs (from 4.9 to 11.5 mg/t). In gauge quartz content of palladium is specified at the level of bulk earth value (4.6 mg/t).

4 Conclusion

The studies of the levels of gold, platinum, and palladium in different types of rocks and metasomatites of Verkhneamylskiy gold and ore district showed relatively high contents in the analyzed specimens of gold and platinum and low contents of palladium. During metasomatic engineering of rocks redistribution of metals takes place that specifically leads to either insignificant increase or decrease of its content in rocks.

Acknowledgements

The study was supported by Government Program «Science» of Tomsk Polytechnic University, grant No. 1.1488.2015.

References

1. I.E. Vasilyeva J. Research and monitoring, **14**, 16 (2010)
2. V.G. Torgov, T.M. Korda, M.G. Demidova, E.A. Gus'kova, G.L. Bukhbinder, J. Anal. At. Spectrom, **24**, 1551 (2009)
3. Y.A. Oskina, E.G. Pakrieva, E.M. Ustinova, A.N. Kryazhov, J. Advanced Materials Research, **1040**, 278 (2014)
4. G.S. Reddi, C.R.M.Rao, J. Analyst., **124**, 1531 (1999)
5. E.M. Ustinova, Y.A. Oskina, E.G. Pakrieva, J. Key Engineering Materials, **685**, 748 (2016)
6. N.A. Kolpakova, Factory laboratory. Diagnosis materials, **80**, 5 (2014)
7. E.V. Gorchakov, E.M. Ustinova, N.A. Kolpakova, J. of Solid State Electrochemistry, **16**, 2425 (2012)
8. E.M. Ustinova, E.V. Gorchakov, N.A. Kolpakova, J. of Solid State Electrochemistry, **16**, 2455 (2012)
9. N.A. Kolpakova, U.A. Oskina, E.N. Dyachenko, A.Y. Pshenichkin, J. Procedia Chemistry, **15**, 335 (2015)
10. E.N. Dyachenko, N.A. Kolpakova, J. Proceedings of the universities. Chemistry and chemical technology, **57**, 10 (2014)
11. A.Y. Pshenichkin, T.V. Timkin, Y.A. Oskina, IOP Conference Series: Earth and Environmental Science, **24**, 012025 (2015)
12. N.A. Kolpakova, E.V. Gorchakov, D.M. Karachakov, J. of Analytical Chemistry, **64**, 52 (2009)
13. E.G. Pakrieva, Y.A. Oskina, E.M. Ustinova, J. IOP Conf. Series, **21**, 012031 (2014)
14. E.V. Gorchakov, *Determination of gold and palladium at a simultaneous presence in natural objects by stripping voltammetry* (Tomsk: Izdatelstvo Tomskogo universiteta, 2006)
15. S.M. Panova, N.A. Kolpakova, *Studying the process of electro-precipitation of palladium-hydrogen electrode with graphite surface* (Ekaterinburg: Ural'skoye otdeleniye Rossiyskoy akademii nauk, 2016)
16. E.M. Ustinova, E.G. Pakrieva, Y.A. Oskina, *The technique (method) for determination of mass concentration of platinum in the samples of ores and rocks by voltammetry* (MU 08-47/361, Tomsk) 22 (2014)

MATEC Web of Conferences **85**, 01013 (2016) DOI: 10.1051/matecconf/20168501013
CCT 2016

Recycling of iron-containing deposits – the main way to increase the efficiency of water-protective measures on the territory of the Tom lower course

Viktor K. Popov[1], *Elena Yu.* Pasechnik[1,*], and *Anna* Karmanova[1]
[1]Tomsk Polytechnic University, 30 Lenin Avenue, Tomsk, 634050, Russia

Abstract. The paper presents the investigation of iron-containing deposits at water treatment in the Tomsk water supply intake. The chemical composition of the waters facilitates the formation of deposits in a large amount on the territory of the Tom lower course. A large ground water deposit is located here, and it is used for Tomsk water supply. Currently, the deposit obtained at water treatment is discharged in the Kislovka, and this practice causes a serious environmental problem. The ways of deposit usage are offered in the paper, which will sufficiently increase environmental protection measures on the territory of the Tom lower course.

1 Introduction

The high level of surface water contamination [1, 2, 3] promotes transition to underground water supply systems. In Tomsk region the use of ground waters is also in priority for drinking water supply. Tomsk is provided with drinking water due to reserves of a major drinking ground water deposit, which is called «Tomskoye». It is located in Tomsk region, in the northern part of the Ob-Tom interstream area [4].

A water supply intake started its operation in December, 1973. Currently, it consists of 198 operating wells arranged in three of its lines. The water supply intake gets out ground waters from the aquifer system in Paleogene sediments of «Tomskoye» deposit in the volume of 149.32 thousand m^3/day [5, 6]. The same aquifer system is exploited by 2 water supply intakes of Seversk located on the bank of the Tom. Their productivity is 60 thousand m^3/day.

However, the use of deep-well water cannot solve the task of providing the population with the water of satisfactory quality.

* Corresponding author: paseyu@yandex.ru

2 Geology and Hydrogeology

The considered area is situated at the joint of two major hydrogeological structures – a Kolyvan-Tomsk bow area and a West-Siberian artesian aquifer basin. It is located in a forest-taiga landscape zone with a normal humidity and heat provision, with favourable physical and geographical conditions for infiltration recharge of ground waters. Within the boundaries of Tomsk and Novosibirsk regions the Kolyvan-Tomsk bow area is called «Tomsk downfold» («Tomsky progib»). It is formed with terrigenous flysh deposits of Devonian-Carbonic periods. Its formation is complicated by the presence of a major folding in the shape of anticlinal and synclinal folds divided by the system of longitudinal thrust and upthrust faults of the early-late folded deposit.

The studied region is included in the outer area of the West-Siberian artesian basin, where all the aquifer systems participate in intense water exchange and contain infiltration waters, which have similar hydrogeochemical characteristics. Only in the northern part of the interstream area the waters of chalk and Paleozoic deposits are located in the zone of delayed water exchange.

Three hydrogeological complexes are distinguished within the boundaries of the Ob-Tom interstream area and the right-bank Tom, according to structural and geologic demarcation of the West-Siberian artesian basin: water drive systems of fault waters on the Paleozoic basement, a water drive system of chalk deposits (low and upper hydrogeological levels), water drive systems of Paleogene and Quaternary deposits (an upper hydrogeological level) [7].

3 Chemical composition of ground waters

The ground waters of the area are divided into four types (according to S.A. Shcherbakov, in descending order): hydrocarbonate calciferous-magnesium, hydrocarbonate sodium, hydrocarbonate-chloride sodium, and chloride sodium one. They are nitrogen-carbonate in composition, with oxygen to 1-3 mg/L [7, 8, 9]. Table 1 shows the chemical composition of ground waters on the territory of the Ob-Tom interstream area. Within the boundaries of a hydrogeochemical section a general zonal sequence is observed. It is caused by the change of water exchange conditions from active flushing to relatively delayed one.

Table 1. Composition and characteristics of ground waters on the territory of the Ob-Tom interstream area

Determined indicator	Units of measu-rement	MAC	Water characteristics			
			Waters of Neogene Quaternary deposits	Waters of Paleogene deposits	Waters of chalk deposits	Waters of Paleozoic deposits
Ammonium	mg/L	2	0.1…2.63	0.5…2.3	0.02…1.50	0.16…1.85
Nitrates	mg/L	45	0.00…4.65	0.00…0.14	0.00…1.00	0.005…4.4
Nitrites	mg/L	3	0.00…0.10	0.00…0.05	0.00…0.04	0.001…0.075
Iron	mg/L	0.3	0.4…16.5	0.9…10	1.5…10	0.93…16.8
Manganese	mg/L	0.1	0.5…1.25	0.2…0.4	0.1…8.0	0.1…30.26
Mineraliztion	g/ dm^3	1	0.15…0.60	0.25…0.50	0.31…0.58	0.13…4.58
Hardness	mmol-eq/ dm^3	7	0.4…7.0	1.3…7.2	3.65…18.4	5.5…8.35
Sulphates	mg/L	350	0.0…14.0	0.00…4.0	0.0…3.5	
Fluorides	mg/L	1.5	0.15…0.50	0.20…0.60	0.45	0.10…0.30
Chlorides	mg/L	500	0.51…16.70	0.72…17.70	3.9…1874	0.47…12.50

Oxidation	mg/L	5	--	2.0…5.0	3.2…3.84	0.7…3.6

According to the data [6], the waters of the Palogene horizon used for water supply change in their chemical composition from South to North from hydrocarbonate calciferous-magnesium with the mineralization of 0.13-0.34 g/dm^3 to chloride sodium with the mineralization of 1.85-4.58 g/dm^3, pH and Eh are within the limits of 6.8…7.6 and +90…+120 mV, respectively. Under the conditions of relatively delayed water exchange, iron is easily decomposed from the rocks, forming Fe^{2+} ions. The waters of a Palogene water-bearing stratum are enriched with iron in the amount of 0.9-10 mg/L.

The following scheme of water treatment from iron is approved at designing the underground water supply intake. The water delivered by artesian wells passes to aerators, where it is enriched with oxygen through water outpouring from the pipe located under the chamber surface. Then, water passes from aerators by gravity to rapid filters, where it is treated from iron, and it transits into the reservoir for pure water via a clarified water pipe line. There out, using the secondary pumps, it is supplied in pure water lines to the third pump stations and, finally, it is pumped into the water supply system of Tomsk. In order to decontaminate this water, its disinfection with sodium hypochlorite is provided.

Several water samples were taken in the Tomsk water supply intake at different stages of water treatment (Table 2). The analyses are performed at the accredited laboratory of Scientific and Research Centre «Water» (TPU). The water is characterized by a high concentration of iron and silicon. In the process of water treatment pH increases, the concentration of chlorine, hydrocarbonate and nitrates grows, and the decrease in the majority of indicators is observed. Besides the above mentioned components (iron and silicon), the other indicators from the operating wells of the Tomsk underground water supply intake do not exceed maximum admissible values.

Table 2. Chemical composition of waters at different stages of water treatment in Tomsk underground water supply intake

Indicators	Units of measu-rement	Norms in SanRaN 2.1.4.1074-01	Ground water	System of reuse	Water passed through filter treatment
pH	pH	6-9	6.8	6.38	7.14
CO_2	mg/L		41.4	92.4	29.5
CO_3^-	mg/L		<3	<3	<3
HCO_3^-	mg/L		329	354	369
SO_4^{2-}	mg/L	500	2.29	3.73	2.07
Cl^-	mg/L	350	7.98	13.45	10.4
Total hardness	mg/L	7.0	5.20	5.60	5.82
Ca^{2+}	mg/L		80	80	87.2
Mg^{2+}	mg/L		14.6	19.56	17.8
Na^+	mg/L		13.9	16	14
K^+	mg/L	200.0	0.72	1.04	0.79
$Fe^{общ}$	mg/L	0.3	2.2	7.5	0.075
Mineralization	mg/L	1000	448.5	487.8	501.3
Permanganate oxygen demand	mgO$_2$/dm^3	5.0	1.5	2.22	0.98
Electric conductivity	mS/cm		0.487	0.51	0.559
NH_4^+	mg/L	2.0	0.85	0.0029	0.22
NO_2^-	mg/L	3.0	<0.02	<0.02	<0.02

NO_3^-	mg/L	45	0.1	2.23	0.98
PO_4^{3-}	mg/L		0.57		0.121
Si	mg/L	10.0	12.06	11.94	12.2
J	mg/L		0.15	0.083	0.04

Also, the results on rare-earth elements in ground waters were obtained. It can be concluded that water treatment gives positive results for such elements as gadolinium, dysprosium, and erbium (Table 3).

Table 3. Content of rare-earth elements in ground waters

Elements	ICP MS 480X, mg/L	Concentration, mg/L		
		Water passed through filter treatment	System of reuse	Ground waters
La	0.00005	0.0000015	0.000015	0.0000017
Ce	0.00005	0.0000007	0.000025	0.0000006
Pr	0.00005	<0.0000005	<0.0000005	<0.0000005
Nd	0.00005	<0.0000005	0.0000088	<0.0000005
Sm	0.00005	0.0000024	0.0000031	0.0000023
Eu	0.00005	<0.0000005	<0.0000005	<0.0000005
Gd	0.00005	<0.0000005	0.0000052	0.0000011
Tb	0.00005	<0.0000005	<0.0000005	<0.0000005
Dy	0.00005	<0.0000005	0.0000051	0.0000014
Ho	0.00005	<0.0000005	<0.0000005	<0.0000005
Er	0.00005	<0.0000005	0.0000041	0.0000011
Tm	0.00005	<0.0000005	<0.0000005	<0.0000005
Yb	0.00005	<0.0000005	<0.0000005	<0.0000005
Lu	0.00005	<0.0000005	<0.0000005	<0.0000005

ICP MS 480X – determination of elemental composition for natural and drinking waters with mineralization to 2000 mg/L.

4 Iron-containing deposits of water treatment

Ferrous iron is stable at lack of dissolved oxygen and other oxidizing agents in ground waters. In ground waters carbon dioxide is dissolved in large amounts, therefore, probably, the main form of iron presence in ground waters is the solution of iron hydrocarbonate (II) – $Fe(HCO_3)_2$.

In the process of ground water aeration a large amount of deposits is precipitated in the iron removal station in Tomsk, 100-150 kg/day. To determine the deposit composition several water samples from filters were taken in the water supply intake. Investigations were conducted in the scientific laboratory of optical and electronic microscopy of International Innovation Research and Education Centre «Uranium geology» based on the Department of Geoecology and geochemistry (TPU). They were carried out using a diffractometer Bruker D2 Phaser. The method of X-ray diffractometry allowed detecting SiO_2 in deposits. It was difficult to identify the other minerals using this method.

The method of atom-emission photocolorimetric analysis as well as flame photometry according to [10] showed the following results, in %: iron – 30.3, manganese – 4.5, silicon – 4.0, calcium – 1.0, magnesium and aluminum – a tenth of per cent, copper, barium, silver, chrome – a hundredth of per cent, boracium, titanium, lead, tin, strontium, lutetium, europium – a thousandth of per cent. In addition, losses on calcination are of 27-29%.

5 Problem of iron-containing deposit recycling

Formation of a larger amount of deposits at water treatment causes a serious environmental problem, which has arisen from the very beginning of water supply intake operation, and it has not been solved yet. Currently, the deposit is discharged together with untreated waste waters in the river Kislovka. This water course is flowing through a specially protected natural reservation in the zone of the Ob-Tom interstream area, and this fact makes the environmental issue of this region even worse.

Deposit recycling through underground disposal seems also to be impossible as far as more and more new territories are required, which need remediation. Dry deposit is characterized by higher dusting, and, if the technology is not maintained, the danger arises to pollute the atmosphere and stuff vast territories with iron-containing dust.

The method of wet deposit briquetting with its further drying and use in metallurgy is economically irrational, in comparison with the use of ore.

In the current settings, it is possible to reprocess iron-containing deposits into chemical agents, and to use them as colouring agents and building materials.

6 Conclusion

Features of the hydrogeological conditions and geological structure in the Ob-Tom interstream area facilitate the formation of ground waters with a high iron content. In the iron removal station the large amount of iron-containing deposits is formed in the process of aeration, and its recycling becomes a serious scientific problem. It is possible to implement some world practices in the Tomsk water supply intake: to use iron-containing deposits for production of chemical agents, colouring agents, and building materials. The solution of the problem concerning the recycling of iron-bearing deposits at water treatment will allow improving the environment on the territory of the Tom lower course and increasing the efficiency of present water-protective measures.

Acknowledgements

The reported study was funded by RFBR according to the research project No. 16-35-00429 мол_а.

References

1. V.A. Zemtsov, O.G. Savichev, *Resources, regime and quality of surface waters in the Ob River basin: history, current state and problems of research. International Journal of Environmental Studies*. **72(3).** 386–396, (2015).
2. S.L. Shvartsev, N.M. Rasskazov, O.G. Savichev, *Contents and migration forms of elements in natural waters of the mid-Tom' basin. Geologiya i Geofizika*. **12** (1997).
3. O.G. Savichev, I.A. Matveenko, E.V. Ivanova, *Conditions of contaminant distribution in the wetland water of Western Siberia (the Russian Federation). IOP Conf. Series: Earth and Environmental Science*. **33**. 1–6, (2016).
4. Ed. by A.F. Poryadin, V.P. Zinoviev, *From a pure spring. Lifetime of Tomsk water pipe-line*. 304 (2005).
5. Geo-environment conditions of Tomsk region territory in 2014. Information

bulletin 20. 84 (2015).
6. V. Lgotin, Y.Makushin, *Groundwater monitoring to assess the influence of injection of liquid radioactive waste on the Tomsk public groundwater supply, Western Siberia,* Russia. *Geological Society Special Publication.* **128**. 255–264, (1998).
7. V.K. Popov, O.D. Lushevich, V.A. Korobkin *et al., Environmental and economic aspects of ground water exploitation in the Ob-Tom interstream area.* 174, (2003).
8. E.Y. Pasechnik, *Ecological and geochemical conditions of natural waters on the territory of Tomsk (the right-bank Tom): extended abstract of a dissertation for PhD in Geology and Mineralogy.* 22 (2010).
9. O. Levina, E.Y. Pasechnik, A.V. Baranova, *Changes in Chemical Composition of Natural Waters in Response to Technogenic Load.* **10**. 137–144, (2014).
10. V.N. Lisetsky, V.N. Bryukhantsev, A.A. Andreichenko, *Water treatment deposit recovery and recycling in Tomsk water supply intakes.*164 (2003).

Potential uses of natural zeolites for the development of new materials: short review

V. Petranovskii[1], F. Chaves-Rivas[2], M.A. Hernandez Espinoza[3], A. Pestryakov[4*], E. Kolobova[4]

[1]CNyN, Universidad Nacional Autonoma de Mexico, Ensenada, 22800, Mexico
[2] Instituto Politechnico Nacional, Mexico Distrito Federal, 07738, Mexico
[3]Benemerita Universidad Autonoma de Puebla, Puebla 72570, Mexico
[4]Tomsk Polytechnic University, Tomsk, 63450, Russia

Abstract. Zeolites stand apart among the many families of porous materials. Due to their structure, they have a number of important practical properties such as a high adsorption capacity and the ability to cation exchange. However, the main feature distinguishing them from other porous bodies and inherent only for zeolites is the ordering of the pores in their size, orientation, and mutual locations. Pore diameters for the crystalline structures of various zeolites may vary from 0.25 to about 15 nanometers. The crystalline arrangement of these pores in space converts zeolites in the naturally existing nanostructured materials, possessing the unique property to present ordered location of stabilized nanoparticles.

1 Introduction

Since the last millennium, zeolites are known as a "magic rock" [1]. The search for methods of synthesis of nanoparticles, investigation of their unique properties, and development of new materials for exploring these properties for applications in advanced technology have undoubted value. Clearly, nanoparticles with sizes in range of a few nanometers can be applied in industrial processes only if supported on solid matrices. Such matrix must be inert in the environment, where the prepared materials are applied. Fortunately, such matrices are created by nature; which are natural zeolites, with a structure possessing channels and cavities of nanometric size, and relatively stable.

In countries such as Cuba, Japan and the USA, many studies have been conducted that have led to new and innovative applications of natural zeolites [1-4]. Other countries, such as Mexico and Russia, where zeolites are found in abundance, have the potential to become an industry leader of zeolitic materials. Innovative research work is needed to find new ways in which zeolites can be used to benefit people and the environment. The aim of the present paper is to estimate potential use of natural zeolites for production of new industrially and environmentally important materials.

2 Available reserves of zeolites

2.1 Estimates of World reserves

Although, there are no certain figures on the total amount of zeolites in the world, it is well known that they are present on all the continents with varying mineral contents and kinds. World production of natural zeolites was estimated in the year 2004 to be between 2.5 and 3 million metric tons (Mt) based on reported production by some countries [5]. Three years later world production of natural zeolites was estimated to be approximately the same [6]. Last year (2015) situation [7] is shown in Figure 1.

2.2 Availability of natural zeolites in Russia

Less than ten deposits are exploited at present in Russia, between others the Holinsk deposit (clinoptilolite) and Mukhor-Tala deposit (mordenite) in the Republic of Buryatia, and “Priargunskoe” Mining and Chemical Association [8]. Evaluation of the Mukhor-Tala deposit was reported in Ref. [9]. Such a small amount of produced zeolite (between

* Corresponding author: pestryakov2005@yandex.ru

3000 and 5000 t for all countries of Russian Commonwealth) contradicts to huge reserves of known deposits, and mainly is due to the few researches dedicated to the use of natural zeolites in industry and agriculture.

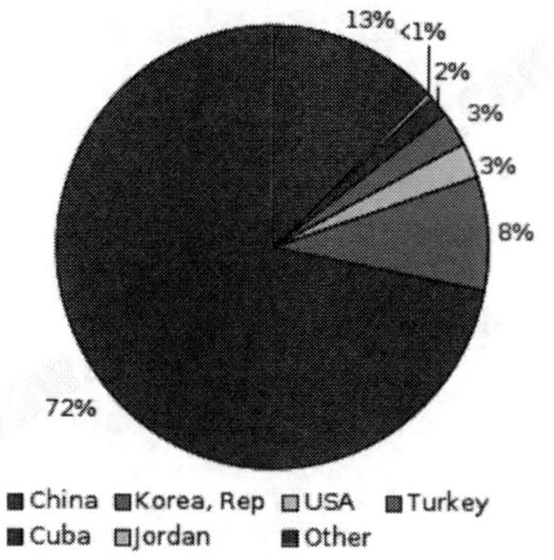

Fig. 1. Where do natural zeolites come from? Estimated world mine production for 2015. China produced almost three quarters of all natural zeolites (2 million tons) [7]. Total world reserves of zeolites are unknown "but are estimated to be large" [7].

3 Zeolites

Zeolites are the aluminosilicate members of the family of microporous solids known as "molecular sieves" (Figure 2). Due to isomorphous substitution of Si^{4+} by Al^{3+}, the zeolite framework get a negative charge and so, need for any cations; those are not the structural units of their crystalline structure, but serve to neutralize the charge of framework only. These charge carriers are located in the voids of zeolite structure, and can easily be exchanged for any other ions.

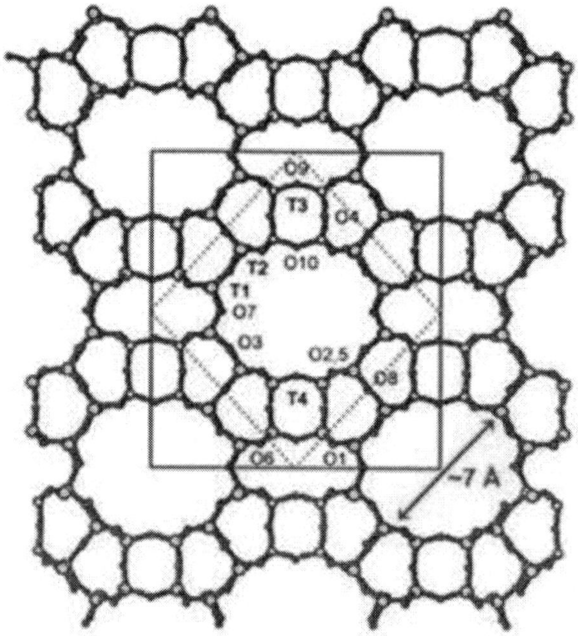

Fig. 2. Projection of mordenite framework along the *c* axis, presenting a view along the channel (twelve-membered ring) and shoving also the side pockets (eight-membered rings). The four inequivalent tetrahedral sites labeled T1 to T4 (T = Si or Al) are shown by the larger light grey spheres. Oxygen atoms are represented by smaller dark spheres and numbered O1 to O10. The solid frame shows the conventional centrosymmetric orthorhombic unit cell, the dashed frame - the primitive unit cell (adapted from Ref. [24]).

3.1 Most important properties of zeolites

The ion exchange property is one of the main features that zeolites exhibit (both natural and synthetic ones). This feature is important from two points of view. First, cations introduced in zeolite matrices by ion exchange treatment allow subsequent development of materials for catalytic, medical and antibacterial applications, between many others [10-17]. Secondly, one of the most important applications is the processes of removing of pollutants such as heavy metals from the wastewater of industrial plants, soil, and other sources [18-23]. Chip and abundant natural zeolites are very suitable material for harmful ions removal, because polymetallic cation mixture inherent to natural zeolitic rock, mineral impurities, and as a rule several zeolitic phases presenting in the row natural mineral do not hinder their use for accumulation of undesired cations in the ambient.

Cation exchange properties are very important applications, for example in environment protection to trap radioactive cations. It is well known that after Chernobyl catastrophe "permeable reactive barriers" made from natural zeolites, especially from clinoptilolites [25] were applied to prevent diffusion of the radionuclides [26,27].

Similar methods were used to tolerate Fukushima nuclear contamination. The Japanese government hopes that the zeolites help reduce radioactive contamination of the sea. Availability and reasonable cost of natural clinoptilolite together with its high selectivity towards Sr^{2+} and Cs^{+} radionuclides make possible the use of a clinoptilolite tuff as a barrier. The potential use of natural zeolites as cation exchangers in environmental protection include scavenging of fall-out products from living hosts [27], ammonium removal from municipal sewage [18], desalination treatment of sea water [28], or remediation of hazardous heavy metal-polluted soils [29] or sediments [30].

3.2 Future applications of zeolites

All of the above-mentioned applications of zeolites are based on the fact what they are naturally existing nanostructured materials. Advances of modern preparative chemistry and materials engineering allow improving existing materials, and to give them new properties. Two main features of zeolites allow synthesize on their base new modern materials - their regular pore structure (Fig. 2) and ion exchange properties. The processes that include formation of complex–cations on zeolites are important to develop industrially significant materials, such as catalysts [10-12] and photocatalysts [13], medicine [14], sensor materials [31,32], feeding additives for farm animals [33], etc.

3.3 Nanozeolites

Very interesting question also arises in connection with the size of the zeolite particles. All known applications were developed while applying zeolites in the regular granulometric scale. What happens if this natural nanostructured material will be prepared in the form of nanoparticles?

Recently, the methods of nanozeolite synthesis were developed [34,35]. Nanozeolites are a type of zeolites, which have narrow particle size distribution with sizes of less than 200 nm [36]. Compared to "ordinary" zeolites of which the particle diameters are of micrometer order, nanozeolites represent very small particle size, and especially, the fact that they are composed of discrete particles (single crystals) rather than aggregates.

One of the advantages of nanozeolites is their extremely high external surface area. The external surface is of vital importance in numerous processes, including adsorption and catalysis. For example, in the fluidized catalytic cracking (FCC) process, the commercial catalysts are manufactured by dispersion of micron-sized zeolites in an amorphous alumina-silica matrix. For cracking to occur, gas oil molecules must pass through the matrix and reach the surface of the zeolite crystals. The molecules then diffuse through the micropores of Y zeolites until they reach an active site. Due to the zeolite structure, molecules larger than 7.4 Å cannot reach active sites located inside the mordenite. This problem can be avoided by replacing the micrometer-sized zeolites with the corresponding nanozeolites. The substitution could lead to a decrease in the diffusional resistance and an increase in the external surface area, hence raising the number of active sites available for large molecules.

3.4 Potential applications of nano-zeolites

Zeolite particles in the 10-100 nm range can bring in new applications of zeolites. The huge surface areas of the nanosized materials dictate that many of the atoms are on the surface, thus allowing good "atom economy" in surface-gas and surface-liquid interface. The ratio of atoms available on the surface increases as the crystal size decreases.

Such phenomenon can be responsible for the observation that clinoptilolite powder can be used as a potential adjuvant in anticancer therapy [37]. It was shown, that anticancer activity of natural clinoptilolite significantly increased after fine milling. Better results were observed if the zeolite used for medical purposes was processed by a special micronization machine that increased the surface area several times over, and ground the material into various sized particles, with about 15% being nanoparticles as small as 100 nm [38]. Many other medical applications of zeolites are actively being explored [39].

4 Hybrid materials based on zeolites

Another method of creating hybrid materials composed of nanoparticles deposited on nanostructured supports also exploits the ion exchange properties of zeolites. In recent years, zeolites have gained importance for their capacity to accommodate clusters of other materials in their void space (regular channels and cavities, as well as irregular mesoporosity existing due to defects of crystalline structure) and on the surface. This facility to insert clusters allows one to create new composites with desirable properties for photocatalysis.

4.1 Intra-zeolite chemistry

General formula of zeolite is $(M^{+1})_x[Al_xSi_yO_{2(x+y)}]$, where M^{+1} usually is Na^{+1} ion. These Na^{+1} ions can be exchanged for any other ions, for example Ag^+, Cd^{2+}, Zn^{2+}, Cu^{2+}, etc. After that, Ag^+ can be reduced, forming Ag^0. Reduced silver atoms can agglomerate inside the cavity, forming small clusters, according to the reaction $nAg^0 = Ag_n$. Similar processes take place with other exchanged transition metal ions [11,40,41].

Another possibility is to realize chemical reactions inside of nano-voids of zeolite structure. Silver ions Ag^+ can react with NaBr solution, precipitating AgBr nanoparticles:

$$Ag[Zeolite] + NaBr = AgBr@Na[Zeolite]$$

Where symbol "@" means "localized inside of zeolite porosity".

If solutions containing S^{2-} anions will be used, CdS or ZnS sulfides dispersed inside of zeolite cavities can be obtained [13,42,43]. In these works, ion exchange by hydrothermal technique was used to synthesize the composites. Nanoclusters with a broad size distribution were formed inside the channels and on the external surface.

4.2 Confinement of nanoparticles in zeolite nanoporosity

New nanostructured materials will provide new interesting physical properties, which are characteristic of electron confinement in curved space with 3D periodicity [44]. Semiconductor nanoparticles exhibit quantum confinement effects when their dimensions become as small as or smaller than the Bohr radius. This quantum confinement causes drastic changes in the electronic and physicochemical properties of the semiconductor-zeolite composites [45-47]. Due to transparency of zeolites in the UV-Vis range, the zeolites can serve as inert supports for the photocatalytically active semiconductors.

Schematically, such a material consisting of nanoparticles deposited on nanostructured carrier is shown in Figure 3. As an example, erionite framework (A) is selected. After chemical reactions, nanoparticles (balls in B (side view) and C (projection view) parts of Fig. 3) are stabilized inside of erionite cavities.

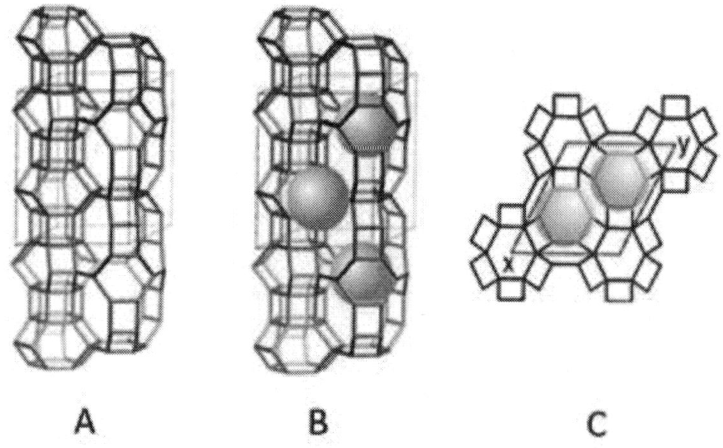

Fig. 3. Schematically represented framework of erionite channels (A), and metal clusters shown as a bolls. Dash frames show the conventional unit cell.

Concluding Remarks

Zeolites represent a class of well-known minerals and materials with a high potential for industrial utilization. Classical fields are detergent additives, catalysts in cracking processes of the petroleum refining industry and as molecular sieves in environmental applications, soil improvement and nuclear waste treatment. Recent innovative applications of zeolites regard their use as host matrices for a large variety of guests. These hybrid materials can be exploited in several research fields, as devices for solar energy harvesting, processing/storing of information, and advanced sensing technology for analytics and diagnostics on the nanoscale. Many microporous heteropolyhedral compounds, containing metals such as Mo, V, Sn, Zr, Ti, Ca and lanthanides, with a wide number of distinct structural types, have been reported and their potential applications in the areas of catalysis, separation of molecular species, ion exchange and optical and magnetic devices have been evaluated. We expect contributions on mineralogy, structure, P-T stability, crystal-chemistry and applications of natural and related synthetic porous phases, and hence from mineralogy and petrology researchers as well as from crystallographers and material scientists.

Acknowledgments

This research was supported by Russian Government Program «Science» of Tomsk Polytechnic University, grant No. 4.1187.2014/K. F.C.R. acknowledge support from COFAA and PIFI-IPN.

References

1. F.A. Mumpton, *Geology, Mineralogy, and Human Welfare*: Proc. of the National Academic of Science of the USA colloquim **96**, 3463 (1999)
2. T. Selvam, W. Schwieger, W. Dathe, Clay Miner. **49,** 501 (2014)
3. V.J. Inglezakis, A.A. Zorpas, *Handbook of natural zeolites* (Potomac: Bentham Books) (2012)
4. C. Laurino, B. Palmieri, Nutr. Hosp. **32,** 573 (2015)
5. Virta R.L. Zeolites *Minerals Year Book* (Reston, VA: USGS) 83.1–83.3 (2004)
6. Virta R.L. Zeolites *Minerals Year Book* (Reston, VA: USGS) 83.1–83.3 (2007)
7. *US Geological Survey: Mineral Commodity Summaries: Zeolites (Natural), January 2016.* http://minerals.usgs.gov/minerals/pubs/commodity/zeolites/index.html; Last accessed at May 24 2016
8. *Overview of market of natural zeolites in the Commonwealth of Independent States*, INFOMINE Research Group, http://www.infomine.ru/files/catalog/76/file_76_eng.pdf; Last accessed at May 24 2016
9. V.I. Prihodko, *New technologies for production and processing of raw materials in terms of environmental restrictions:* Proc. Russian Scientific Conference (Ulan-Ude) 72 (2004)
10. N.M. Xavier, S.D. Lucas, Rauter A.P., J. Mol. Catal. A: Chem. **305,** 84 (2009)
11. V. Petranovskii, A. Pestryakov, L. Kazantseva, J. Cruz, A. Kryazhov, Stud. Surf. Sci. Catal. **174,** 513 (2008)
12. V. Petranovskii, A. Pestryakov, L. Kazantseva, F.F. Castillon Barraza, M.H. Farias, Int. J. Modern Phys. B **19,** 2333 (2005)
13 A. Nezamzadeh-Ejhieh N. Moazzeni, J. Ind. Eng. Chem. **19,** 1433 (2013)
14. J. Hrenovic, J. Milenkovic, T. Ivankovic, N. Rajic, J. Hazard. Mater. **201–202,** 260 (2012)
15 A.N. Pestryakov, N. Bogdanchikova, A. Simakov, I. Tuzovskaya, F. Jentoft, M. Farias, A. Diaz, Surf. Sci. **601,** 3792 (2007)
16. E. Smolentseva, N. Bogdanchikova, A. Simakov, A. Pestryakov, M. Avalos, M.H. Farias, A. Tompos V. Gurin, J. Nanosci. Nanotechnol. **7,** 1882 (2007)
17. A. Pestryakov, I. Tuzovskaya, E. Smolentseva, N. Bogdanchikova, F. Jentoft, A. Knop-Gericke, Int. J. Modern Phys. **B 19,** 2321 (2005)
18. C. Colella, *Natural microporous materials in environmental technology:* Proc. NATO Advanced Research Workgroup (NATO Advanced Study Institute, Series E: Applied Sciences vol **362**) (Boston: Kluwer Academic Publishers) 207 (1999)
19. X. Wanjing, Y.L. Loretta, J.R. Grace, G. Hébrard, Appl. Clay Sci. **80–81,** 31 (2013)
20. M.K. Doula, K. Elaiopoulos, V.A. Kavvadias, V. Mavraganis, J. Hazard. Mater. **207–208,** 103 (2012)
21. G.S. Sultanbayeva, R. Holze, R.M. Chernyakova, Micropor. Mesopor. Mat. **170,** 173 (2013)
22. S. Babel, T.A. Kurniawan, J. Hazard. Mater. **97,** 219 (2003)
23. A. Nezamzadeh-Ejhieh , M. Kabiri-Samani, J. Hazard. Mater. **260,** 339 (2013)
24. T. Demuth, J. Hafner, L. Benco, H. Toulhoat, J. Phys. Chem. B **104,** 4593 (2000)
25. V. Nikashina, I. Serova, E. Kats, N.A. Tikhonov, M.G. Tokmachev, Clay Miner. **46,** 233 (2011)
26. A. Nesterenko, V. Nesterenko, A.Yablokov, Ann. NY Acad. Sci. **1181,** 287 (2009)
27. A. Dyer, *Environmental mineralogy: Microbial interactions, anthropogenic influences, contaminated land and waste management:* Proc. Mineralogical Society's winter meeting (London: Mineralogical Society) 319 (2000)
28. T. Wajima, Anal. Sci. **29,** 139 (2013)
29. W.Y. Shi, H.B. Shao, H. Li, M.A. Shao, J. Hazard. Mater. **170,** 1 (2009)
30. Y. Chiang, R.M. Santos, K. Ghyselbrecht, V. Cappuyns, J.A. Martens, R. Swennen, T. Van Gerven, B.Meesschaert, J. Environ. Manage **105** (2012)
31. Y.G. Zheng, X.G. Li, Sensors **12,** 5170 (2012)
32. O.Y. Saiapina, V. Pyeshkova, O. Soldatkin, V. Melnik, B. Akata Kurç, A. Walcarius, N. Dzyadych Jaffrezic-Renault, Mat. Sci. Eng. C **31,** 1490 (2011)
33. L. Nimatsyrenova S. Zhamsaranova, Bull. Buryat State Agricultural Acad. V.R. Filippov **4** , 106 (2011)
34. D. Karami, S. Rohani, Rev. Chem. Eng. **23,** 1 (2007)
35. T. Tago, T. Masuda, *Nanocrystals* (Sciyo: [S.I.]) 191 (2010)
36. L. Tosheva, V.P. Valchev, Chem. Mater. **17,** 2494 (2005)
37. K. Pavelić, M. Hadzija, L. Bedrica, J. Pavelić, I. Dikić, M. Katić, M. Kralj, M.H. Bosnar, S. Kapitanović, M. Poljak-Blazi, S. Krizanac, R. Stojković, B. Subotić , M. Colić, J. Mol. Med. **78,** 708 (2001)
38. N. Zarkovic, K. Zarkovic, M. Kralj, S. Borovic, S. Sabolovic, M.P. Blazi, A. Cipak, K. Pavelic, Anticancer Res. **23,** 1589 (2003)
39. S.M. Auerbach, K. Carrado, P.K. Dutta, *Handbook of zeolite science and technology* (New York: M. Dekker) 1184 (2003)
40. Yu. A. Alekseev, V.N. Bogomolov, T.B. Zhukova, V.P. Petranovskii, S.G. Romanov, S.V. Kholodkevich, Bull. Acad. Sci. USSR Phys. Series **50,** 1 (1986)
41. Y. Meteleva, F. Roessner, G.F. Novikov, J. Photochem. Photobiol. Chem. A **196,** 154 (2008)
42. M.S. Sadjadi, A. Pourahmad, S. Sohrabnezhad, K. Zare, Mater. Lett. **61,** 2923 (2007)
43. O. Jaime-Acuña, H. Villavicencio, J. Díaz-Hernández, V. Petranovskii, M. Herrera, O. Raymond, Chem. Mater. **26,** 6152 (2014)
44. J. Cejka, H. Bekkum, A. Corma, F. Schueth, *Introduction to zeolite molecular sieves* (3rd ed. *Studies in Surface Science and Catalysis* vol. **168**) (Amsterdam: Elsevier) 800 (2007)
45. F. Iacomi, Surf. Sci. **532–535,** 816 (2003)
46. P.V. Kamat, J. Phys. Chem. C **111,** 2834 (2007)
47. P.V. Kamat, J. Phys. Chem. C **112,** 18737 (2008)

APPLICATION OF COMBINED SORBENT FOR PROCESS OF WATER PURIFICATION FROM MICROBIAL CONTAMINATION

Irina Martemianova [1,*], *Sergey* Zhuravkov[1], *Evgenii* Plotnikov[1] , *Dmitrii* Martemianov[1], *Elena* Korotkova[1], *Bohumil* Kratochvil[1], *Olesya* Voronova[1] and *Svetlana* Antonova[1]

[1]Tomsk Polytechnic University, Lenin av. 30, Tomsk, 634050, Russia

Abstract. The results of sorption efficiency of filter material which was obtained on the base of river sand and modified by aluminium oxyhydroxide with the sol-gel process are presented. All components were tested as sorbents, and their properties were compared. New filter material was found effective in the process of water treatment from bacterial contamination.

1 Introduction

Microbiological contaminant is a special challenge among the various pollutants that are in the hydrosphere. Different microorganisms can cause a lot of negative consequences both in industrial and home use, and it take a lot of effort to treat consequences of that including health loss, necessity of antibacterial drugs development, etc [1-4]. Obviously, it is an important task to eliminate bacteria from water. Various water treatment methods, including chemical treatment (chlorination, ozonation), membrane cleaning, UV sterilization, modified filter materials are used to prevent the negative effects associated with the use of contaminated water [5]. In recent years, a lot of attention is paid to development and spread of modified sorptive filter materials in combination with silver and aluminum nanoparticles, which have several advantages over other methods of treatment [6-10]. Therefore, the creation of new modified materials with improved sorption, filtration properties, and low cost, has relevance.

The aim of this work is to study some physical and chemical properties of the samples of the combined filter material and its components (in the form of aluminum oxyhydroxide and sand), as well as to determine the extraction degree of the bacteria Escherichia Coli (E.coli) from the model contaminated solution by the test samples of the modified materials.

2 Materials and methods

* Corresponding author: martemiv@yandex.ru

The objects of the present study were the sand from the Ob river (Kireevsk, Tomsk region) with the particle size of 0.1-0.7 mm, aluminum oxyhydroxide (AlO(OH)) synthesized from aluminum food grade AK7 (the particle size is less than 0.1 mm) and the modified filter material from the specially treated river sand and aluminum oxyhydroxide.

Surface morphology of all the samples of aluminum oxyhydroxide and river sand was studied with the transmission electron microscopy (the transmission electron microscope JEM-2100F (JEOL, Japan) with the system of sample preparation EM-09100IS "Ion Slicer".

The specific surface area (S_{sp}) and pore volume (P_{sp}) in all the studied samples were determined using the analyzer "SORBTOMETR M" with the method of thermal desorption of nitrogen (BET).

The bacteria Escherichia Coli (strain Escherichia coli ATCC25922) were used as the contaminating microbial agent for the experiment. The bacteria E.coli were filtered under dynamic conditions using the peristaltic pump for water suspension. A suspension of bacteria passed through a tube filled with the test sample of the material. The following samples were prepared for the filtration test: the river sand, aluminum oxyhydroxide, and three samples of the combined sorbent with different contents of the active ingredient, were obtained by mixing. All the test materials were placed in the glass tubes (length - 150 mm, internal diameter - 8 mm) to determine the filtration capacity. The weight and composition of the materials in the tubes were: sand - 10.3 g; aluminum oxyhydroxide - 7 g; sorbent # 1 (sand 90%/10% aluminum oxo-hydroxide - 9.94 g); sorbent # 2 (sand 80%/20% aluminum oxo-hydroxide - 9.64 g); sorbent # 3 (sand 70%/30% aluminum oxo-hydroxide - 9.3 g). The test bacterial suspension was prepared in tap water, it was settling for 1 day. Then, the water was added to the 24-hour culture of Escherichia Coli in the meat-peptone broth. The final concentration of the bacterial suspension was 2.3×10^7 CFU/cm^3. The total number of microorganisms (TMC) in the sample bacterial suspension was determined according to the Methodological Instructive Regulations MUK 4.2.1018-01. After filtering, the samples of the suspension were placed in the petri dishes. The results were analyzed in 24 hours by counting the number of colonies expressed in the colony forming units (CFU) per 1 cm^3 of the sample.

3 Results

Figure 1 shows the agglomeration of the aluminum oxyhydroxide nanoparticles (19000×). The size of the separate nanoparticle is about 50-200 nm in length and 2 nm in diameter. Figure 2 shows the particle of the river sand with a magnification of 5000 times.

Fig. 1. Agglomeration of the aluminum oxyhydroxide nanoparticles (19000×).

DOI: 10.1051/matecconf/20168501015

Fig. 2. River sand particle with the transmission electron microscopy (5000×).

As shown, the sand particle surface is very smooth opposite to the aluminum nanoparticles with the clearly visible needles on the surface. According to this image, the low specific surface value and possible absent or very low porosity of the sand material and, consequently, low adsorbent properties were predicted. It was completely proved by the method of thermal desorption of nitrogen (Table 1). Presumably, sand could be used as very cheap and available mineral base for modified adsorbent production. Table 1 summarizes the specific surface area and pore volume of all the tested material samples.

Table 1. The specific surface area and pore volume of the initial and combined materials

Sample	**S_{sp}, m^2/g**	**P, cm^3/g**
River sand	0.444	0
Aluminum oxyhydroxide	192.5	0.083
Sorbent # 1	20.37	0.009
Sorbent # 2	40.41	0.017
Sorbent # 3	62.8	0.025

The results revealed the smallest value of the specific surface and absence of pores in the river sand samples. Aluminum oxyhydroxide has the highest values of all the studied parameters. The values of the specific surface and porosity of the modified sorbents increase with a rise in the active component of the sample.

Figure 3 shows the sorption properties of the material samples for extraction of the Escherichia Coli culture from the model solution under dynamic conditions.

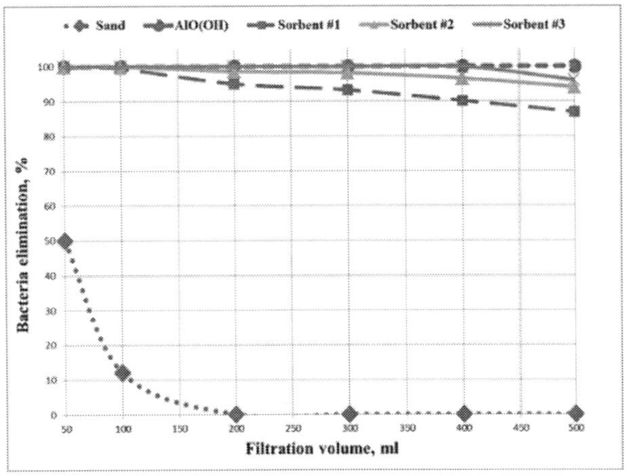

Figure 3. The degree of bacteria elimination vs. different filtration volumes.

The data showed (Figure 3) that the river sand does not remove notable number of microorganisms from the aqueous suspension but has a low flow resistance, therefore, the process has high performance (250 cm^3/h). The lowest effectiveness revealed in this test makes the non-modified sand inappropriate for microbiological filtration.

Opposite to the sand, aluminum oxyhydroxide completely removed bacteria Escherichia Coli via the filtration process. However, a very high flow resistance of the material was determined, so the filtration performance was 5 cm^3/h.

Among the combined sorbents, the best filtering properties showed the sorbent #3, the worst one was the sorbent #1. Filtration performance of the sorbents was as follows: the sorbent #1 - 223 cm^3/h; the sorbent #2 - 197 cm^3/h; the sorbent #3 - 152 cm^3/h. So, the flow resistance of the material decreased with fall in aluminum in the sorbent.

4 Conclusion

The surface morphology of the river sand and aluminum oxyhydroxide nanofibers was investigated, the sizes of aluminum oxyhydroxide nanofibers were determined - length of 50-200 nm; width of 2 nm. It was found that the bicomponent composite sorbents exhibit good extraction rate of the bacteria Escherichia Coli from the test suspension at a high flow rate. Increased quantities of the active component in the samples of the combined sorbent lead to the improved filtration properties of the material. The river sand was shown as an ineffective agent for microbiological water treatment. In turn, the aluminum oxyhydroxide completely removes the bacteria from the contaminated water but has very low hydrodynamic properties.

Thus, a prospective way of production and effective use of the combined sorbent based on the river sand mixed with aluminum oxyhydroxide to extract the microbiological contaminants from aqueous media was shown.

This work was supported by the grant of the President of the Russian Federation for young Russian scientists MK-5939.2016.8

References

1. Y. Chen et al, Acta Scientiae Circumstantiae **35**, 337-351, (2015)

2. M. Telitchenko, S. Ostroumov, Introduction to Problems of Biochemical Ecology: Biotechnology, agriculture, protection of the environment. (Moscow: Science, 1990)
3. E. Plotnikov, V. Pehenko, V. Plotnikov, Physiology and Pharmacology **19**, 247-252, (2015) https://www.scopus.com/record/display.uri?eid=2-s2.0-84960958582&origin=inward&txGid=2A01D10CFF2B47942E9AAD44C40053BD.wsnAw8kcdt7IPYLO0V48gA%3a1
4. E. Plotnikov, V. Silnikov, A. Gapeyev, V. Plotnikov, Advanced Pharmaceutical Bulletin **6**, 71-74, (2016) http://dx.doi.org/10.15171/apb.2016.011
5. V. Kosoy, A. Bazhenov, Water and Ecology: Problems and Solutions (Rus). **1**, 40-45, (2001)
6. S. Zhuravkov, E. Plotnikov, D. Martemiyanov, N. Yavorovsky, U. Hasse, S. Zander, Advanced material research **872**, 74-78, (2014) http://dx.doi.org/10.4028/www.scientific.net/AMR.872.74
7. D. Martemianov, E. Korotkova, A. Galanov, Bulletin of the Karaganda University **3**, 61-65, (2012)
8. E. Plotnikov, I. Martemianova, D. Martemianov, S. Zhuravkov, O. Voronova, E. Korotkova, V. Silnikov, Procedia Chemistry **15**, 219-224, (2015) http://www.sciencedirect.com/science/article/pii/S187661961500114X
9. E. Plotnikov, S. Zhuravkov, A. Gapeyev, V. Plotnikov, D. Martemiyanov, Advanced material research **1040**, 65-70, (2015) http://dx.doi.org/10.4028/www.scientific.net/AMR.1040.65
10. E. Plotnikov, I. Martemianova, D. Martemianov, S. Zhuravkov, T. Kan, O. Voronova, Journal of Materials and Environmental Science **7 (11)**, 3944-3948, (2016) http://www.jmaterenvironsci.com/Document/vol7/vol7_N11/423-JMES-2509-Plotnikov.pdf

Technological characteristics of glauconite rocks in Bakchar deposit (Western Siberia)

M.A. Rudmin[1, a], A.K. Mazurov[1], B.I. Makarov[2], I.V. Reva[1]
[1] National Research Tomsk Polytechnic University, Institute of Natural Resources, Department of Geology and Mineral Exploration, 634050 Tomsk, *Russia*
[2] Tomsk State University, Biological Institute, Department of Plant Physiology and Biotechnology, 634050 Tomsk, *Russia*

Abstract Glauconite-containing rocks found within the boundaries of the Bakchar deposit include the following: glauconite ores (glauconitolite), glauconite sandstone, hydrogoethite-chlorite, iron ores with glauconite. Glauconite content in glauconite ores is 50…70 %, in glauconite sandstone – 20…40 %, in hydrogoethite-chlorite ores- 10…20 %. This paper describes the technological characteristics of glauconite rocks. The authors proposed the optimal technological scheme of producing glauconite concentrate. It has been defined that the most highly-qualified product with commercial components 90…100 % is recovered from glauconite ores. It has been evaluated that Bakchar glauconite is applicable as a favorable mineral fertilizer based on the concentrate produced from enriched hydrogoethite-chlorite ores. Experimentally, it has been proved that this concentrate has a positive effect on the growth of oats (Avéna sativa). This is a direct indicator of its practical applicability. Obtained results indicate the fact that the extraction and application of Bakchar glauconite could be widely used in agriculture.

1. Introduction

The present research is focused on the development of the technological production scheme of glauconite concentrate from enclosing rocks and its possible application in agriculture to enhance the investment potential of Bakchar deposit. Bakchar iron ore deposit in S-E area of Western Siberia (fig.1) and 200km. north-westward of Tomsk. This deposit is related to Western-Siberian iron ore basin which is a wide correlative belt from Turukhan river in the north to the Tobol river head in the south-west. The following large iron ore areas are within this basin (fig.1): Elogy-Turukhansky, Kargasokin, Narym- Parabelsky, Kolpashevo and well-developed Bakchar deposit.

Iron ore deposits are located within Cretaceous-Paleogene sediments at the depth of 150…250 meters (fig.2).Ore-hosting sediments are typical sandstones of the coastal-marine genesis: sandstones, sands, aleurolites, aleurites, clays with horizontal oolitic iron ores [1-3]. Currently, this deposit is not being developed due to unfavorable mining conditions, i.e. increased water content in overlying loose horizon. The detailed investigation of ore-hosting series indicated the fact that besides the iron in Upper Cretaceous Galkin suite there is enriched glauconite deposits (fig.2).

Glauconite group includes stratified low-temperature magnesian-ferrous hydromica, conditional formula (K, Na, Ca)(Fe, Al, Mg, Mn)$_2$(Si, Al)$_4$O$_{10}$(OH)$_2$, having mica texture and diverse chemical composition [4]. The minerals in this group include random and /or rarely non-random dioctahedral lattice of subtype 1M and 1 Md, as well as uniform micro-concretion forms of aggregates which formed in marine sediments [5]. The criterion proving that glauconite is an indicator of ancient sedimentation conditions has been discussed in such publications as [6-13]. However, the issue concerning the diversified application of such a non-traditional mineral source has only been recently reflected in different publications.

Due to its specific properties – coloring oxides, active potassium cations, laminated structure-glauconite is a valuable source applicable in agriculture. These minerals are used as compound mineral fertilizers, in livestock farming and fishery, production of emulsion and oil paints, water waste treatment etc [14-18]. Glauconite fraction in agriculture concentrates should be more than 40 % (average 60…70 %).

[a] Corresponding author: rudminma@tpu.ru

DOI: 10.1051/matecconf/20168501016

Excluding the commercial value of glauconite rocks in Bakchar deposit, most probably, these rocks would be either extracted into disposal areas (during pit mining) or be concentration tailings. This fact itself significantly underestimates the true value of this deposit.

Consumers of glauconite products could be regional agrarian enterprises and petroleum refining sector, as well as chemical enterprises.

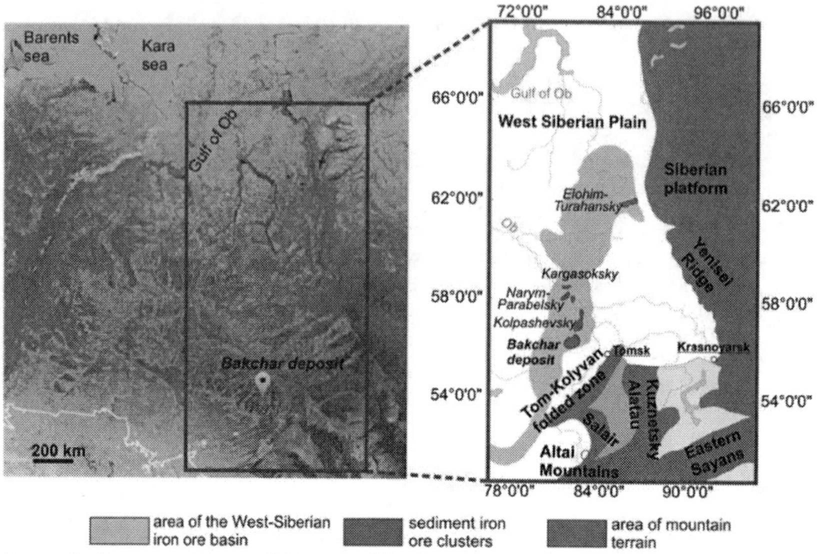

Figure 1. Overview map of Western-Siberian iron ore basin and Bakchar deposit

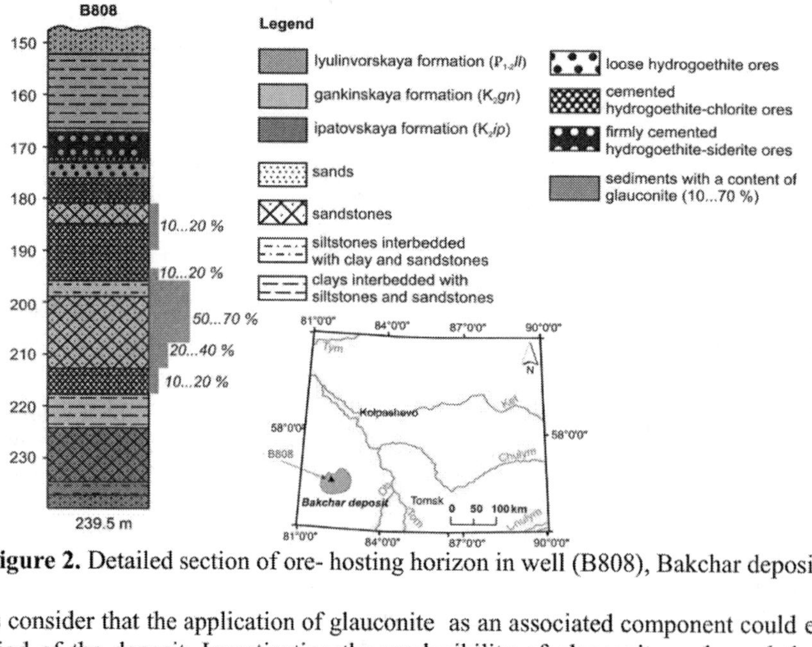

Figure 2. Detailed section of ore- hosting horizon in well (B808), Bakchar deposit

The authors consider that the application of glauconite as an associated component could enhance the pay-off period of the deposit. Investigating the producibility of glauconite rocks and their possible application in different economic sectors is a challenging advanced research task.

2. Factual material and research methods

The research material include core samples. All the glauconite rock samples were studied under petrographic microscope before developing the processing route. Glauconite liberation is based on "wet" screening to split the samples into granulometric classes: more than 1 mm, 1…0.5 mm, 0.5…0.2 мм, 0.2…0.1 mm, less than 0.1mm. Granulometric classes 0.5…0.2 mm and 0.2…0.1 mm were split in electromagnetic ore separator (EVS 10/5) at current 3.5…1.8 A, on unmagnetic (more than 3.5 A), submagnetic (3.5…1.8 A) and magnetic (less than 1.8 A) fractions. In each electromagnetic fraction the glauconite content is calculated by gravimetric method with monofraction selection under binocular microscope. The optimal processing route was developed on the basis of a small technological sample of 2 kg.

Glauconite concentrate produced from the small technological sample was tested to evaluate its possible application in agriculture. Soil glauconite concentrate inoculation was performed in the following three steps. Soil dry glauconite inoculation included concentrations of 30,60 and 90 kg/hectar, and further distilled watering. Seed steeping in glauconite solution concentration of 0.0015, 0.005 %, 0.01 %, 0.05 %, 0.1 % for 30 minutes before sowing and succeeding sowing and distilled watering. The last step – soil watering by 20mliter glauconite solution concentration of 0.03 %, 0.06 %, 0.09 %. All three- variant experiments were conducted in Petrie dish. Oat (Avéna sativa) germinating energy was determined on the 4th experiment day, while germination – 7th day. Plant height, herbage and dry mass die out on the11^{th} day.

3. Results and discussion

Comparable to the Cauvery glauconite basin (India) [8]. Bakchar deposits include glauconite pellets, formed as a result of altered fecal pellets and glauconite infilling, developed within pores, cracks of fossil bioclasts. Glauconite pellets have isometric rounded (globular, nodular) morphological forms [19]; infilling – elongated globular, bunchy biomorphic aggregate compounds. The two major color variations were pistachio-green, green and greenish-yellow, greenish-brown. There are also grains of bluish-green, practically black, as well as yellow with light greenish hue. In this paper the terms glauconite "globule", "granule", "grain" and "ooid" are considered to be synonyms.

Crystal-chemical formula has an averaged form: $K_{0.51}(Fe_{1.83}Mg_{0.24}Ca_{0.05})_{2.12}[Si_{3.27}Al_{0.69}O_{10}](OH)_2nH_2O$. Content of basic oxides in glauconite ranges within the following limits: K_2O – 2.5…7.7 %, $Fe_2O_{3(total)}$ – 20.2…69.4 %, MgO – 1.7…2.8 %, CaO – 0.5…0.9 %, Al_2O_3 – 5.5…11.5 %, SiO_2 – 32.5…52.5 %, H_2O – 8.3…12.3 % (difference). Internal structure of glauconite globules is randomly distributed pallets and flakes with jagged contours.

It is well-known that [20] the producibility of raw glauconite is determined by the degree of glauconite-containing rock cementation and grained glauconite amount in the rock. In the Bakchar deposit cross-section glauconite rocks are confined to Gankin suite and extends to a depth of 180…230 meters (fig.2). Also glauconite rocks occur in the Upper Jurassic deposits of Western Siberia [21]. Sediments containing more than 10% glauconite and could be divided into three types, different in composition, texture and technological properties.

Glauconite ores (glauconitolite) (fig.3-a) includes 50…70 % granulated glauconite with insignificant impurities of terrigenous material. Rocks are bright green, rarely, dark green and tobacco green, usually nonstratified, sometimes, indistinctly stratified due to sunhorizontal pelitomorphic clots. The texture is aleuritic-(arenaceous) psammitic (rarely aleuritic), fine-grained, weakly-cemented. Cement is basal-porous, fine-grained, hydromica composition. Sometimes, one can observe micro-grained siderite impurities or thin veinlets, as well as phosphate and pyrite nodules in the cement. The fraction content of 0.2…0.1 mm includes 37…40 %; 0.5…0.2 – 25…34 % and less than 0.1 – 25…31 %. Glauconites are predominating green, tobacco, rarely black of less than 0.2 mm. Globules have blade-like form. There is detritus in the rocks, whereas in some areas- coating sediment components. Glauconite ores possess the most favorable phase, morphological and technological criteria for industrial application. Applying these rocks as mineral fertilizers, nutritional environmental components (during reclamation etc.) and effective sorbents could be used without preliminary enrichment. Experimentally electromagnetic separation exhibited that magnetic fraction is "rich"

DOI: 10.1051/matecconf/20168501016

glauconite concentrates with commercial component content of 90…100 %, which, in its turn, could be used as a coloring pigment.

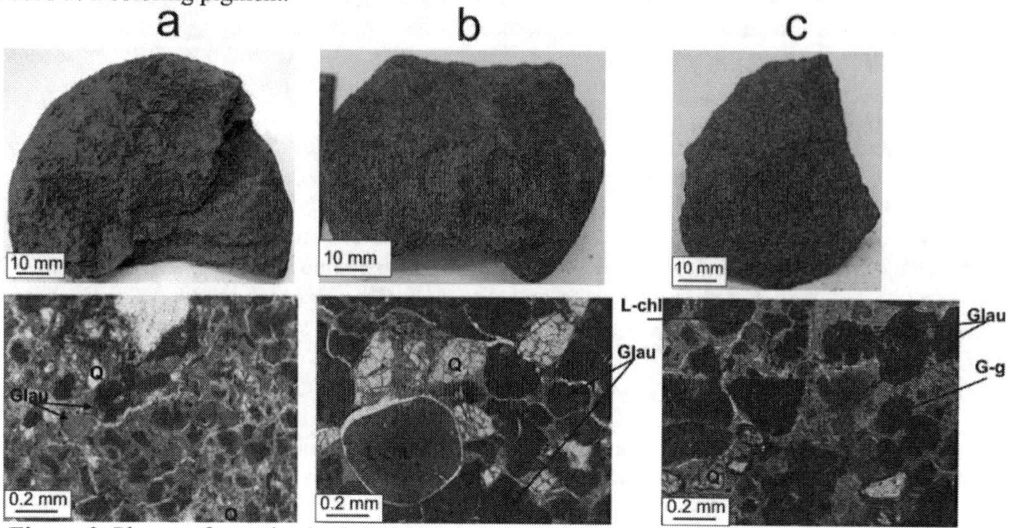

Figure 3. Photos of samples (top) and their images in transmission (bottom): a – glauconite ores (glauconites), b –glauconite sandstone, c – hydrogoethite-chlorite ore with glauconite, Q – quartz, Glau – glauconite, L-chl – leptochlorite ooid, G-g – hydrogoethite oolites

Table 1. Technological indicators of different glauconite rock categories

		Glauconite ores		Glauconite sandstone		Hydrogoethite-chlorite ore with glauconite	
Fraction, mm		0.5…0.2	0.2…0.1	0.5…0.2	0.2…0.1	0.5…0.2	0.2…0.1
Output, %		25.8	42.5	55.4	18.1	58.7	30.7
Output, %	magnetic	47.1	92.6	66.6	72.6	25.9	19.0
	submagnetic	51.6	5.3	6.8	5.5	9.9	9.7
	unmagnetic	1.3	2.1	26.6	21.9	64.2	71.4
Glauconite content , %		98.7	95.8	33.4	50.9	21.3	16.5
Glauconite content , %	magnetic	100.0	100.0	50.0	70.0	65.0	70.0
	submagnetic	100.0	60.0	1	2	45.0	33.0
	unmagnetic	0	0	0	0	0	0

Glauconite sandstone (fig.3-b) includes 20…40 and up to 50 % of grained glauconite, remaining is terrigenous matter, leptochlorite ooids and cement. The rocks are yellowish, greenish, dull tobacco, non-stratified. Texture- psammitic, medium- grained, weakly cemented. Cement – clay (hydromica), porous, micro-grained. The most significant output is granulometric fraction 0.5…0.2 mm– 45…60 %. Tobacco and brownish glauconite has the following sizes 0.4…0.1 mm, whereas grain size is predominately less than 0.2 mm. Through enrichment these rocks include the magnetic fraction of such concentrates as glauconite – 60…70 %, hydrogoethite oolites and leptochlorite ooids – 10…20 %, and clay particles up to 30 %. Technological properties indicate that this concentrate is applicable in agriculture, in livestock farming and fishery, as well as land reclamation.

Hydrogoethite-chlorite ores with glauconite (fig.3-c) include hydrogoethite oolites of 20…25 %, leptochlorite ooid– 15…20 %, glauconite – 10…20%, quartz – 10 %. The cement content in these rocks changes from clay to leptochlorite and in some cases, hydrogoethite. The cement is fine-crystalline and basal. Rocks are of tobacco color, non-stratified. Texture- psammitic, medium- grained to fine-grained, cemented sometimes, loose. Fraction content for 0.5…0.2 mm is 53…59 %, 0.2…0.1 – 30…38 % and less than 0.1 – 9…13 %. Iron oxide content in these ores changes from 26 to 36 %, which could be considered to be "poor" within this deposit. Glauconite grains are rounded, blade-like, micro-flaked of pistachio-green and tobacco color with numerous syneresis cracks. Different oxidized globules with brown crust of hydrooxide iron are predominate. Magnetic fraction in these rocks is glauconite of 50…70 %, remaining being hydrogoethite oolites, leptochlorite ooids and aleuritic-clay clots of cementing matter. According to the technological criterion and composition this concentrate could be used in agriculture.

It should be noted that glauconite is always found in hydrogoethite-chlorite ores with a fraction of less than 10 %. Glauconite concentrates could be produced even if there is an insignificant content of this mineral under electromagnetic enrichment of this ore.

An integrated optimal scheme of producing glauconite concentrates was developed on the basis of the above-described glauconite rocks (fig. 4). Agrarian analysis involved the concentrate obtained from a small technological sample. Small technological sample including combined samples from three wells at the depth of 184…186 m are "poor" hydrogoethite-chlorite iron ores with glauconite. Glauconite fraction in this sample is 18 %. The authors used this rock type to evaluate the possible associated extraction and application of glauconite in agriculture, being obtained from "poor" glauconite-containing deposits. As a result of enriching small technological sample according to described scheme (fig.4) glauconite concentrate was produced. This concentrate (fig.5) includes granulometric glauconite of 71 % and 29 % from hydrogoethite oolites, leptochlorite ooids and aleuritic-clay clots of cementing matter.

The results of agrarian analysis proved the positive effect of glauconite concentrate on the growth of plants, especially oats (*Avéna satíva*). Soil dry glauconite inoculation where the concentrations are 3…9 gr/m^2 for 2.6…8.6 % increased germination energy up to 3.4…10 %, at small and medium concentrations (3 and 6 gr/m^2) increases herbage up to 8.2 and 11.7 %, respectively. If seed steeping in glauconite solution with concentrations of 0.001…0.1% for 1.4…7.4 % increased plant growth and at 2…4 % – germination energy of oats. If watering the soil with glauconite solution of 0.03…0.09 % concentration the growth and development of plants insignificantly change.

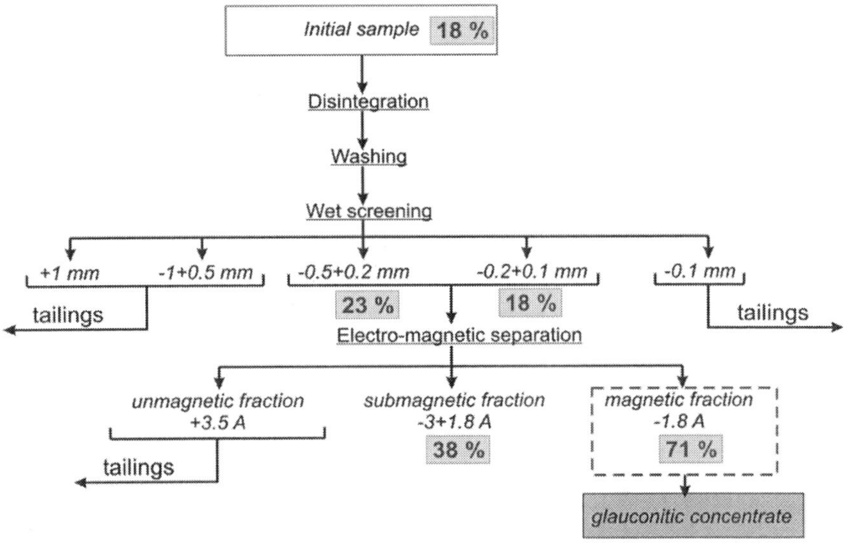

Figure 4. Technological scheme of producing glauconite concentrate from Bakchar deposit rocks. Green numbers refer to the glauconite content in corresponding products (fraction) from small technological sample of hydrogoethite-chlorite ore with glauconite.

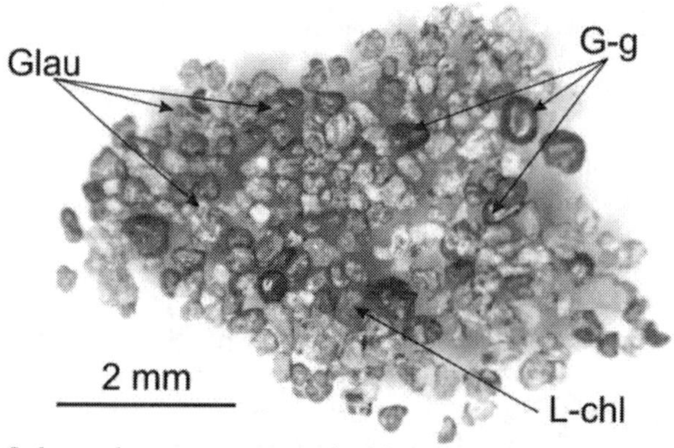

Figure 5. Photo of glauconite concentrate under binocular mircoscope, produced from small technological sample: Glau – glauconite, L-chl – leptochlorite ooids, G-g – hydrogoethite oolites

4. Conclusion

Several types of glauconite containing rocks were identified in the ore-hosting horizons of Bakchar deposit: glauconite ores (or glauconitolites), glauconite sandstones, hydrogoethite -chlorite iron ores with glauconite. Based on the research results the optimal processing route in producing glauconite concentrates from identified rocks was developed. Depending on the glauconite content in initial samples (rock type) concentrate enrichment was possible for further application. A product containing commercial component of 90…100 % was produced from glauconite ores. According to technological criteris this product is applicable in all economic segments where there is a demand for raw glauconite. Glauconite sandstones and iron ores with glauconite content of more than 10% under enrichment are especially applicable in agriculture.

Agrarian analysis showed that adding glauconite concentrate produced from enriched "poor" iron ores could increase oat germination. Soil glauconite concentrate (3…9 gr/m^2) inoculation results in increasing germinating energy, germination and herbage.

Obtained results prove the fact that it is more profitable and efficient to use Bakchar glauconite in agriculture. As a secondary by-product raw glauconite could further the development of this deposit.

Acknowledgements

This study was supported by the Russian Foundation for Basic Research and Tomsk Region Administration (project no. 16-45-700090). The authors express their thanks to employees of the Siberian Research Institute of Agriculture and Peat Zmeeva O.N., Proskurina L.D. and Petrova L.V. for the performance of agricultural research.

References

1. M.A. Rudmin, A.K. Mazurov, A.S. Ruban. *IOP Conf. Ser. Earth Environ. Sci.* **24.** 012028 (2015)
2. M.A. Rudmin, I.V. Reva, A.P. Gunko, A.K. Mazurov, R.N. Abramova. *IOP Conf. Ser. Earth Environ. Sci.* **27.** 012026 (2015)
3. West Siberian iron ore basin (1964)
4. I. Nikolaeva. Minerals of Glauconite Group in Sedimentary Formations 319 (1977) (in Russian)
5. V.A. Drits, A.G. Kossovskaya. *Lithology and mineral resources.* **5.** 563-580 (1986)

6. S. Banerjee, S.L. Chattoraj, P.K. Saraswati, S. Dasgupta, U. Sarkar. *Marine and Petroleum Geology.* **30.** 144–160 (2012)

7. S. Banerjee, S.L. Chattoraj, P.K. Saraswati, S. Dasgupta, U. Sarkar, A. Bumby. *India Geological Journal.* **47.** 357–371 (2012)

8. S. Banerjee, U. Bansal, K. Pande, S.S. Meena. *Sedimentary Geology.* **331.** 12–29 (2016)

9. S. Banerjee, U. Bansal, A.V.Thorat. *Journal of Palaeogeography.* **5 (1).** 43-71 (2016)

10. A.G. Kossovskaya, V.A. Drits. *Sedimentology.* **15.** 83–101 (1970)

11. A. Meunier, A. El Albani. *Terra Nova.* **19.** 95–104 (2007)

12. G.S. Odin, A. Matter. *Sedimentology.* **28.** 611–41 (1981)

13. R. Wigley, J.S. Compton. *Research II.* **54.** 1375–1395 (2007)

14. E. Dutova, I. Vologdina, D. Pokrovsky, N. Nalivaiko, K. Kuzevanov, V. Pokrovsky. *IOP Conference Series: Earth and Environmental Science.* **33 (1).** 012017 (2016)

15. V. Lyashko, V. Zabilov, I. Dobrovolski. *AIC Russia.* **72(2).** 141-148 (2015)

16. S.S. Mandzhieva, T.M. Minkina, S.N. Sushkova, G.V. Motuzova, T.V. Bauer, V.A. Chapligin *American Journal of Agricultural and Biological Science.* **9 (1).** 19-24 (2013)

17. N. Rahimzadeh, F. Khormali, M. Olamaee, A. Amini, E. Dordipour. *Biology and Fertility of Soils.* **51 (8).** 973-981 (2015)

18. A.K. Yapparov, L.M. Bikkinina, I.A. Yapparov, S.A. Aliev, A.M. Ezhkova, V.O. Ezhkov, R.R Gazizov. *Eurasian Soil Science.* **48 (10).** 1149-1158 (2015)

19. M.A. Rudmin, A.K. Mazurov, L.M. Bolsunovskaya. *IOP Conf. Ser. Earth Environ. Sci.* **21.** 012003 (2014)

20. E. Levchenko, N. Ptyk-kara, M. LEvchenko. *International Geological Congress* (2008)

21. L. Krasnoschekova, D Cherdansteva, R. Abramova. *IOP Conf. Ser. Earth Environ. Sci.* **24.** 012018 (2015)

MATEC Web of Conferences 85, 01017 (2016) DOI: 10.1051/matecconf/20168501017
CCT 2016

Adsorption of inorganic ions from aqueous solutions using mineral sorbent - tripoli

Tatyana Yurmazova[1], *Nina* Shakhova[1,*] , and Hoang Tran Tuan[1]

[1]Tomsk Polytechnic University, Tomsk, Russia

Abstract. The present research considers the adsorption of $H_2AsO_4^-$, CrO_4^{2-}, Ni^{2+}, Fe^{3+} ions from their aqueous solutions by mineral sorbent - tripoli. Tripoli was characterized by different physico-chemical methods such as X-ray phase analysis, inductively coupled plasma atomic emission spectrometry, method of thermal desorption of nitrogen. Contact time, specific surface area, specific pore volume and surface charge of tripoli have been determined. The effect of tripoli surface area modification by iron (III) oxide-hydroxide on tripoli sorption capacity for the arsenic anions has been investigated. The maximum adsorption was found to occur within 30 minutes of contact time. Different models including the pseudo-first-order kinetic and the pseudo-second-order kinetic equations were used to analyse kinetic data. All the models being considered, it has been stated that the pseudo-second-order kinetic model is the most appropriate to describe the adsorption behaviour of Fe^{3+} ions on tripoli. The adsorption has been explained in terms of Langmuir and Freundlich isotherms. Based on values of correlation coefficients, $H_2AsO_4^-$, CrO_4^{2-}, Ni^{2+}, Fe^{3+} sorption isotherm data were better fitted by Langmuir model. It has been detected that the modification of mineral sorbent (tripoli) leads to the increase in $H_2AsO_4^-$ adsorbing capacity of tripoli. In general, the results indicated that tripoli can be an efficient low-cost sorbent for removing $H_2AsO_4^-$, Ni^{2+}, Fe^{3+} ions from aqueous solutions.

1 Introduction

Existing methods for water purification from $H_2AsO_4^-$, CrO_4^{2-}, Ni^{2+}, Fe^{3+} ions are based on processes of oxidation, sedimentation, coagulation and membrane technologies. Nevertheless, they cannot be used to reduce concentrations of these ions to permissible levels. Nowadays sorption method is generally given preference over other water purification techniques due to its high efficiency, easy handling; besides, it is environmentally friendly [1-12]. Literature review [1-7] showed that mineral sorbents are cheap and available material capable to remove heavy metal ions from industrial wastewater.

Since mineral sorbents hardly ever have appropriate adsorption properties, they are to be chemically or physically modified [12-15]. Due to modification the maximum adsorption capacity of mineral sorbent increases severalfold. Thus, the search for low-cost

* Corresponding author: nina.b.shakhova@gmail.com

and effective adsorbents has become the focus of attention for many studies. Bannova et al. [12] studied the conditions and parameters of modification of tripoli in order to obtain a hydrophobic sorbent. Anisimov et al. [13] determined that synthetic chemically modified terra silica (complex sorbent) effectively decreases water hardness. In the papers [14, 15] the influence of ultrasonic treatment as well as thermal and microwave modification of tripoli were discussed. However, mechanism of adsorption has not been the subject of a detailed analysis.

The present study is focused upon adsorption of $H_2AsO_4^-$, CrO_4^{2-}, Ni^{2+}, Fe^{3+} ions from their aqueous solutions by mineral sorbent tripoli. The effect of various parameters affecting sorption behaviour such as contact time, specific surface area, tripoli surface charge, modification of the surface area of sorbent has been investigated to better understand the adsorption process and ways of possible usage of the mineral sorbent (tripoli) for wastewater purification.

2 Experimental studies

2.1 Sorbent

Low-cost locally available tripoli (Zikeevsk deposit, Russia) was selected as a natural mineral sorbent. The chemical composition of tripoli was defined by inductively coupled plasma atomic emission spectrometry using iCAP 6300 Duo ThermoScientific (USA). X-ray phase analysis with a Shimadzu XRD-7000 CuK_α-radiation diffractometer (Germany) was used to identify phase composition of tripoli. The specific surface area (S) of tripoli was determined by thermal desorption of nitrogen using specific surface analyser Sorbi-3M (Russia).

2.2 Adsorption studies

All chemicals used in this research were of analytical grade. To determine the optimal time required for reaching the sorption equilibrium, static adsorption was induced, with different contact time being considered. Equal quantity of 0.05g of dry adsorbent was mixed with 50 mL of the solution with equal initial concentration of Fe^{3+} ions 20 $mg \cdot L^{-1}$ over the periods corresponding to different contact time; and then the mixture was centrifuged with the centrifuge «Allegra 64R» (USA) for 10 min at 10000 rpm. The concentration of the residual Fe^{3+} ions in the solution was measured using spectrophotometer «ApelPD-303UV» (Japan).

The static adsorption of Fe^{3+}, Ni^{2+}, CrO_4^{2-}, $H_2AsO_4^-$ ions on tripoli was investigated using 100 mL flasks where 50 mL of Fe^{3+}, Ni^{2+}, CrO_4^{2-}, $H_2AsO_4^-$ ions solution of different initial concentrations and 0.05 g of tripoli were added at temperature of 25°C. The initial concentration of adsorbate ions in the solutions was in the range of 10-100 $mg \cdot L^{-1}$.

In 30 min, after the sorption equilibrium having been reached, the adsorbent was separated from the solution by spinning for 10 min at 10000 rpm and the equilibrium concentration was measured. The equilibrium concentration of Fe^{3+}, Ni^{2+}, CrO_4^{2-}, $H_2AsO_4^-$ ions was determined by chemical and physico-chemical methods. The equilibrium concentration of Fe^{3+}, Ni^{2+}, CrO_4^{2-}, $H_2AsO_4^-$ ions was measured using UV/visible spectrometer with appropriate reagents.

Equilibrium adsorption capacity of the adsorbent A_e ($mg \cdot g^{-1}$) was calculated by the formula:

$$A_e = \frac{(C_i - C_e)\cdot V}{m}$$

Where, A_e is the adsorption capacity, mg·g^{-1}; C_i is the initial concentration of ions in the solution, mg·L^{-1}; C_e is the equilibrium concentration of ions, mg·L^{-1}; V is the volume of ions solution used, mL; m is the mass of adsorbent used, g.

2.3 Effect of tripoli surface charge on adsorption

Indicator method was applied to detect the influence of the tripoli surface charge on the adsorption properties. Methylene blue (monovalent cationic dye) and eosin (monovalent an anionic dye) were used as indicators of tripoli surface charge. The static adsorption of methylene blue and eosin was analysed.

$C_{16}H_{18}ClN_3S\cdot HCl \rightarrow [C_{16}H_{18}ClN_3S]H^+ + Cl^-$ - methylene blue, a cationic dye; (Russia)

$Na_2[C_{20}H_6O_5Br_4] \rightarrow 2Na^+ + [C_{20}H_6O_5Br_4]^{2-}$ - eosin, an anionic dye (Russia).

Using Zetasizer Nano ZS instrument (Germany), the value of initial tripoli surface ζ - potential was measured as well as the value of surface ζ-potential of tripoli, with different amounts of adsorbed methylene blue being added.

2.4 Effect of chemical modification of tripoli surface area on adsorptive capacity

To enhance the adsorption properties of mineral sorbent the surface area of tripoli was subjected to chemical modification by iron (III) oxide-hydroxide (FeO(OH)).

The experiment on chemical modification of tripoli was carried out, with tripoli being treated with the ferric chloride (III) solution and then sodium hydroxide solution at temperature of 50°C. In each test 10 g of tripoli were added to the solution of ferric chloride (III) of different concentration (20 – 100 mg·L^{-1}), with the contact time being 2 hours. Then the solution was decanted and tripoli with adsorbed Fe^{3+} ions was treated with sodium hydroxide at temperature of 50°C. After that the solution was decanted again, and tripoli modified by iron (III) oxide-hydroxide (FeO(OH)) was washed with distilled water to reach pH 6.5 – 7.2. Finally, the samples were filtered and dried.

3 Results and discussion

Based on the results of inductively coupled plasma atomic emission spectrometry, the tripoli is composed of SiO_2 (89 %), Al_2O_3 (3.9 %), Fe_2O_3 (2.38 %), CaO (0.69 %), MgO (0.44 %), other components account for 3.59 %. X-ray diffractogram of tripoli is given in figure 1.

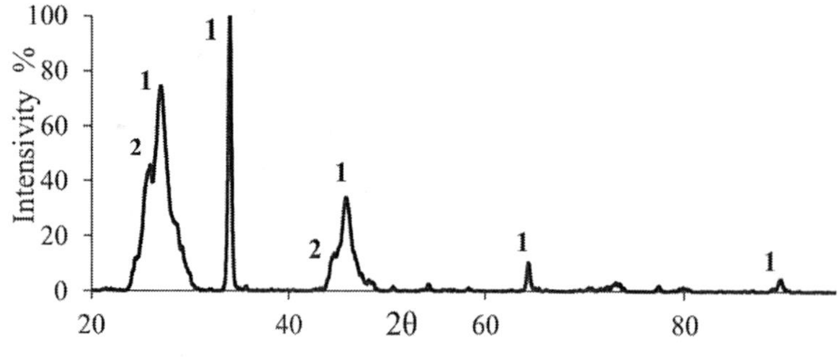

Fig. 1. XRD patterns of tripoli 1 – β quartz, 2 – β cristobalite.

 DOI: 10.1051/matecconf/20168501017

The diffractogram of tripoli showed that the dominant phase is SiO_2 (β quartz), with β cristobalite being present. The specific surface area and specific pore volume of tripoli were 94.6 $m^2 \cdot g^{-1}$ and 0.011 $sm^3 \cdot g^{-1}$, respectively.

The static adsorption of Fe^{3+} ions on tripoli was studied to determine the optimal time required for reaching the sorption equilibrium. The effect of contact time on the adsorption of Fe^{3+} ions by tripoli can be observed in figure 2. It has been indicated that the adsorption of Fe^{3+} ions tends to increase with time and attains the maximum value over 30 min and then remains constant. Therefore, all further adsorption experiments were carried, with this reference time being considered. The kinetics of the adsorption of Fe^{3+} ions on tripoli was analyzed by using pseudo-first-order and pseudo-second-order kinetic models [4, 5]. The linear plot of kinetics based on pseudo-second-order model and the correlation coefficient are shown in figure 3.

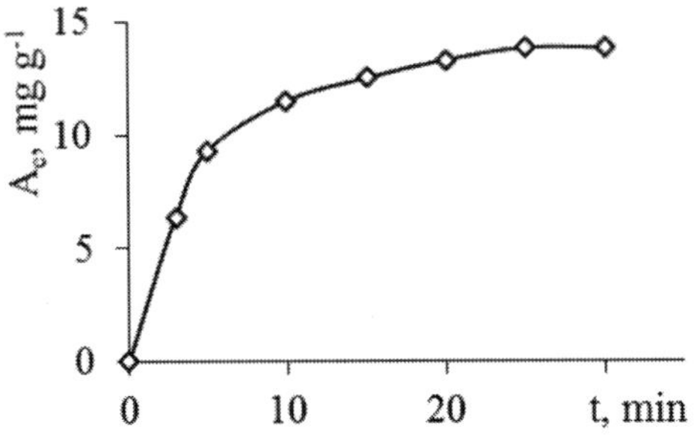

Fig. 2. The kinetics of Fe^{3+} ion adsorption on tripoli.

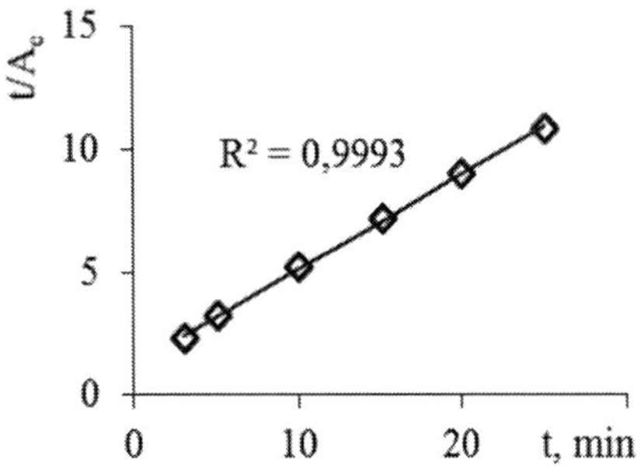

Fig. 3. Pseudo-second-order kinetic plot for the adsorption of Fe^{3+} on tripoli.

It can be observed that the values of the correlation coefficient for both the pseudo-first-order and pseudo-second order models are close to unity. However, the curve-fitting plots of t/A_e versus t give a straight line, with the correlation coefficient close to 1 (R^2=0.999) confirming the applicability of the pseudo-second-order equation. It can be assumed that the adsorption rate is determined not only by the speed of diffusion processes, but also by the rate of chemical interaction between the sorbate and the sorbent. The adsorption of $H_2AsO_4^-$, CrO_4^{2-}, Ni^{2+}, Fe^{3+} ions using tripoli is shown in figure 4.

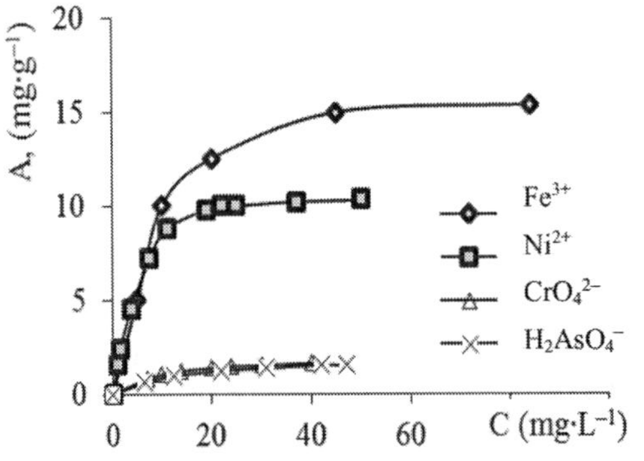

Fig. 4. Adsorption isotherm of $H_2AsO_4^-$, CrO_4^{2-}, Ni^{2+}, Fe^{3+} on tripoli.

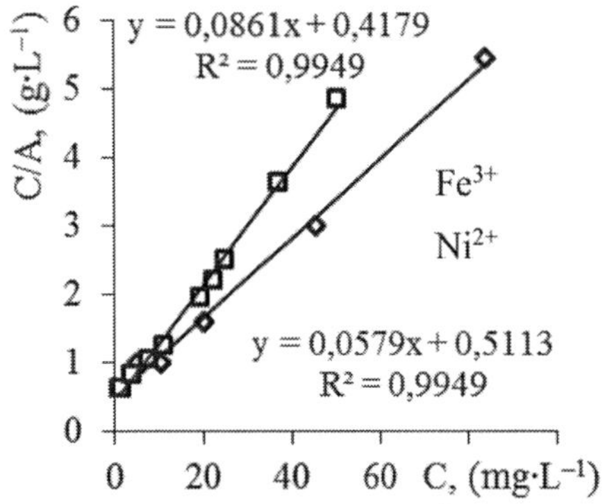

Fig. 5. Lengmuir plots for the adsorption of Ni^{2+}, Fe^{3+} on tripoli.

From figure 4, it can be observed that the uptake of Ni^{2+}, Fe^{3+} ions on tripoli is much better than adsorption of $H_2AsO_4^-$, CrO_4^{2-}. The capacity of tripoli to adsorb $H_2AsO_4^-$, CrO_4^{2-}, Ni^{2+}, Fe^{3+} ion is shown in table 1.

The adsorption isotherm data were analyzed in terms of Langmuir and Freundlich isotherms (figure 5). Langmuir and Freundlich isotherms are the most widely used models for solid-liquid adsorption [4-6]. Based on the values of the correlation coefficients, it can be inferred that the equilibrium adsorption data were better described by the Langmuir adsorption model. The Langmuir model assumes that each adsorbate molecule is located at specific homogenous sites within the adsorbent. Thus, it predicts the formation of a monolayer of the adsorbate on the homogenous adsorbent surface and does not consider surface heterogeneity of the adsorbent.

Table 1. Parameters for ion adsorption on tripoli.

Ions	Fe^{3+}	Ni^{2+}	CrO_4^{2-}	$H_2AsO_4^-$
A_∞ ($mg \cdot g^{-1}$)	17.27	11.614	1.9988	1.97
K ($L \cdot mg^{-1}$)	0.113	0.206	0.102	0.07

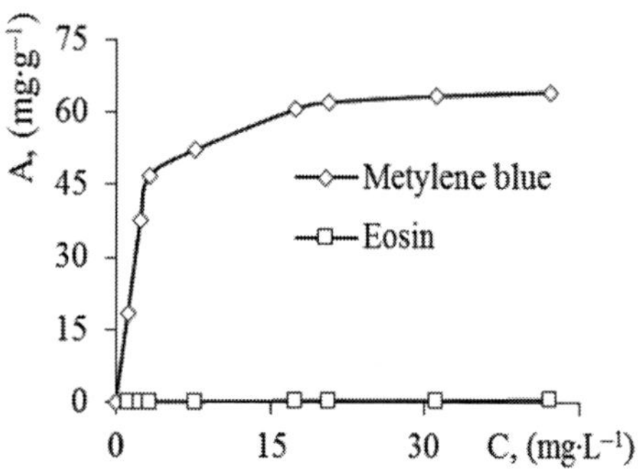

Fig. 6. Adsorption isotherm of methylene blue and eosin on tripoli.

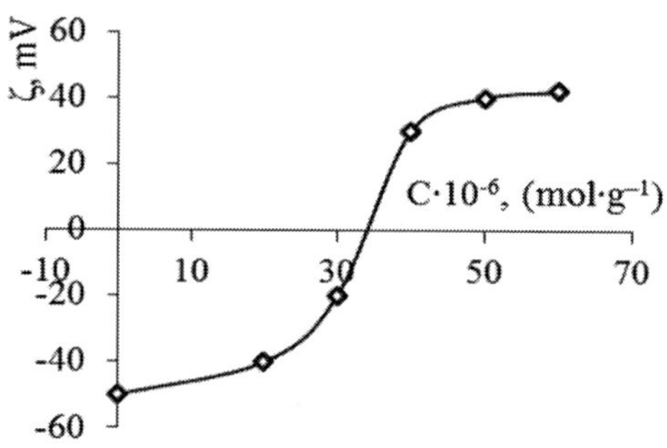

Fig. 7. The dependence of the tripoli ζ-potential on the amount of adsorbed methylene blue.

Figure 6 shows the effect of the tripoli surface charge on the adsorption properties with the use of indicator method. Based on these data, it can be seen that methylene blue uptake by tripoli is much better than eosin uptake. This can be explained by the fact that the tripoli active sites are negatively charged.

The investigation of the value of initial tripoli surface ζ-potential, as well as the value of surface ζ-potential of tripoli with different amounts of adsorbed methylene blue (figure 7) indicates that ζ-potential behavior is described by a curve passing through a zero value. When methylene blue being adsorbed, both the value and ζ-potential charge change. This proves that charge interactions between the adsorbent and adsorbate are of great significance.

In this study, the chemical modification of mineral sorbent (tripoli) is performed to enhance the capacity of tripoli to adsorb $H_2AsO_4^-$ ions. Arsenic concentration above acceptable standards (0.05 $mg \cdot L^{-1}$) in drinking water has been detected in many countries and, thus, the removal of arsenic from drinking water is a global issue [1-3]. Iron (III) oxide-hydroxide (FeO(OH)) was used to modify the tripoli surface area due to low solubility of ferric arsenate (K_{sp} ($FeAsO_4$)= $5.8 \cdot 10^{-21}$) and non-toxicity of Fe^{3+} ions.

The values of specific surface area and specific pore volume of initial tripoli and modified tripoli are given in table 2. As shown in table 2, the studied adsorbents are characterized by low pore volume and high specific surface area. Therefore, we can assume that in case of water treatment with tripoli no intraparticle diffusion processes occur, which, in its turn, will speed up sorption equilibrium attainment.

Table 2. Characterization of adsorbent.

Adsorbent	**Concentration of iron ions, (%)**	**Specific surface area, ($m^2\ g^{-1}$)**	**Specific pore volume, ($sm^3\ g^{-1}$)**
Tripoli	0.38	94.6	0.011
Tripoli modified by FeO(OH) (sample 1)	0.96	98.8	0.042
Tripoli modified by FeO(OH) (sample 2)	2.8	114	0.052

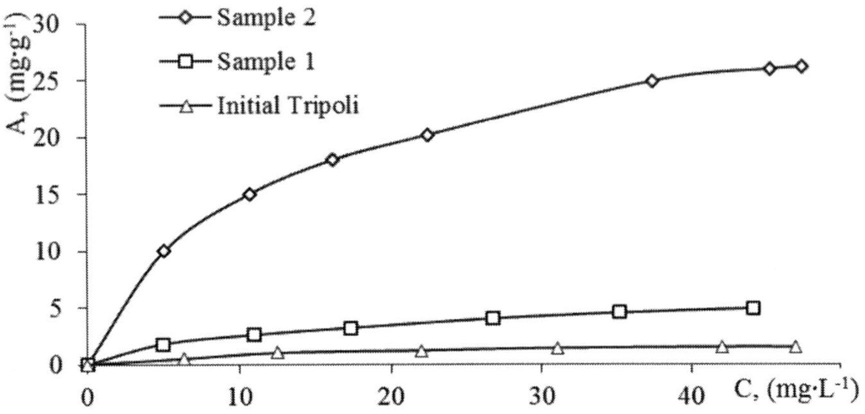

Fig. 8. Adsorption isotherms of $H_2AsO_4^-$ ions on initial tripoli and tripoli modified by FeO(OH).

Adsorption isotherms of $H_2AsO_4^-$ ions on modified tripoli are shown in figure 8. From this figure, it can be observed that static absorption of $H_2AsO_4^-$ ions on modified tripoli is much better than the uptake of $H_2AsO_4^-$ ions on initial tripoli. An increase in the adsorption capacity of tripoli modified by FeO(OH) is due to formation of insoluble ferric arsenate with extremely low value of the solubility product constant ($K_{sp} = 5.8\cdot10^{-21}$). From figure 8, it can be seen that maximum adsorption capacity of $H_2AsO_4^-$ ions increases severalfold. The maximum adsorption capacity of $H_2AsO_4^-$ ions increased with the increase in the amount of iron (III) oxide-hydroxide in modified tripoli. The maximum adsorption capacity of tripoly for $H_2AsO_4^-$ ions is A_∞=33.33 $mg\cdot g^{-1}$ which is equivalently high level.

To assess the stability of the compounds formed as a result of arsenic ions sorption on tripoli the desorption process of arsenic ions in distilled water was carried out. Periodical sampling during a month showed that the arsenic concentration does not exceed the level of trace concentrations. Sediments obtained as a result of arsenic ions sorption on tripoli belong to the class of low-solubility compounds which are suitable for disposal [11].

4 Conclusion

1. The adsorption of $H_2AsO_4^-$, CrO_4^{2-}, Ni^{2+}, Fe^{3+} ions from their aqueous solutions by mineral sorbent tripoli has been studied. It has been stated that the maximum adsorption occurs within 30 min of contact time. Based on high correlation coefficient value, the sorption equilibrium data for Fe^{3+} ions were better fitted to the pseudo-second-order kinetic model.

2. The maximum adsorption capacity of tripoli for $H_2AsO_4^-$, CrO_4^{2-}, Ni^{2+}, Fe^{3+} ions has been determined. It has been deduced that the sorption of $H_2AsO_4^-$, CrO_4^{2-}, Ni^{2+}, Fe^{3+} ions was better described by the Langmuir absorption model. This means that each adsorbate

molecule is located at specific homogenous sites within the adsorbent, with a monolayer of the adsorbate on the adsorbent surface being formed.

3. The study has indicated that tripoli can be an effective adsorbent for the removal of Ni^{2+}, Fe^{3+} ions from aqueous solutions. This is due to the fact that surface charge of tripoli is negative, which is confirmed by recharging of the surface in course of methylene blue sorption by tripoli.

4. This research suggests that the mineral sorbent – tripoli can be effectively used as an adsorbent for the removal of Ni^{2+}, Fe^{3+} ions from aqueous solutions as well as tripoli modified by FeO(OH) can be effectively used in the removal of $H_2AsO_4^-$ ions in different concentrations. The research can be carried on further to study application of tripoli to water treatment.

References

1. G. Saveliev, T. Yurmasova, N. Shakhova Bulletin of Higher Sch. Chem. and chem. technology **53,** 36–39 (2011) (in Russian)
2. M. Dinesh, U. Charles, Jr. Pittman J. of Hazard. Mater. **142,** 1–53 (2007).
3. A. Malik, Z. Khan, Q. Mahmood, S. Nasreen et al. J. of Hazard. Mater. **168**, 1–12 (2009)
4. K. Nebaghaa, K. Ziata, L. Rghiouib, M. Khayetd, M. Saidia, K. Aboumariaf, A. Hourchg, S. Sebtih J. Mater. Environ. Sci. **6**, 3022–33 (2015)
5. K. Nebaghaa, K. Ziata, L. Rghiouib, M. Khayetd, A. Najif, M. Saidia J. Mater. Environ. Sci. **6**, 2694–2702 (2015)
6. P. Shah, N. Raval, N. Shah J. Mater. Environ. Sci. **6**, 2573–82 (2015)
7. A. Alzaydien American J. of Environm. Scien. **5** (3), 197–208 (2009)
8. M. Zankuli, H. Devlin, N. Silikas Dental Materials **30**, 324–329 (2014)
9. M. Avdeev, N. Blagoveshchenskii, V. Garamus, A. Novikov, A. Puchkov Crystallography Reports **56**, 1090–95 (2011)
10. D. Martemiyanov, A. Galanov, T. Yurmazova Fundamental research **8**, 666–670 (2013) (in Russian)
11. G. Pushkareva, K. Kovalenko Mining Informational and Analytical Bulletin (scientific and technical journal) **9**, 294–298 (in Russian) (2012)
12. E. Bannova, E. Zaloznaya, N. Kitaeva, S. Merkov, M. Muchkina, A. Chaban, A. Alekseev J. Water: chem. and ecology **11**, 73–78 (2012) (in Russian)
13. V. Anisimov, K. Petrov, A. Ratnikov, P. Martynov, E. Podzorova, A. Chaban J. Water: chem. and ecology **10**, 11–16 (2009) (in Russian)
14. A. Shilina, V. Milinchuk, E. Podzorova, P. Martinov, A. Chaban J. Water: chem. and ecology **11,** 25–29 (2009) (in Russian)
15. Y. Pyatko, R. Akhmetova, A. Khatsrinov, V. Fakhrutdinova, A. Akhmetova, A. Gubaydullina Fundamental research **12**, 320–324 (2015) (in Russian)

INFLUENCE OF HOMOGENIZATION OF CLAY BATCH FOR LABORATORY TEST ON THE STRUCTURE OF THE CERAMIC SAMPLES

Nikita Toropkov [1,*], *Alexander Abakumov*[1]

[1]Tomsk Polytechnic University, Lenin av. 30, Tomsk, 634050, Russia

Abstract. In this paper, we study the chemical and mineralogical composition of the Krasnoyarsk Kray clay used in the production of coarse ceramics. It is found that the fire –strength depends of method of preparation of clay paste. It is shown that the forced dissolution of the raw clay material effectively homogenizes clay paste. As a result, strength of the samples increased during firing. It is also justified by the influence of their mineralogical composition and there method of the preparation. It is proved that conditions of preparation and mineralogical composition affect equally.

1 Introduction

Batch preparation is an important technological step in the production of ceramic construction materials, which significantly determine the stability of the parameters of technological processes at the stages of forming, drying and firing.

Changes in moisture, mineralogical and fractional composition of batch lead to variability in product quality [1, 2]. Therefore, one way to improve the quality of the product is to optimize characteristics of batch raw material with respect to homogeneity criteria. Homogeneity of batch composition can be achieved at the stages of dosing, and during processing of clay paste by department of mass-preparation and forming [1, 3].

Special attention should be paid to the technological methods of humidity correction of clay paste, which determine average moisture content of the total volume of batch [4, 5]. The technological effect is a change of rheological properties of clay –water system, according to the amount of correction additives and the method of their distribution due to the batch volume. In addition, interaction of water and dispersed particles of a solid phase plays a specific role [6]. The technological properties of the clay –water system depends on the amount and the method of mechanical processing, thermal regime during mass processing, as well as the time course of physical and chemical processes of water and particles of a dispersed solid phase [7].

Thus, batch of the same composition may represent different technological parameters depending on the particular implementation of mass processing [8]. The objectives of this study are to determine the effect of different methods of mass processing on the final

* Corresponding author: zerogooff@gmail.com

properties of samples of coarse ceramics in laboratory testing [3, 9]. One of the tasks of laboratory research is to produce prototypes

Based on the results it is possible to determine not only the e ways to improve quality of the final product by varying the mass processing, but also the feasibility of introducing additional operations into the existing technological schemes.

2 Materials and methods

Investigation of properties of clay raw materials was carried out on three types of clay loam, which is typical for clay deposits in Krasnoyarsk Kray. GL1 is an average sample of low-melting red-burning clay, GL2 and GL3 are average samples of Kantatsky deposits: fireproof and refractory clay, respectively. These clay deposits are used as a basic raw material component in the current production of ceramic wall products in accordance with the GOST 530–2012. The properties of clay raw materials are shown in Table 1.

Table 1. Characteristics of clay raw materials of Krasnoyarsk Kray

Index	**Types**	**Plasticity**	**Grain size distribution by Rutkowski**	**Smallest capillary moisture**
GL1	Red-burning, fusible	moderately plastic	Clay loam	12,1%
GL2	White-burning, fireproof	moderately plastic	Clay loam	12,7%
GL3	White-burning, refractory	moderately plastic	Clay loam	11,9%

Granulation composition was determined by the Rutkowski method [2]. The lowest capillary moisture was determined by the procedure of excess moisture extraction on a press at constant pressure [10].

Table 2 shows the chemical composition of the oxide raw clay determined by an electron microscope with analyzing attachment JEOL–9000. It should be noted that the electron microscopy data differ from those obtained by the other methods of analysis.

Table 2. Oxide chemical composition of clay raw materials of Krasnoyarsk Kray

Oxide	**GL1**	**GL2**	**GL3**
SiO_2	58,09%	61,71%	57,4%
Al_2O_3	16,82%	27,66%	28,41%
Fe_2O_3	11,92%	5,33%	7,58%
MgO	3,09%	1,48%	0,75%
CaO	6,25%	2,15%	1,28%
Na_2O	1,44%	0,02%	0,33%
K_2O	2,39%	1,65%	1,85%
TiO_2	–	–	2,39%

Also, X–ray analysis using a DRON–3M device was performed. The samples were irradiated with monochromatic Cuα radiation with a wavelength of 1.54056 Å due to the following parameters: accelerating voltage 90 κW, beam current 30 mA, scanning angle range 10–80°, scanning step 0,03°, set signal time 1s. Synchronous thermal analysis was carried out on the STA 449 F3 Jupiter.

Clay raw materials samples were dried to the air humidity and crushed by a jaw crusher to an average particle size of 1 mm passing through a sieve of 1 mm mesh. The further sample preparation was a different way of plastic mass preparation. The first method was the implementation of a traditional method of mass preparation in a laboratory.

 DOI: 10.1051/matecconf/20168501018

20% of water was added to the dry mixture of raw clay powder, and after stirring the mass was kept in a sealed container.

The second method involves forming an aqueous slurry by dispersing coarse suspension in a centrifugal homogenizer of BASKEY company. Dehydration to a residual moisture content of 20% was carried out in a centrifuge with membrane filters.

Table 3. Mixture compositions

Index	**Contents of GL1**	**Contents of GL2**	**Contents of GL3**
1	100%	–	–
2	85%	15%	–
3	70%	30%	–
4	50%	50%	–
5	85%	–	15%
6	70%	–	30%
7	50%	–	50%

3 Results

To predict raw clay material behaviour during drying and burning processes, it is necessary to know a compound composition shown in Figure 1.

As seen in Figure 1, all clay minerals consist of minerals, but GL –3 contains a large amount of montmorillonite and white feldspar, mainly represented by albite. The sample of GL –2 feldspar contains less quantity as shown in Table 2 and Figure 1, and more quantity of kaolinite, which explains its size distribution. The sample GL –1 is also represented mostly by kaolinite, quartz and albite. High content of ferric oxide gives red color (table 2). These assumptions are confirmed by synchronic analysis of three samples – figure 2, 3 and 4.

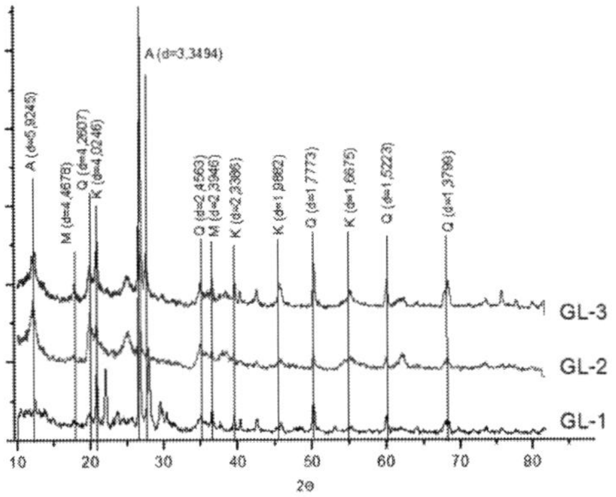

Fig. 1. X-ray pattern of Krasnoyarsk Kray clay used in the production of coarse ceramics. K – kaolinite, M – montmorillonite, A – albite, Q – quartz.

Endothermic effects observed in the temperature range of 20 – 200°C are accompanied by weight loss up to 5 %. It is mainly caused by loss of the adsorption and interlayer water due to clay minerals – hydromica and kaolinite. Exothermic effect in the range of 200 – 400 ^{O}C, accompanied by mass loss of 2.5% is connected to the processes of burning of fine organic impurities. Thermodestruction process of kaolinitic clay minerals occurs at low endothermic effect with a minimum at 455 ^{O}C, and is accompanied by a weight loss of 2.5% in the temperature range of 400 – 600 ^{O}C.

High endothermic effect with a minimum at 700 ^{O}C and the weight loss of 3% in the temperature range of 600 –800 ^{O}C is due to dehydration of hydromicaceous clay minerals. The negative heat effect above 890 ^{O}C is caused by the silicate materials formation in the melt system.

Post-firing compression strength of the molded test cubes with the 25x25x25 mm size of the raw material baked at a temperature of 850 –1000 ^{0}C with the step of 50 ^{0}C was investigated to determine the effect of batch preparation on the basis of clay raw materials. [3, 10].

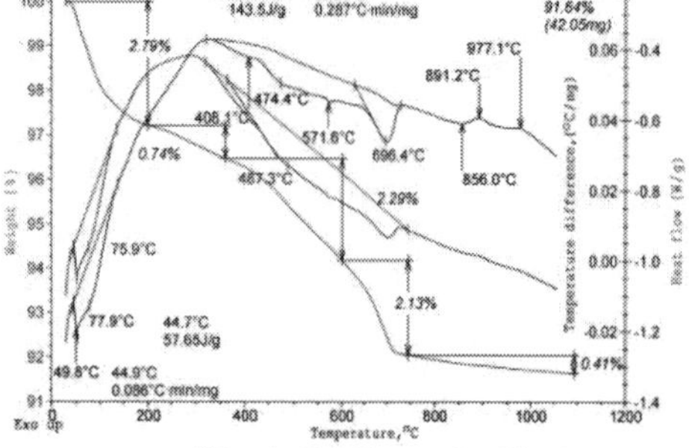

Fig. 2. DSC sample GL–1.

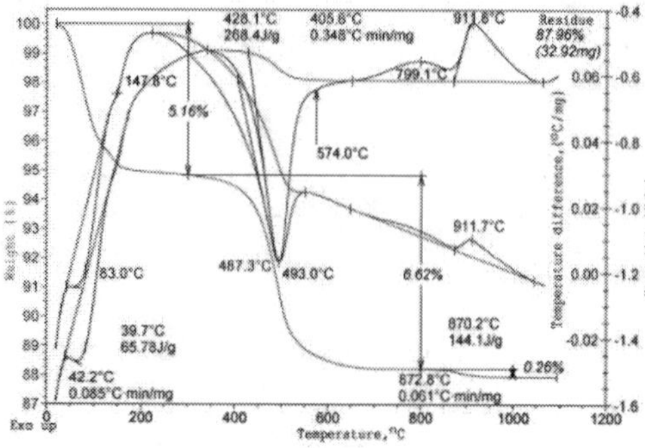

Fig. 3. DSC sample GL–2

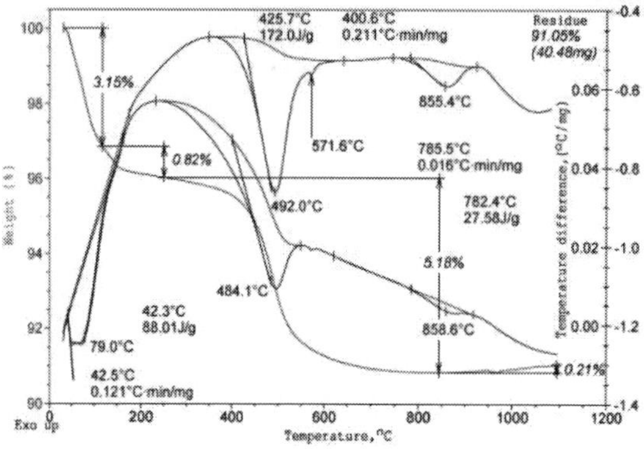

Fig. 4. DSC sample GL–3

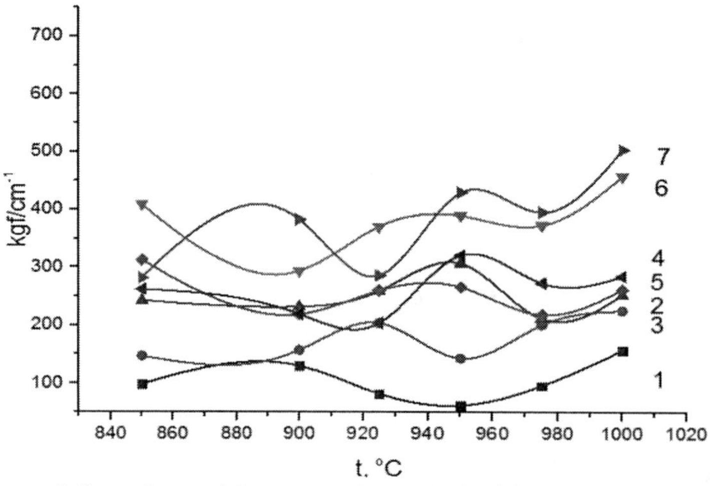

Figure 5. Dependence of the compressive strength of the final firing temperature for the samples prepared by method 1 prepared by the method 1.

Comparing the graphs shown in Figure 5 and 6 it can be concluded that the moisture averaging by method 1 gives a strong fluctuation effect of the strength values and, consequently, the marks of bricks. This deviation is related to averaging the chemical composition of the material in the volume of the material. It is worth noting that the sample 1 is pure clay, and the same fluctuations of the strength are presented on the graph. Due to the graphs the forced dissolution of clay raw materials both in pure form and in a mixture gives the best strength values of the baked samples. According to the sample 4, the maximum strength reaches 508 kgf/cm^2, while at the stock preparation of raw materials subjected to the centrifugal forced dissolution it reaches 609 kgf/cm^2. The other samples have the same analogy.

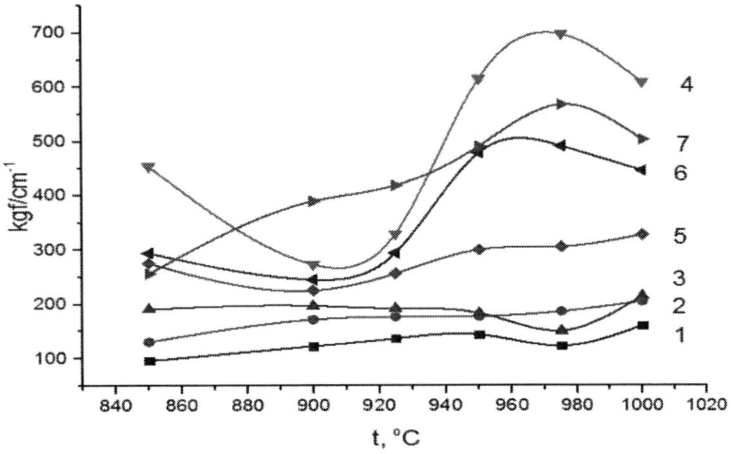

Figure 6. Dependence of the compressive strength of the final firing temperature for the samples prepared by method 2.

These deviations were repeated on a series of 10 experiments. Apparently, it is due to the fact that at dissolution the clay particles have lower between each other, and at drying and further firing small conglomerates become more active to sinter. It explains high strength values of the samples obtained by the method 2.

4 Conclusion

The aim of the work was to study changes of the technological properties of the batch depending on the composition and mass preparation. It was shown that the forced dissolution of raw clay materials in the centrifugal mixer provides a smaller variance in laboratory tests. It is discovered that the final properties of the ceramic body depended not only on the composition of the batch, but also the method of forming and mass processing.

References

1. T. Vakalova, V. Pogrebenkov *Refractories and Industrial Ceramics* **50** 17 (2009)
2. G. Cultrone, E. Sebastián, I. Sidraba *Applied Clay Science* **28** 297 (2005)
3. V. Lotov *Glass and Ceramics* **57** 164 (1999)
4. V. Lotov *Bulletin of the Tomsk Polytechnic University* **311** 77 (2007)
5. V. Pogrebenkov, K. Kostikov *Advanced Materials Research* **1040** 367 (2014)
6. A. Stolboushkin, A. Ivanov, O. Fomina, A. Fomin, G. Storozhenko, *Advanced Materials and Structural Engineering Proceedings of the International Conference on Advanced Materials and Engineering Structural Technology* 87 (2015)
7. G. Ogandzhanyan, D. Ogandzhanyan *Theoretical & Applied Science* **5 (25)** 144 (2015)
8. A. Tatarin, R. Vogt *Ziegelindustrie International/Brick and Tile Industry International* **4** 20 (2013)
9. N. Dhami, M. Reddy, A. Mukherjee *Ecological Engineering* **39** 31 (2012)
10. V. Lotov *Glass and Ceramics* **55** 123 (1998

MATEC Web of Conferences **85**, 01019 (2016) DOI: 10.1051/matecconf/20168501019
CCT 2016

ANALYSIS OF PRESSURE REDUCING EFFECTIVENESS IN THE REFORMING INDUSTRIAL UNIT WITH MATHEMATICAL MODELLING METHOD USING

Inna Yakupova [1,*], *Ekaterina* Chernyakova[1], *Anton* Koksharov[2]

[1]Tomsk Polytechnic University, Lenin av. 30, Tomsk, 634050, Russia
[2]Ltd "Kinef"

Abstract. In this paper the effect of pressure on the product yield and quality in industrial catalytic reforming unit was done with mathematical modelling method using. It was proved that the work at reduced pressure will increase the product yield by 2% mass, and also will enhance the yield of hydrogen and aromatic hydrocarbons.

1 Introduction

Catalytic naphtha reforming is one of the important and the most effective processes in the petroleum refining and petrochemical industries. Hydrogen and valuable aromatics are being produced as by-product. In most cases, this industrial process occurs in three or four adiabatic reactors mounted in series at a temperature of 480-520°C and an operating pressure of 0.3 - 2 MPa. Petroleum fraction in the range of 85-180°C is being used as raw material. This complex mixture consists of hydrocarbon molecules with 5–12 carbon atoms, mainly including paraffins, olefins, naphthenes, and aromatics [1-7].

Catalyst is the most important constituent of the process. Naphtha reforming catalyst is a bifunctional catalyst consists of a metal function, mainly platinum, and an acid function, usually chloride alumina. The metal function catalyzes the hydrogenation and dehydrogenation reactions and the acid function promote the isomerization and cyclization reactions. In order to achieve an optimum performance of the naphtha reforming catalyst, adequate balance between these functions is needed [1].

The yield of catalytic naphtha reforming process depends strongly on the catalyst properties. During operation, the catalyst undergoes physiochemical changes, which contribute to decrease in the activity for aromatic production.

Improving the stability and selectivity of the catalyst as well as reducing catalyst deactivation is a vital issue for enhancing the efficiency and yield of the process[1]. This practice could be achieved by mathematical modeling method application.

* Corresponding author: yakupovaiv@tpu.ru

2 Materials and methods

Catalytic reforming unit L-35-11/450K with a preliminary hydro-treating is an important subject of "Komsomolsk Refinery" JSC in the Far East of Russia [8].

Foreign catalyst charge PR-9 («Criterion» company) has proven to be appropriate in operation and ensure high specified quality yield. Results of catalyst work were modeled on the mathematical model, and the results of the pressure effect on the main industrial process indicators were predicted.

In the catalytic naphtha reforming process, coke formation is the most important cause of the catalyst deactivation. It's showed on the Figure1 that the existence of formation and hydrogenation of unsaturated intermediate products of compaction is possible. Therefore, under certain conditions coke formation doesn't occur, because resins and asphaltenes (previous substances of coke) can be hydrogenated to hydrocarbons or be in equilibrium with gas-phase reaction medium. Thus, controlling of feed temperature in reactor, it could be possible to provide process operation mode with the existence of equilibrium of formation and hydrogenation of coke structures [9, 10].

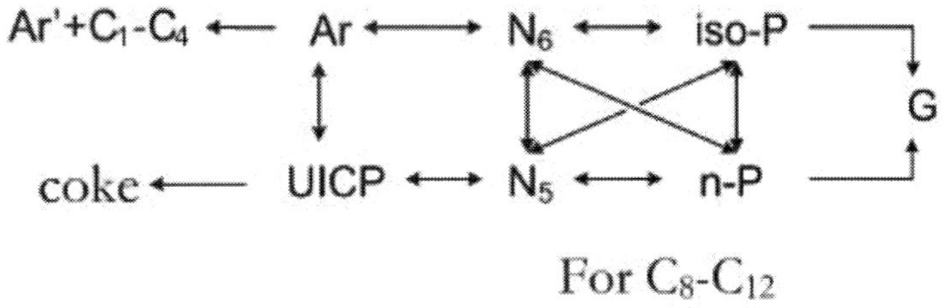

Fig.2. Formalized scheme hydrocarbon conversions C8 – C12: G – gas; n-P – normal paraffins; iso-P – izoparaffins; N_6 – cyclohexanes; N_5 – cyclopentanes; Ar – aromatic hydrocarbons; UICP - unsaturated intermediate products of compaction

But it's difficult to realize in practice, because it's required to obtain the final high-octane product, so there is an objective need for deviations from thermodynamic equilibrium. As a result, accumulation of unsaturated intermediate products of compaction occurs. However, the solution of very important industrial problem – how to find desired catalyst – is based on this contradiction [7].

The studies have been carried out using a mathematical model, which enables to consider physical and chemical laws of hydrocarbons conversion on the catalyst surface [12-14], and changes in the composition feedstock. The results of chromatographic analysis of the hydrocarbon feedstock composition, technological modes of production unit operation were used as initial data.

$$\begin{cases} G\left(\dfrac{\partial C_i}{\partial z} + \dfrac{\partial C_i}{\partial V}\right) = \sum\limits_{j=1}^{m} W_j \\ G\left(\dfrac{\partial T}{\partial z} + \dfrac{\partial T}{\partial V}\right) = -\dfrac{1}{C_p^{mix}} \sum\limits_{j=1}^{m} Q_j W_i \end{cases}$$

The initial conditions: Z=0, Ci=0, T=0, V=0, Ci=Cen (at the reactor entrance), if Z=0, T=Ten, where Ci – a concentration of the i-th component, mol/m3; T – temperature; Z – a raw material volume, m³; Wj – j-th reaction heat, mol/(m³·m); V – a volume of the catalyst layer, m3; G – a raw material flow rate, m3/h; Qj – j-th reaction heat, J/mol; Cpmix – a heat capacity of mixture, J/mol.

 DOI: 10.1051/matecconf/20168501019

3 Results

Using the mathematical model described above results of previously catalyst charge PR-9 were analyzed (Fig. 3). Also the catalyst activity changes during 2 previous work cycles are performed.

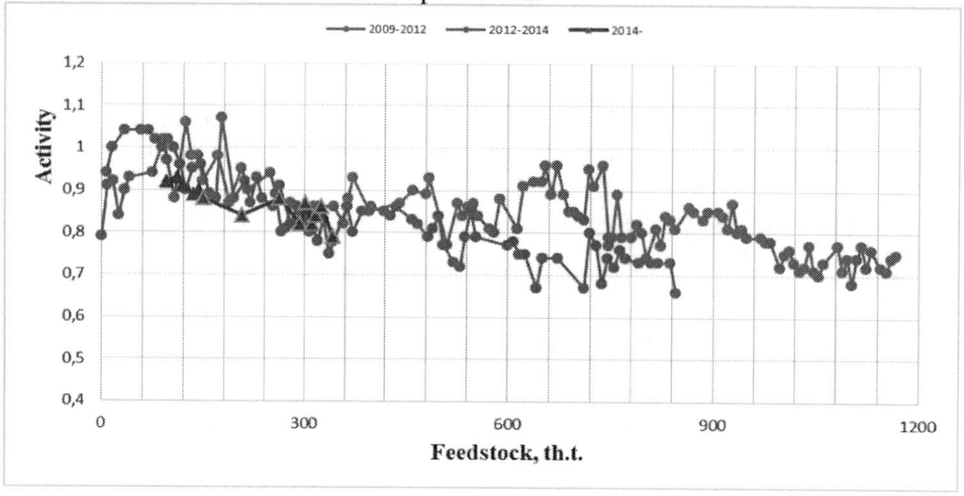

Fig.3. The comparison of catalyst current activity in different work periods

The rate of the catalyst activity change in previous cycle is slightly different on the dynamics of the current what could be compared with changes in the feedstock composition and catalyst aging.

The difference in the rate of coke accumulation (Fig.4) in the cycles is directly related to differences in technological modes of industrial plant operation.

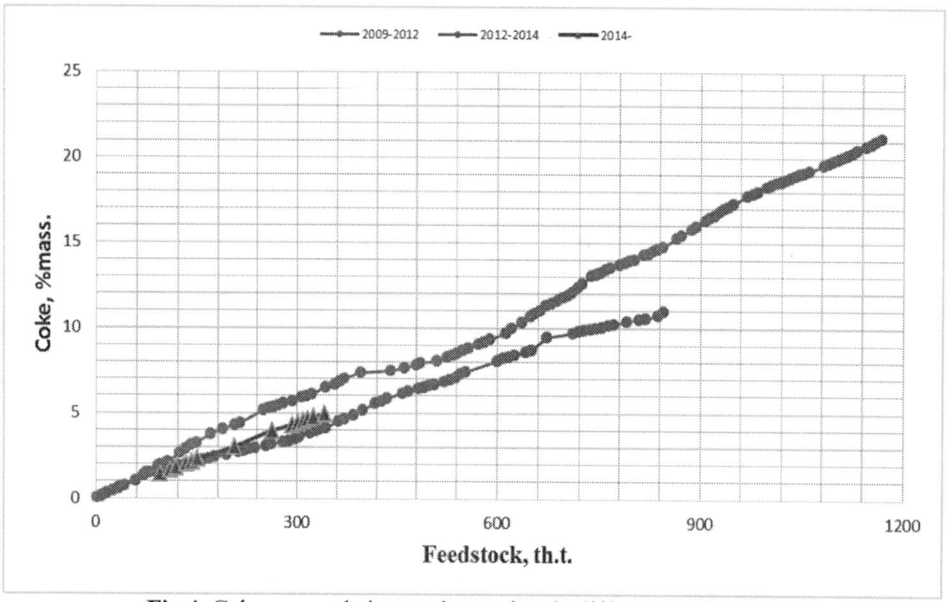

Fig.4. Coke accumulation on the catalyst in different work periods

Product output (Fig. 5) during all cycles remains quite high and ranges from 80 to 90% mass. The juddering changes product output confirm changes in the composition of the feedstock in this work period.

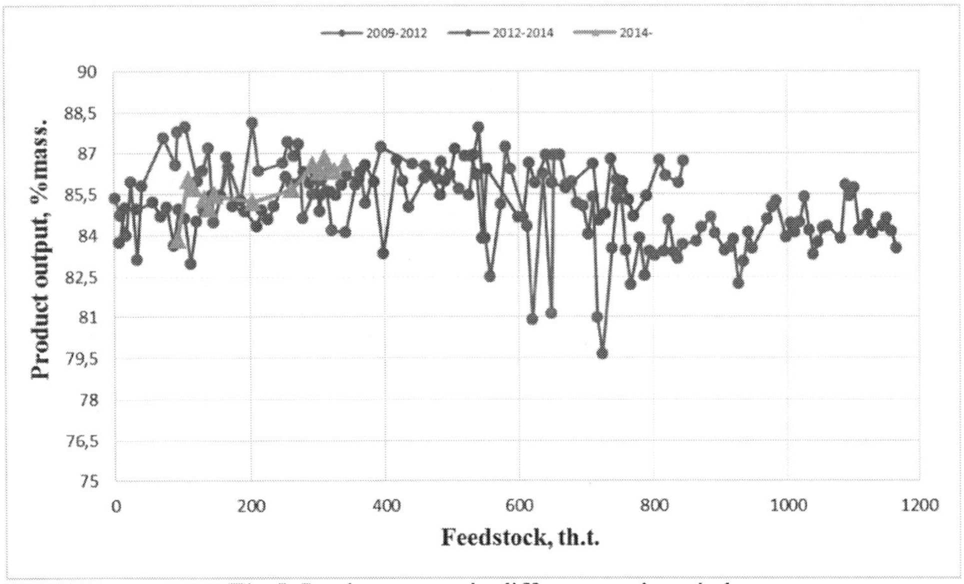

Fig.5. Product output in different work periods

The strategy of reforming technology improving goes hand in hand with a pressure reduction. However, with the pressure decreasing the coke deposition on the catalyst and the catalyst deactivation rate increase. For this reason, the reforming process with a fixed bed catalyst is not carried out at a pressure lower than 1.4 - 1.5 MPa. With a mathematical model using the pressure effect on the product yield and quality was analyzed. The results obtained are performed in Table 1.

Table 1. Pressure effect on the product yield and quality with PR-9 using

Pressure, Мпа	**P=1,7**	**P=1,6**	**P=1,5**	**P=1,4**	**P=1,3**	**P=1,2**
Octane number (research method)	92,9	92,9	93	93,1	93,1	93,2
Aromatics, %mass.	60,32	60,27	60,22	60,22	60,25	60,18
Hydrogen output, %	1,67	1,63	1,59	1,55	1,51	1,47
Coke, %mass.	4,03	4,05	4,08	4,18	4,26	4,31
Product yield, %mass.	86,68	87,06	87,44	87,79	88,13	88,53

The operating pressure deacreasing of the reforming process promotes its selectivity, favors the naphthenes dehydrogenation and paraffins dehydrocyclization reactions, and hydrocracking and hydrogenolysis side reactions inhibiting. Wherein pressure decreasing increases coke formation in the reactor according to the Le Chatelier's principle (Table 1).

4 Conclusion

With the mathematical modeling method using the calculations of PR-9 industrial characteristics have been done, as well as the effect of pressure on the quality of the product has been carried out.

From the results of studies it can be concluded that the catalyst over several cycles provides a high yield of quality product. Reduced pressure promotes the process target reaction, which leads to selectivity increasing, which in its turn promotes quality product, hydrogen and aromatics increasing. With the pressure reduction the rate of coke formation will increase, so the pressure is not recommended to drop below 1.4 MPa.

 DOI: 10.1051/matecconf/20168501019

References

1. Mohammad Reza Rahimpour, Mitra Jafari and Davood Iranshahi 2013 Applied Energy 109C 79
2. R.B. Smith 1959 Chem.Eng.Prog. 55(6) 76
3. Lapinski, M.L., Baird L., James, 2014 The McGraw Hill Companies R. 4.32004
4. D. Bommannan, R.D. Strivastava, D.N. Saraf, 1989 Chem. Eng. 67 405
5. H.G. Krane, A.B. Groh, B.L. Shulman, J.H. Sinfelt World Petroleum Congress 1960 39
6. A.S. Belyi 2005 Kinetica and Catalysis 46(5) 684
7. A.V. Kravtsov, E.D. Ivanchina, S.A. Galushin, D.S. Poluboyartsev 2004 TPU
8. Inna V. Yakupova, E.S. Chernjakova (Sharova), Je.D. Ivanchina, A.S. Belyj, M.D. Smolikov 2015 Procedia Engineering 113 51
9. Emilia D. Ivanchina, Ekaterina S. Sharova, Inna V. Yakupova 2014 Procedia Chemistry 10 197
10. Inna V. Yakupova, Ekaterina S. Chernyakova, Emilia D. Ivanchina, Anton G. Koksharov 2015 Procedia Chemistry 15C 378
11. Zagoruiko A., Belyi A., Smolikov M., Noskov A. 2014 Catalysis today 220-222 168
12. Tataurshikov, A.; Ivanchina, E.; Krivtcova, N.; Krivtsov, E.; Syskina, A. 2015 IOP Conference Series-Earth and Environmental Science 27 1
13. Zagoruiko A., Belyi A., Smolikov M., Noskov A. Unstedy-state kinetic simulation of oil reforming and coke combustion processes in the fixed and moving catalyst beds 2014 Catalysis today 220 168

MATEC Web of Conferences **85**, 01020 (2016) DOI: 10.1051/matecconf/20168501020
CCT 2016

Rheological properties of inverse emulsions stabilized by ethanolamides of tall oil fatty acids

V. A. Yanovsky[1,*], M. O. Andropov[1,2], R. S. Fakhrislamova[1,3], R. A. Churkin[1,3], K. M. Minaev[3] and O. S. Ulyanova[3]

[1]Siberian Physical-Technical Institute of Tomsk State University, 1 Novosobornaya sq., Tomsk, 634050, Russia
[2]Seversk Technological Institute of National Research Nuclear University "MEPhI", 65 Kommunisticheski avenue, Seversk, 636036, Russia
[3]National Research Tomsk Polytechnic University, 30 Lenin Ave., Tomsk, 634050, Russia

Abstract. The paper presents the comparative analysis of inverse emulsion rheological behavior stabilized by monoethanolamide and diethanolamide of tall oil fatty acids, which are promising reagents for petroleum engineering. Flow curves of emulsions and volume ratio of dispersed phase have been obtained. The analysis of the dependence of Herschel–Bulkley model parameters on concentration of emulsifiers has been carried out. It is proved that emulsifiers affect weakly on the rheological parameters in high water phase content emulsions, while in cut-back emulsions their impact is registered to be stronger. It has been found that rheological parameters of the emulsions stabilized by monoethanolamide throughout the range of concentration are considerably higher than the samples on the base of diethanolamide. The experimental data proved the key impact of viscosity of dispersion medium on rheological behavior of cut-back emulsions

1 Introduction

Inverse emulsions also known as invert emulsions or water-in-oil emulsions are widely used in various branches of industry, for example, food, cosmetic and pharmacy, as well as building and chemistry [1]. Also they are used for petroleum engineering [2-4]. Oil-base muds (OBM) or the inverse emulsions of the mineralized water solutions in hydrocarbon phase due to their unique technological parameters are very essential for drilling in complex environments, producing formations completion and as well-killing fluids.

In recent decades inverse emulsions are becoming increasingly important to enhance oil recovery [5-7]. Their efficiency is defined by their nature: they are viscous water-proofing fluids, which firstly, enhance swabbing needed for an oil drive, and secondly, get remaining oil in the pipes. In addition, in high-permeability watered layer water is sucked by the emulsion, which increases its viscosity, puts watertight barrier and as a consequence, decreases the water relative permeability. Such parameter as viscosity of inverse emulsions

* Corresponding author: mailto:yavatpu@yandex.ru

can be easily adjusted in various ways, the simplest of which is to change the ratio of the water phase to hydrocarbon phase. Furthermore, the viscosity of such emulsions can be controlled the type of emulsifier, its concentration, and addictives, stabilized the emulsions and structured within a dispersion medium. Inverse emulsion provides additional 5 – 20 tons of oil per 1 ton of drilling mud [8].

Rheological parameters of inverse emulsions define their efficiency, when they are used as drilling muds, for example, OBM or to enhance an oil drive in various canals (wellbore, pores and reservoir-scale fractures) as well as in the process of cuttings.

In previous works [9, 10] the authors carried out the synthesis of such emulsifiers as ethanolamides of tall oil fatty acid, and the stability of these emulsions was studied. This paper is a continuation of these studies and its purpose is to study of rheological parameters of inverse emulsions stabilized by ethanolamides of tall oil fatty acids.

2 Methods and materials

2.1 Reagents and materials

Monoethanolamide (MATO) and Diethanolamide (DATO) of tall oil fatty acid were produced by direct amidation of distilled tall oil (Monoethanolamide and Diethanolamine, respectively) at 150–160 °C under vacuum (60-70 mm Hg). The synthesis was carried out to full dewatering of the reaction mixtures (approximately 3-4 hours) for MATO – 2.5 mg KOH/g, DATO – 4.0 mg KOH/g. The following materials are used such as the distilled tall oil, produced by OAO "Selenginskij cellyulozno-kartonnyj kombinat", diesel fuel (DT-EVRO), class 3, type I-III (winter) according to GOST R 52368-2005 with 2.13 mm^2/s kinematic viscosity (GOST 33-2000) at 40 °C, and such p.a. reagents as chloride of calcium anhydrous ($CaCl_2$) and chloride of sodium (NaCl).

2.2 Preparation of emulsions

Hydrocarbon phase was preliminary prepared by estimated amount of emulsifier solution in diesel fuel, while water phase was prepared by solution of $CaCl_2$ and NaCl into a delivery flask. The emulsions were prepared by mechanical dispersion of water phase in hydrocarbon phase by Triple-Spindle Drink Mixer HAMILTON BEACH HMD 400 with 14000 rpm at room temperature. Estimated amount of emulsifier in diesel fuel was put into a cup of the Drink Mixer. Then it was mixed with addition of some portions of required amount of water phase with their full emulsionizing. When the last portion of water phase was added, the emulsions continued to be mixed for 5 min. Then the emulsions were put under vacuum in a vacuum rotary evaporator at room temperature for about 20 min to full air elimination.

2.3 Rheological measurements

Rheological measurements of emulsion were made immediately after preparation and degassing by viscosimeter Brookfield DV3T LV with Brookfield Thermosel® System at speed range of 10 – 250 rpm. Before measuring each sample was thermostated at specified temperature for 30 min. The accuracy of temperature control is ± 0.2 °C. Brookfield Rheocalc T 1.1.13 Software is used for data reduction process.

Kinematic viscosity of emulsifier solutions was registered according to ASTM D445 with accuracy ±1.0 % by Ubbelohde viscometer where a capillary diameter is 0.56 mm.

2.4 Microscopy

Microphotography of the emulsions were taken by Nikon unveils Coolscope II microscope.

3 Results and discussion

The authors have investigated the rheological parameters of inverse emulsions, prepared by using two nonionic emulsifiers of alkanolamide class: Monoethanolamide (MATO) and Diethanolamide (DATO) of tall oil fatty acids. Emulsifier range of concentrations, volume ratio and contain of the dispersed phase were chosen in compliance with real emulsions, used as drilling muds in petroleum engineering. Emulsifier concentration (c) was varied in the range of 0.75 – 15 % (wt) to hydrocarbon phase or 0.192 – 6.764 % (wt) to afforded emulsion. The high-concentrated emulsions with volume ratio of dispersed phase (φ) corresponded to 0.7, 0.6 and 0.5 were studied, where the last was in the form of the mineralized water solution with 32 g/dm^3 of $CaCl_2$ and 16 g/dm^3 NaCl.

The Herschel–Bulkley model was used to describe the rheological behavior of the inverse emulsions. This model demonstrates well the rheological parameters of non-Newtonian fluids and is often used to analyze high-concentrated emulsion behavior [11, 12]. The Herschel-Bulkley model is commonly written as

$$\tau = \tau_T + K \cdot \dot{\gamma}^n ,$$

where τ is the shear stress, Pa; $\dot{\gamma}$ the shear rate, s^{-1}; τ_T the yield stress, Pa; K the consistency index, and n the flow index.

The flow curves of inverse emulsions at 25 °C are shown in figure 1 with different volume ratios (φ) of dispersed phase and different emulsifier concentration *(c)*.

In 70 % water phase emulsions (fig.1a), the flow curves of the studied emulsifiers are arranged in pairs, depending on their concentration. When (c) index is high, DATO provides more viscous emulsion than MATO. When (c) index is lower than 5%, inversion of behavior is observed: MATO provides more viscous emulsions. In general, the impact of chemical nature of emulsifiers on rheological behavior of high-concentrated emulsion can be considered negligible. However, the situation radically changes with solution of the emulsion by dispersed medium (fig. 1b and 1c). The samples stabilized by MATO, demonstrate much higher viscosity than the samples with DATO. In addition, the higher emulsifier concentration is, the more differences in viscosity are. The viscosity of the samples on the base of MATO is higher than in the entire range of emulsifier concentration and shear rate. For example, according to Figure 1b, the emulsion with 10 % of MATO has practically the same rheological parameters as the emulsion with 15 % of DATO.

Then the dependence of the Herschel–Bulkley model parameters on emulsifier concentration in the emulsions was analyzed. The registered data are shown in Figure 2.

The dependences of the yield stress (τ_T) on emulsifier concentration (c) (fig. 2a) are predictably considered to have a form of rising curves. More intense ascend of τ_T index is expectedly observed in higher water phase emulsions, which can be explained by increasing contacts between droplets at increasing dispersed phase content [11]. It was registered the weak impact of chemical nature of emulsifiers on the yield stress for high-concentrated emulsions (φ = 0.7). MATO provided higher τ_T index in cut-back emulsions than DATO, and the difference increased with increasing c index. Therefore, the dependence $\tau_T - c$ with the same φ are considered to be divergent curves for each pair of emulsifiers.

The dependence of the consistency index curves (K) on emulsifier concentration (c) index shown in Figure 2b is similar to curves $\tau_T - c$ mentioned-above. The one difference is the higher K index for DATO, observed in high-concentrated emulsions. This fact naturally follows from the flow curves of the emulsions, shown in Figure 1a.

 DOI: 10.1051/matecconf/20168501020

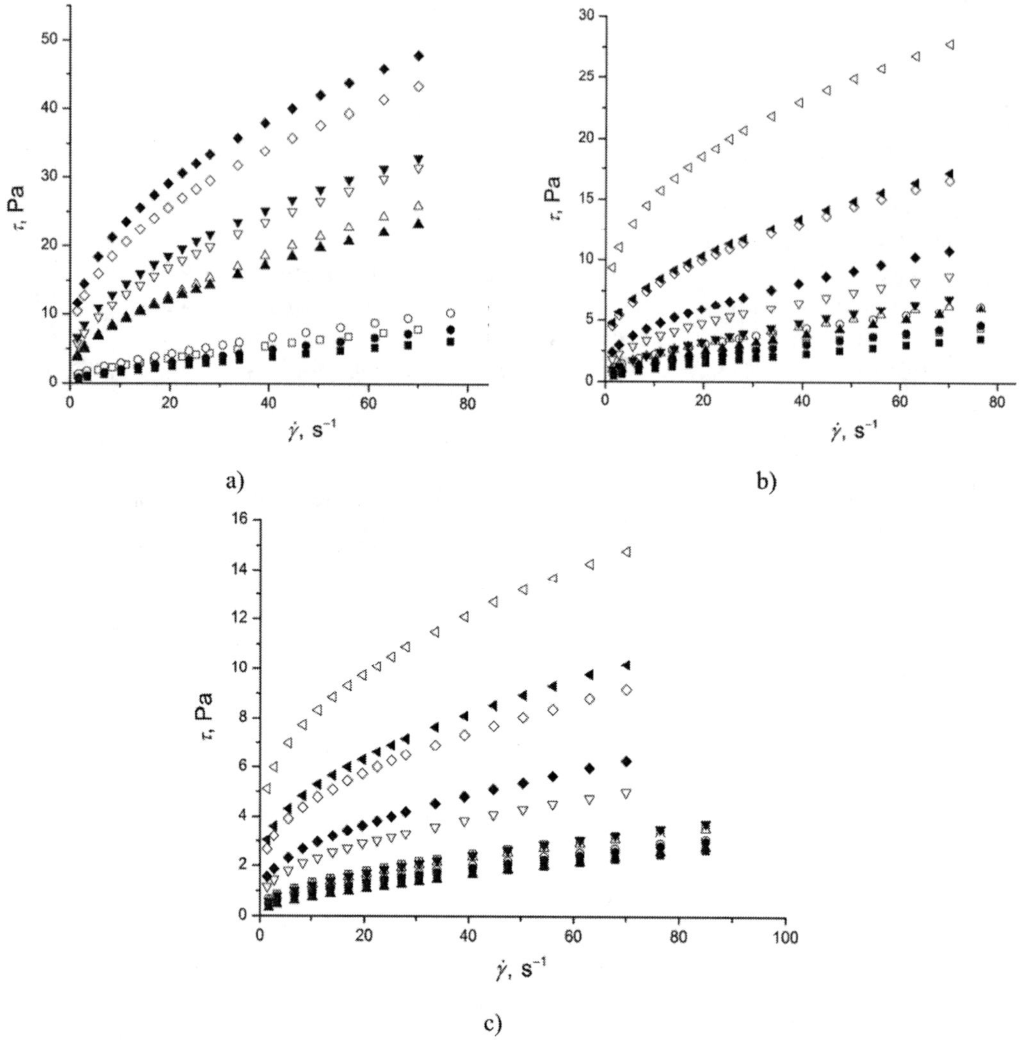

Fig. 1. The flow curves of inverse emulsions (25 °C): a) $\varphi = 0.7$; b) $\varphi = 0.6$; c) $\varphi = 0.5$ (□ –MATO; ■– DATO; □ – 0.75 %; ○ – 1 %; Δ – 2 %; ∇ – 5 %; ◇ – 10 %; ◁ – 15 %; the emulsifier concentration in hydrocarbon phase is shown)

A special feature of flow index dependence on emulsifier concentration (fig. 2c) is maximum value at low *c* index (not more than 1% in the emulsion). And increasing of water phase in emulsion gives maximum values at lower *c* index. Further behavior of the curves also depends on φ: with an increase of *c* index the flow index sharply decreases from ≈0.8 at $c = 0.257$ to 0.55–0.65 at $c = 0.514$ in high-concentrated emulsions. In cut-back emulsions a decrease of *n* index from maximum point is smooth. With further increasing *c* index the curves of all emulsions reaches some maximum *n* index and then, *c* index becomes independent from the flow index (*n*).

It should be noted that under equal conditions DATO provides maximum flow index in more concentrated emulsions, while MATO – in less concentrated emulsions.

DOI: 10.1051/matecconf/20168501020

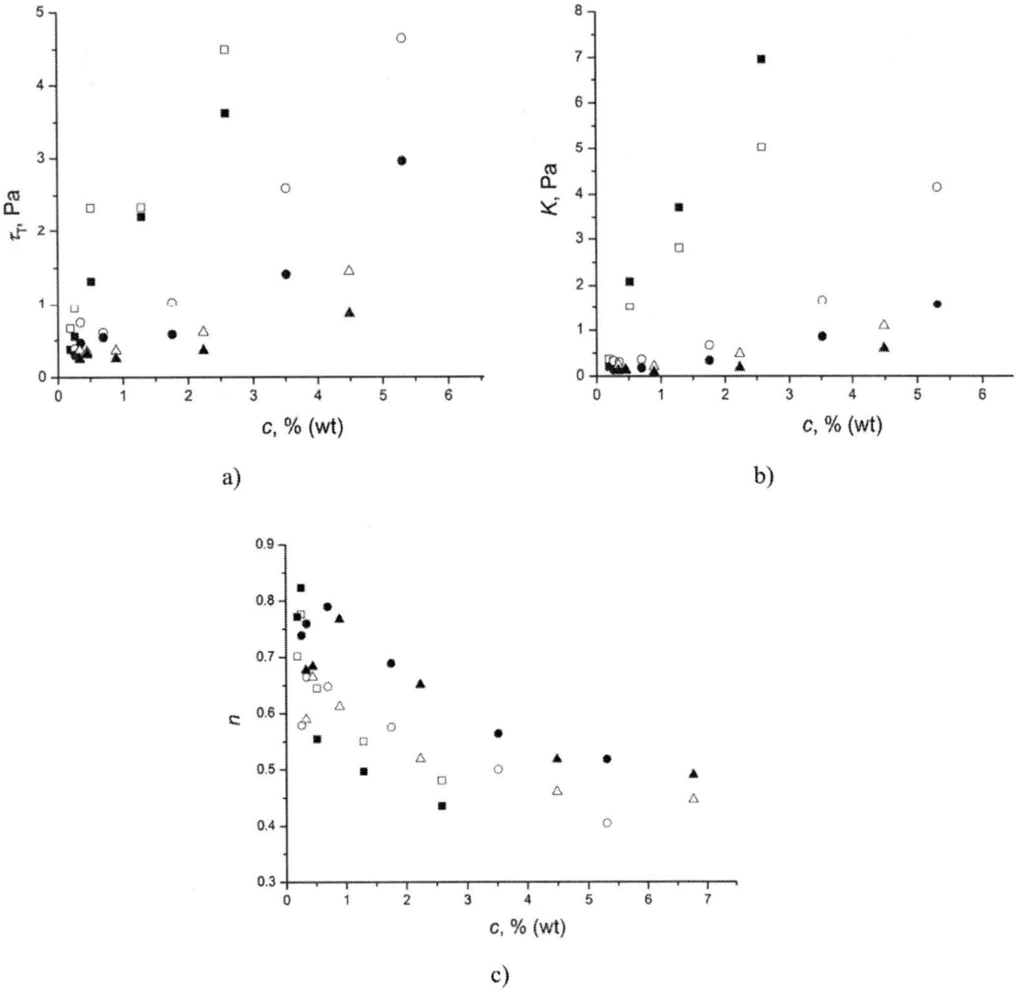

Fig. 2. The dependence of rheological parameters on the emulsifier concentration at 25^0 C
a) the yield stress (τ_T); b) the consistency index (K); c) the flow index (n) (□ –MATO; ■ – DATO; □ – $\varphi = 0.7$; ○ – $\varphi = 0.6$; Δ – $\varphi = 0.5$; the emulsifier concentration in emulsion is shown)

When analyzing the rheological behavior of the emulsions, it is necessary to remember about the simultaneous impact of complex factors. For example, *Sherman* [13] defined approximately two dozens of these factors. Concerning studied emulsions it is possible to distinguish the following factors: the average particle diameter, polydispersity index, continuous-phase viscosity, and droplet flocculation, as well as surface tension, thickness and rheological parameters of adsorption layer.

A microstructure of the emulsions was studied to evaluate the impact of droplets dimension and their polydispersity index. To compare these parameters, the samples of the emulsions, stabilized by MATO and DATO in the same concentration (при $\varphi = 0.6$) which had maximum differences in rheological behavior were selected (fig. 1b).

Fig. 3. The microphotographs of emulsions at $c = 10$ % (hydrocarbon phase), $\varphi = 0.6$: a) MATO; b) DATO

According to microphotographs shown in Figure 3 there are no significant differences in microstructure of studied emulsions. Therefore, the factor of particle diameter, in our opinion, is not considered to be important for studied emulsions. The impact of rheological parameters of dispersion phase is assumed to be a more important factor. These assumptions are confirmed by experimental data. Table 1 shows kinematic viscosity index of emulsifier solution in diesel fuel of different concentration. According to Table 1, low concentrated solutions (up to 5 %) have similar viscosity indexes generally correlate to rheological parameters of corresponding emulsions. With further increasing emulsifier concentration, nonlinear growth of viscosity is observed, and the viscosity of emulsions with MATO increases faster than DATO emulsions. A similar dependence was registered in cut-back emulsions ($\varphi = 0.5$–0.6). It should be noted that this dependence is not observed in high-concentrated emulsions ($\varphi = 0.7$). As mentioned above, at high values of c DATO forms slightly more viscous emulsions than MATO. Consequently, other factors are predominating in high-concentrated emulsions. Taking into account the more density of droplets there, it might be assumed that parameters of adsorption layer of emulsifier at phase boundary is, probably, more important, but this hypothesis needs further confirmation.

Table 1. The kinematic viscosity index of emulsifier solution in diesel fuel at 25 °C.

c, % (wt)	MATO	DATO
0	2.91	–
0.75	2.98	3.00
1	3.00	3.02
2	3.16	3.11
5	3.45	3.52
10	4.25	4.15
15	5.70	5.27

4 Conclusions

To sum up, the rheological behavior of inverse emulsions, stabilized by monoethanolamide and diethanolamide of tall oil fatty acids have been analyzed in the paper. The flow curve of different concentrated emulsions and volume ratio of dispersed phase have been described, as well as the dependence of Herschel–Bulkley model parameters on concentration of emulsifiers. It has been proved that chemical nature of emulsifier affects weakly on the rheological parameters in high-concentrated emulsions ($\varphi = 0.7$). In such emulsions DATO (more than 5 % content) provides higher viscosity and minimum flow index. In cut-back emulsions the impact of rheological parameters on the structure of the emulsifier varies significantly. The rheological parameters of cut-back emulsions, stabilized by MATO in the entire range of concentration are appreciably higher than parameters of DATO samples. The experimental data have proved the key impact of viscosity of dispersion medium on rheological behavior of cut-back emulsions.

This work was supported by Federal Target Program «Research and development on priority directions of scientific-technological complex of Russia for 2014-2020» (RFMEFI57814X0068).

References

1. T. F. Tadros, *Emulsion Science and Technology* (Wiley-VCH Verlag GmbH & Co. KGaA, Weinheim, 2009)
2. V. N. Glushchenko *The inverse emulsion and suspension in the oil and gas industry [Obratnye ehmul'sii i suspenzii v neftegazovoj promyshlennosti]* (Intercontact Nauka, Moscow, 2008) (in Russian)
3. V. I. Tokunov, I. B. Heyfets *Hydrophobically-emulsion drilling muds [Gidrofobno-ehmul'sionnye burovye rastvory]* (Nedra, Moscow, 1983) (in Russian)
4. A. Ya. Soloviev, V. A. Dokichev, G. V. Konesev, Neftegazovoe delo. **2**, 67 (2004) (in Russian)
5. M. K. Rogachev, D. V. Mardashov, A. R. Mavliev , Neftegazovoe delo. **3**, 180 (2011) (in Russian)
6. A. N. Beregovoj, M. I. Amerkhanov, S. G. Rakhimova, E. P. Vasiljev, Neftyanoe khozyaystvo – Oil Industry. **8**, 116 (2010)
7. V. A. Kotelnikov, M. F. Romantsev, V. B. Zavolzhskiy, Neftyanoe khozyaystvo – Oil Industry. **6**, 110 (2002)
8. I. M. Araslanov, Inzhenernaya praktika. **7**, 75 (2011) (in Russian)
9. V. A. Yanovsky, R. A. Churkin, M. O. Andropov, N. I. Kosova, Vestnik Tomskogo gosudarstvennogo universiteta. **370**, 194 (2013) (in Russian)
10. K. Minaev, A. Epikhin, D. Novoseltsev, M. Andropov, V. Yanovsky, O. Ulyanova, IOP Conf. Ser.: Earth Environ. Sci. **21**, 012033 (2014)
11. S. R. Derkach *Rheology of emulsions [Reologiya emulsiy]* (Nauka, St. Petersburg, 2012) (in Russian)
12. T. A. Dyakina, S. R. Derkatch, S. M. Levachev, Vestnik Moskovskogo Universiteta. Khimija. vol. 59, **1**, 42 (2004) (in Russian)
13. R. V. Petersen *Emulsion science*, ed Philip Sherman (Academic Press, London, 1968)

MATEC Web of Conferences **85**, 01021 (2016) DOI: 10.1051/matecconf/20168501021
CCT 2016

Development of a purification technology for treatment of medium- and low-activity radioactive waste of radiochemical production from Co-60 and Cs-137

Gleb Apalkov[1,2,*], *Alexander* Dyachenko[1], *Andrey* Zhabin[2], *Sergey* Smirnov[2]

[1] Tomsk Polytechnic University, Tomsk, Russia
[2] Mining and Chemical Combine, Zheleznogorsk, Russia

Abstract. The technological flowchart of purification of medium- and low-activity waste from Co-60 and Cs-137 is developed and introduced. The developed purification scheme has been successfully tested using genuine medium- and low-level liquid radioactive waste of radiochemical production containing complexing and colloid forming components (complexons, surfactants). The optimal conditions of the presented method of purification ensure reduction of the residual specific activity of 60Co and 137Cs radionuclides in the solution to less than 0,9 Bq per litre.

1 Introduction

A critical task for providing radiation safety at radiochemical plants is solution to the problem of treatment and reprocessing of liquid radioactive waste (LRW), generated from the operation of power installations, various technological procedures with solutions containing radionuclides, during decontamination of equipment, etc. Liquid radioactive wastes are collections of wastes from drains collectors and vacuum drainage of floor, decantates from clarifiers, liquid wastes from specialized laundries. In case of considerable volumes of LRW not only decontamination factor (index), but also residue levels of radionuclides in solutions after their treatment are critical, that enables reuse of water resources without boiling and utilization of stillage residue. The main radionuclides which LRW are purified from are 137Cs, 60Co. Ion-selective methods of purification from 60Co used in the technology suggests preliminary acid treatment of solutions aimed at destroying complexing agents and subsequent stages of radionuclide release, e.g., use of inorganic sorbent agents and cation exchange resins used for desalinization. The employed methods ensure efficient decontamination from137Cs with decontamination factor of 102÷104. However, the decontamination factor for 60Co is usually less than 102, which does not allow reducing general activity of solutions after treatment of less than10÷50 Bq/L.

The process solutions of LRW, generally, contain not only such components fed into solutions as aminopolycarboxylic acids; ethane diacid, citric acid, tartaric acid and their salts; sodium salts of ethylenediaminotetraacetic (EDTA) and diethylene triamine pentaacetic acids (DTPA), but also organic components with variable composition, formed as a result of radiolysis on long storage of LRW which form with 60Co complex chelated compounds with high stability constants [1,2]. Therefore, a prerequisite for increasing decontamination factor for 60Co is total destruction (oxidation) of all organic components of LRW regardless of their composition. Surfactants contained in the solutions (anionic, ampholytic, and nonionic SAS), high-molecular compounds, biocolloids also require both techniques intensifying oxidation decomposition and methods for special destruction of a colloid, containing 60Co.

Well-known methods of extracting the radionuclide 60Co from LRW at nuclear power plants are introducing Fe (III) and Ni (II) cations into LRW upon contact with potassium ferricyanide with molecular ratio from 2:1 to 4:1, allowing the solution to stand at room temperature and centrifugal separation of residue [3]. However, since there is no stage of oxidation of organic components of LRW, reagent consumption and sludge formation increase. Moreover, the technological adaptation of these methods requires expensive equipment, necessary for removing fine-grained sludge – a cascade of filtering centrifuges with the separation factor not less than 15000.

Treatment of LRW generated at nuclear power stations involves their preliminary vaporization resulting in formation of condensate and stillage residue, ozonation of stillage residue, removal of radioactive slurry and concentrating the filtrate by deep evaporation [4,5]. The main drawbacks of this method are: insufficiently effectiveness of a special stage

* Corresponding author: apalkov@tpu.ru

of 60Co extraction from the solution after ozonation and, as a result, inefficient removal of 60Co, that requires concentrating the filtrate by evaporation; high power consumption of the process; increased reagent consumption (ozone); need to keep constant pH level; as well as exclusive fire- and explosion safety requirements during the process. LRW stillage residue is also purified from radioactive cobalt and cesium by oxidation of stillage residue with hard ultra-violet rays (UF) and removal of corrosion products by filtration [6,7]. We must also note low efficiency and technological complexity of arranging LRW oxidation in dynamic regime, technological complexity of arranging LRW oxidation in the static regime due to the process time (over 6 hours), reduced efficiency of UV-lamp when the salt content is over 10 g/l and increased turbidity leading to inefficient oxidative destruction of organic components of LRW and removal of radionuclides.

2 Experimental

The aim of the research is to develop a technologically applicable method for purification of medium- (MAW) and low-activity waste (LAW) from ^{60}Co with simultaneous purification from ^{137}Cs at radiochemical plants. The task was to reduce remaining specific activity of MAW and LAW solutions in terms of ^{137}Cs and ^{60}Co content to the level, sufficient to classify LRW as non-radioactive waste according to the requirements of OSPORB-99/2010 [8] (allowing subsequent use of purified solutions).

A comprehensive approach aimed at increasing the decontamination index of medium- (MAW) and low-activity (LAW) LRW from ^{60}Co, ^{137}Cs and ensuring the required level of radionuclides removal involves catalytic oxidation of organic components of LRW and coagulative and precipitative extraction of radionuclides with mixed ferricyanides of transition metals followed by subsequent treatment of filtrate with cation exchange resin. The technological flow-chart of the developed method of purification is given in figure 1. The critical factor is the adequacy of destruction of complexing and colloid forming (organic) agents of LRW.

As stock solution for purification genuine slurry decantates after precipitation of drainage and desorbing solutions, formed after decontamination of production equipment at the spent fuel storehouse WWER-1000 and equipment in the independent spent-fuel storage installation room RBMK-1000 were used. The composition of stock solution (used during the research) is given in Table 1.

Table 1. Chemical and radionuclide composition of stock solution.

Component or criteria	Dimension	Value
Total salt content	g/l	20÷70
Oxalate-ion	g/l	2
Carbonate-ion	g/l	2
EDTA	mg/l	0÷100
ASs (anionic surfactants)	mg/l	1,5÷50
pH		8,9÷9,2
Cs^{137}	Bq/L	up to $4{,}56\times10^4$
Co^{60}	Bq/L	up to $1{,}02\times10^3$

Well-known catalysts for oxidation destruction of organic compounds in nitric acid media are platinum catalysts, applied to anion exchange resins (VP-1AP and other), silica gels, carbons, etc. [9-11]. Despite the advantages of heterogeneous catalytic processes, their industrial application in radiochemical technology is limited, firstly, due to the high cost of the (platinum) catalyst and non-recoverable losses of a platinum-group metal in the course of operation. As solid state catalyst of oxidation destruction of organic components of LRW in nitric acid media we use: platinum catalyst (Pt/VP-1AP), applied to anion exchange resin VP-1AP (GOST 20301-74) with platinum weight content of up to 2%, and two new types of solid state catalysts developed by the authors – nickel-ferricyanide catalyst [12] on VP-1AP carrier and bimetal platinum-zirconia catalysts (carriers – silica gel, aluminum oxide) [13].

Characteristics of the prepared solid state catalysts used during the process are given in Table 2. Total surface area of catalysts is determined using the method of thermal desorption of nitrogen. Chemical activation of the obtained catalysts is carried out in the dynamic regime with 20÷55 column volumes of 1,5 mol/L solution of nitric acid at the speed of 7÷12 column volume per hour.

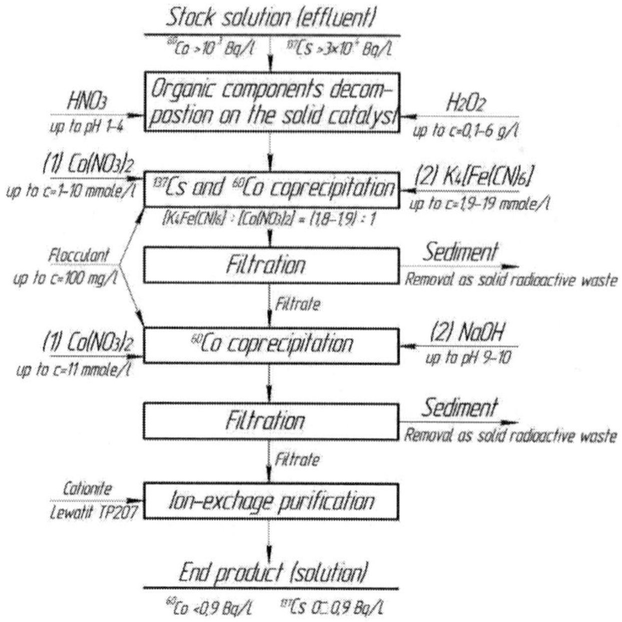

Fig. 1. Flowchart of purification of medium- and low-activity waste from ^{137}Cs and ^{60}Co.

Table 2. Characteristics of the catalysts used during the research.

Catalyst	Catalyst carrier	Pt, %	Particle size distribution, mm	Surface area, m²/g
Pt / VP-1AP	Anion exchange resin	1,97	0,3 ÷ 0,5	18,47
Pt / Zr / SiO_2 (Al_2O_3)	Silica gel (Aluminum oxide)	0,08	0,25 ÷ 0,35	70,31 (10,47)
R-Ni[Fe(CN)$_6$]	Anion exchange resin	-	0,3 ÷ 0,5	18,36

Testing of regimes and optimization of the developed method of purification of MAW and LAW from radionuclides ^{60}Co and ^{137}Cs is assessed on the results of experiments (research) on the laboratory installation given in Fig. 2.

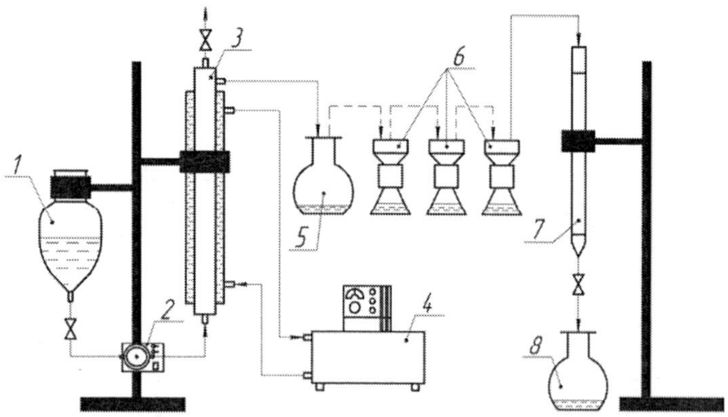

Fig. 2. Laboratory installation diagram of LRW purification from ^{60}Co and ^{137}Cs.

Laboratory installation includes: a stock bottle with stock solution (1), a peristaltic pump (2), a catalytic oxidation column (3), a thermostat (4), a stock bottle for receiving solution from the catalytic column (5), receiving filters with removable fluoroplastic cartridges (6), an ion-exchange column (7), a receiver stock bottle for purified solution (8).

LRW is purified in the following sequence: the LRW solution to be treated, regardless of its conditions, is corrected with hydrogen nitrate up to pH 1÷4. Next, hydrogen peroxide is added to reach concentration of hydrogen peroxide in the solution of up to 0,1-3,3 g/L. The corrected solution passes through the first stage of LRW purification in the catalytic oxidation column, operating at 47,5÷80°C, having the flow speed of 3÷15 column volume/per hour, that ensures contact time of the solution with the catalyst in the interval of 50÷130 sec.

The developed method and catalytic oxidation mode ensure 99,85-99,91 % of decomposition of organic components of LRW (EDTA, oxalate-ion, ASs), which can be achieved when using the indicated types of solid state catalysts, given in Table 2, and depends on the concentration of an organic component in the stock solution. The residual value of organic components does not exceed 100 μg/l.

The laboratory catalytic column is a device made from quartz glass with bottom feed of solution, a laminar flow zone, a catalysis zone, a gas separation zone. The catalysis zone is divided by meshed partitions. The prepared catalyst is loaded through the upper gate and squeezed by the upper meshed partition. In this case, the pore volume of the catalyst granular bed is 5,1÷12,5% of the bulk volume. The catalyst zone is a vertical cylindrical column filled with catalyst with «diameter/height» ratio of 1:3÷1:15. The catalyst is a uniformly graded inorganic porous carrier, silica gel or anion exchange resin with the grain size of 0,25÷0,50 mm, an active surface area of 18÷71 m^2/g with a layer of ultrafine homogeneous chemically inactive in weak solutions of nitric acid metal (salt) composition with ratio to the carrier of 0,08-7,5% mass, which are both oxidation catalysts of organic components of LRW and decomposition catalysts of hydrogen peroxide followed by formation of a hydroxyl radical.

The solution which has gone through the phase of catalytic oxidative destruction proceeds to the second stage – flocculation and precipitation. At the first operation a transition metal nitrate (Co, Ni, Zn) is added to the solution up to a concentration of 1,0-12,6 mmol/L. Then, while stirring intensively for 15÷20 minutes, solution of Besfloc K-051C flocculant is added up to a concentration of 10-100 mg/l, followed by addition of solution of an alkali metal ferricyanide up to a concentration of 1,9-19,0 mmol/L. In this case, the molecular ration between a transition metal nitrate (Co, Ni, Zn) and an alkali metal ferricyanide is from 1,0:1,5 to 1,0:1,9. Upon the total sedimentation of residue after 0,5÷2,2 hours the clarified part is decantated, i.e. intermediate filtration is carried out at the speed of 5,0-15,0 $mL/min \cdot cm^2$ through MFFK-3G membrane (0.45μm pore size), MFFK-4G (0,65μm pore size) or TRUMEM membrane with 0,45-0,60 μm pore size. Dense precipitate from the reactor bottom is deposited on the removable cartridge of the receiving filter and filtrated at the speed of 0,2÷7,0 $mL/min \cdot cm^2$ through MFFK-1G membrane (0,15μm pore size), MFFK-2G (0,25μm pore size) or TRUMEM membrane with 0,15÷0,25μm pore size. Decantate and filtrate are combined and proceed to the next operation of the flocculation and precipitation stage.

Here, Besfloc K-051C flocculate is added, while being stirred intensively, to the solution which has a residual concentration of alkali metal ferricyanide of 0,9-6,4 mmol/L up to a concentration of 10-100 mg/l. Then, diethyldithiocarbamate of an alkali metal is added up to a concentration of 10-100 mg/l. Further, the solution is mixed during 10÷15 minutes and a transition metal nitrate (Co, Ni, Zn) is added up to a concentration of 1,8-12,8 mmol/L. It ensures excess of a transition metal nitrate (Co, Ni, Zn) as compared to ferricyanide. At the second operation precipitate detachment is similar to one at the first operation. The obtained decantate and filtrate are combined and proceed to the third operation of the flocculation and precipitation stage.

Then, Besfloc K-4000 flocculate is added, while being stirred intensively, to the solution up to a concentration of 10-100 mg/l, followed by addition of sodium hydrate solution up to pH 9-10. At the third operation precipitate detachment is similar to one at the first operation. The second and the third operations can be combined without intermediate detachment of sediment. All three operations can also run subsequently in one device if to use an intermediate container for filtrates and decantates.

Final purification from ^{60}Co is carried out with cation exchange resin, selective to ions of transition and heavy metals, e.g., Lewatit MonoPlus TP 207, which is a weakly basic macroporous cation exchange resin with chelate groups of iminodiacetic acid. The solution passes through the ion-exchange column at the speed of 4÷10 column volume/per hour, with «diameter/height» ratio of the filling layer geometry of 1:10÷1:20. The third stage – ion-exchanging – can be excluded, in case the salt content of filtrate after coagulation and precipitation stage does not exceed 10 g/l.

3 Results and discussion

The developed method has significant advantages over the known (indicated earlier [3]-[7]) methods of purification of LRW generated at radiochemical plants and nuclear power stations due to a more complete oxidation of organic components of LRW in the dynamic mode of the column operation during considerably less time available for contact (50÷130 sec) combined with high efficiency of the column (3÷15 column volume/per hour), at lower temperature of the process (47,5÷80°C). A column with the filling layer of a catalyst and a dynamic mode of oxidation decomposition stage condition simple hardware arrangement of the process. Process variables do not depend on turbidity and salt composition of the solution and allow a wide range of the flow rate and temperature setting without sacrificing oxidation efficiency of organic components of LRW, which enables treatment of cloudy solutions (e.g., decantates) without preliminary clarification. Process variables (contact time, efficiency) can be easily changed in terms of both increase/decrease of the height of the catalyst filling layer and flow rate. Application of a catalytic mechanism of oxidation decomposition allows significantly reducing the amount of hydrogen peroxide added to LRW up to 3% volume (unlike UV-treatment of stillage residue) at the oxidation stage. Oxidation decomposition is carried out in acidic medium at pH 1÷4, which allows taking an advantage of synergism resulting from addition of low concentration of the second oxidizing agent – hydrogen nitrate during catalyst oxidation. The solution is mixed in the contact zone due to intensive spontaneous gas release in the pore space of the catalyst filling layer that increases the flow turbulence and decreases the contact time. The suggested method excludes sludge formation and precipitation of products of oxidation decomposition at the stage of oxidation decomposition. Thus, extraction of solid phase prior to coagulation and precipitation can be neglected.

An alkali metal diethyldithiocarbamate is added at the second operation of the coagulation and precipitation stage (when the base quantity of ^{60}Co (65,1÷95,4%) has already precipitated with mixed ferricyanides). It is added into dilute acidic solution (pH 1÷4), containing ferricyanide of an alkali metal and flocculant, which ensures precipitation of ^{60}Co in ion form (the most suitable for fixation of cobalt with diethyldithiocarbamate). The sequence of addition of an alkali metal diethyldithiocarbamate allows, unlike the method described in [4], initial binding of radioactive cobalt with diethyldithiocarbamate (resulting in formation of finely dispersed flocculated cloud, detected by turbidity change of the LRW solution). After 10-15 minutes of contact a transition metal nitrate (Co, Ni, Zn) can be added followed by coprecipitation of ^{60}Co, remaining in the solution, both with residue of a transition metal diethyldithiocarbamate (Co, Ni, Zn) and with residue of a transition metal ferricyanide (Co, Ni, Zn). Since an excessive amount of transition metal nitrate (Co, Ni, Zn) is used, the whole diethyldithiocarbamate is removed from the LRW solution with a precipitate, without polluting LRW with the added organic components. In case of simultaneous coprecipitation of diethyldithiocarbamate of a transition metal (Co, Ni, Zn) and ferricyanide of a transition metal (Co, Ni, Zn) dense precipitate is formed which is easily removed by microfiltration. The suggested method ensures reducing the amount of the introduced diethyldithiocarbamate of an alkali metal more than 100 times, when using it in concentration of 10-100 mg/l (considerably lower as compared to [6]).

When analyzing the results of the experiments conducted on the laboratory installation and improvement of the method of LRW (MAW, LAW) purification from radionuclides, efficiency of the developed method was estimated. The data are given in Table 3.

Table 3. Estimate of efficiency of the developed method of purification from ^{60}Co and Cs^{137}.

№ of treatment stage	Solution activity after treatment, Bq/L		Decontamination index	
	Co^{60}	Cs^{137}	Co^{60}	Cs^{137}
<u>I stage:</u> Catalytic oxidation of the solution	$1{,}02\times10^3$	$4{,}56\times10^4$	-	-
<u>II stage:</u> Coprecipitation of ^{60}Co and ^{137}Cs radionuclides with mixed ferricyanides of transient metals (Co, Ni, Zn)	6	16	170	2850
Coprecipitation of ^{60}Co with Na-diethyldithiocarbamate mixed with ferricyanides of transient metals (Co, Ni, Zn)	3,5	5	1,72	3,20
Coprecipitation of ^{60}Co with hydroxides of transient metals (Co, Ni, Zn)	2	3,6	1,75	1,39
<u>III stage:</u> Ion-exchange purification	< 0,9	< 0,9	> 2,20	> 4
Result			> 1133	> 51667

Low decontamination indexes during after-purification stages are conditioned by low content of radionuclides in the solution (after the first operation of coprecipitation) versus content of an added stable transition metal. The aim of the final ion-exchanging purification in the developed scheme is a barrier protection from slip of removed cobalt-60 radionuclides. At this stage it is important not to ensure high decontamination indexes, but a stable maximum low content of both radionuclides in the flowing solution and added salts of transient metals.

After extracting ^{60}Co and ^{137}Cs radionuclides (with activity lower than 0,9 Bq/L), the solutions, according to [8], do not belong to radioactive waste and do not require any measures on reducing radioactivity which allows their later use for different technological operations.

As a result of a successful laboratory testing of the developed method a technological flowchart for purification of MAW and LAW from ^{60}Co and ^{137}Cs (see fig 3) was suggested. The installation includes: a stock bottle with stock solution of LRW (1), membrane pumps (2,6,9,12,15), a catalytic oxidation column [14] (3), a thermostat (4), a stock bottle for receiving solution from the catalytic column (5), «mixer tank-settling tank» reactors (for every operation of coagulation and precipitation stage, and which have sedimentary residue zones and decantate removal zones) equipped with an agitator (7,10,13), receiving filters with remotely-removable fluoroplastic cartridges (8,11,14), ion-exchanging column (16), control valves (B01÷B33).

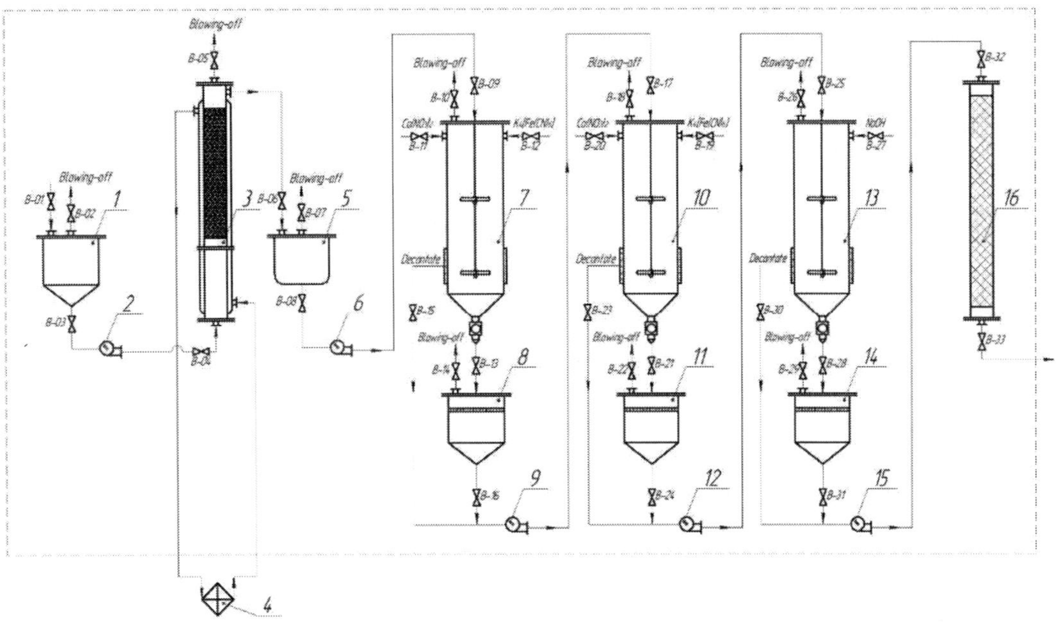

Fig. 3. Technological flowchart of the purification process of medium- and low-activity waste from ^{60}Co and ^{137}Cs.

4 Conclusions

The developed process of LRW purification from ^{60}Co and ^{137}Cs has been successfully tested using real medium- and low-level liquid radioactive waste of radiochemical production containing complexing and colloid forming components (complexons, SAS).

The optimized parameters of the process of purification allowed reducing residual specific activity of ^{60}Co and ^{137}Cs in the solution up to less than 0,9 Bq/L.

A technological flowchart for purification of MAW and LAW from ^{60}Co and ^{137}Cs was developed and suggested for application.

The analysis of the results allows considering the developed method to be technologically applicable for treatment of decontamination fluids, drain water, wastes from specialized laundries at radiochemical enterprises, containing different concentrations of complexons and SAS.

References

1. *Management of Low and Intermediate Level Radioactive Wastes with Regard to Their Chemical Toxicity* 2002 IAEA-TECDOC-1325 IAEA: Vienna, Austria
2. *Handling and Processing of Radioactive Waste from Nuclear Applications* 2001 TRS No. 402 IAEA: Vienna, Austria
3. Kulyuhin S A, Konovalova N A, Gorbacheva M P 2013 *Radiochemistry* 2013 vol **55** № **1** pp 242-248
4. Seliverstov A F, Lagunova Yu O, Ershov B G, Gelis V M, Basiev A G 2009 *Radiochemistry* vol **51** № **3** pp 326-328
5. Kudryavcev B K, e.a. *Patent RF* 2004 № 2226726
6. Shmakov L V, e.a. *Patent RF* 2012 № 2467419
7. Seliverstov A F, Lagunova Yu O, Ershov B G, Morozov P A, Kamrukov A S, Shashkovskii S G 2008 *Radiochemistry* vol **50** № **1** pp 70-74
8. *Main Sanitary Regulations to Ensure Radiation Safety* (Osborn - 99/2010). SP 2.6.1.2612
9. Ananyev A V 2005 *PhD dissertation* Frumkin Institute of Physical Chemistry and Electrochemistry of Russian Academy of Sciences Moscow 391 p (in Russian)
10. Tyumentsev M S 2013 *PhD dissertation* Frumkin Institute of Physical Chemistry and Electrochemistry of Russian Academy of Sciences Moscow 173 p (in Russian)
11. Shilov V P, Dzuybenko V I, Astafurova L N, Krot N N 1990 *J. applied chemistry* vol **2** pp 348-353
12. Apalkov G A 2014 *IOP Conf. Ser.: Mater. Sci. Eng.* vol **66** 012033
13. Zhabin A Yu, Apalkov G A, Smirnov S I *Patent RF* 2016 № 2598944
14. Apalkov G A, Zhabin A Yu, Smirnov S I *Patent RF* 2016 № 158504

MATEC Web of Conferences **85**, 01022 (2016) DOI: 10.1051/matecconf/20168501022
CCT 2016

Microbiological synthesis of methyl ethyl ketone

Anna Astashkina[1,*], *Yuliya* Kolbysheva[1], *Alena* Nikiforova[1], and *Abdigali* Bakibayev[2]

[1] Tomsk Polytechnic University, Institute of natural resources, Tomsk, Russia
[2] Tomsk State University, 634050 Tomsk, Russia

Abstract. Optimal conditions for the microbiological synthesis of methyl ethyl ketone (MEK) on the mineral medium with butanol-1, butanol-2, 2-methylpropanol-2 by hydrocarbon-oxidizing bacteria Pseudomonas spp. were chosen. Optimal conditions for microbiological synthesis are liquid mineral Adkins' medium containing 3% (vol.) of the substrate, stirring speed – 80-100 rev/min, temperature – 37° C, and synthesis time 24-48 hours. The specific growth rate and time of culture generation during the microbial synthesis were determined. It was found that methyl ethyl ketone is formed in the reaction mixture after 8, 48 and 72 hours when using 2-methylpropanol-2, butanol-2, butanol-1 as substrates, respectively. Maximum methyl ethyl ketone yield in the reaction mixture is observed after 12 hours in the case of using 2-methylpropanol-2 as a carbon source and accounts for 10%.

1 Introduction

Currently, there is interest in biotechnology worldwide. This leads to the development of methods enabling to synthesize valuable organic compounds [1-2]. Methyl ethyl ketone is a valuable selective solvent at the dewaxing units in the petrochemical industry, in the woodworking industries to produce a variety of adhesives, and other industries as well [3-5]. Its world production is 1.2 million tons, and the demand for it is constantly increasing on average by 7% per year. Currently, Japan, Western Europe, China, and the USA are major producers of methyl ethyl ketone. More than 10 chemical methods of producing MEK based on the catalytic oxidation of butene, dehydrogenation or oxidation of butanol, sec-butylbenzene or n-butane are known [6-20]. In industry MEK is produced by a liquid phase free radical oxidation of n-butane with the target product yield not more than 23%. It should be noted that the main disadvantages of the chemical synthesis process of MEK consists in the high energy consumption, the presence of by-products, greatly complicating and increasing the cost of MEK selection process. Besides it is a long and laborious process [12, 16-20]. Therefore, the development of environmentally safe and cost-effective process for the synthesis of methyl ethyl ketone with the use of microorganisms is important. Thus, the goal of the work consisted in the choice of optimal conditions for microbial synthesis of methyl ethyl ketone from butyl alcohols.

* Corresponding author: apa2004@mail.ru

2 Experimental

In the work the hydrocarbon-oxidizing microorganisms of the genus Pseudomonas spp., isolated from soil and water reservoirs of Zapadno-Katylginskoe field in Strezhevoy (Tomsk region) by the company "Ecoil" (Tomsk) during the remediation of oil-contaminated land and oil, were examined. The taxonomy of pure isolated cultures was determined on the basis of the sequence analysis of nucleotides in 16S pRNA, their modification and primary bioinformatical analysis (laboratory "Genotech" Ltd., Moscow). To study the bacteria viability, the bacterium culture in the amount of 5% with the concentration of CFU 108 was inoculated into liquid mineral Adkins' medium with the addition of butyl alcohols: butanol-1 (puriss p.a., GOST 6006-78), butanol-2 (puriss p.a., cas № 78-92-2), 2-methyl-2-propanol (puriss p.a., cas № 75-65-0) in the amount of 1-3 % (vol.). The cultivation was carried out at the speed of 80-1000 rev/min and temperature of 28-30° C within 48 hours. The number and growth dynamics of hydrocarbon-oxidizing microorganisms during the biotransformation were determined by the McCready [2] and Koch's [2] methods, using the culture media of the meat-and-peptone broth (MPB) and meat-and-peptone agar (MPA) as well. Microbiological synthesis of methyl ethyl ketone was carried out in Erlenmeyer flasks on mineral Adkins' medium with the addition of butyl alcohols as a substrate in the amount of 1% (vol.), inoculum with the concentration of CFU 108 in the amount of 5% (vol.) within 72 hours at the temperature of 28-30° C and constant stirring 80 rev/min. The research of residual hydrocarbons during the biotransformation of butyl alcohols was carried out by gas chromatography. For chromatographic analysis GOST R 52714-2007 was used [21]. Changes in the hydrocarbon composition of reaction mixture was measured by gas chromatography "Chromatec-Crystal 5000" with a flame ionization detector (FID) under the following conditions: column length – 100 m, column diameter – 0.25 mm, evaporator temperature – 2900 C, oven temperature – from 0.2 to 2900 C, carrier gas – helium, carrier gas flow rate – 315 ml/min; injected sample volume – 0.2 μl; split ratio – 200:1, total analysis time – 20 min.

3 Results and discussion

Hydrocarbon-oxidizing bacteria of the genus Pseudomonas spp. are of the greatest interest [22-25]. A variety of biosynthetic and catabolic reactions, high growth rate on the simple composition media, opportunities for genetically engineered manipulation can be considered as a promising object for works in the field of biotechnology [23, 26].
Schematically the process of hydrocarbons biotransformation by bacteria [2, 25] can be represented as follows:

alkane, arene ⟹ alkene, isoalkane, isoalkene, alkadiene, cycloalkane, cycloalkene ⟹ alcohol ⟹ ketone ⟹ carboxylic ester

According to this scheme, simple hydrocarbons under the influence of the enzymic complex are converted to alcohols, aldehydes, ketones and etc. Therefore, it was important to examine the viability of bacteria on the mineral medium using only alcohols as a carbon source. According to published data [22-23], representatives of the genus Pseudomonas are able to assimilate ethanol as the only source of carbon nutrition in the conditions of

 DOI: 10.1051/matecconf/20168501022

submerged cultivation. At the same time they accumulate biomass, which is rich in amino acids and vitamins. Some strains such as P. aeruginosa, P. fluorescens, P. putida, P. aurantiasa, P. mendocina, and P. rathonis grow well on the synthetic medium with 1% ethanol as the only carbon source. However, only 41 strains grow well on the same medium with 2.5% ethanol. As the goal of the work was to obtain methyl ethyl ketone, so butanol-1, butanol-2, 2-methylpropanol-2 were selected as the substrate. The number of colony forming units (CFU) of bacteria on the nutrient medium with butyl alcohols was determined by limiting dilution analysis after every 8 hours. On the basis of the results of microbiological analysis, the kinetic growth curves of bacteria were made according to the amount of substrate (1-3% (vol.)) in the reaction mixture (fig. 1-3).

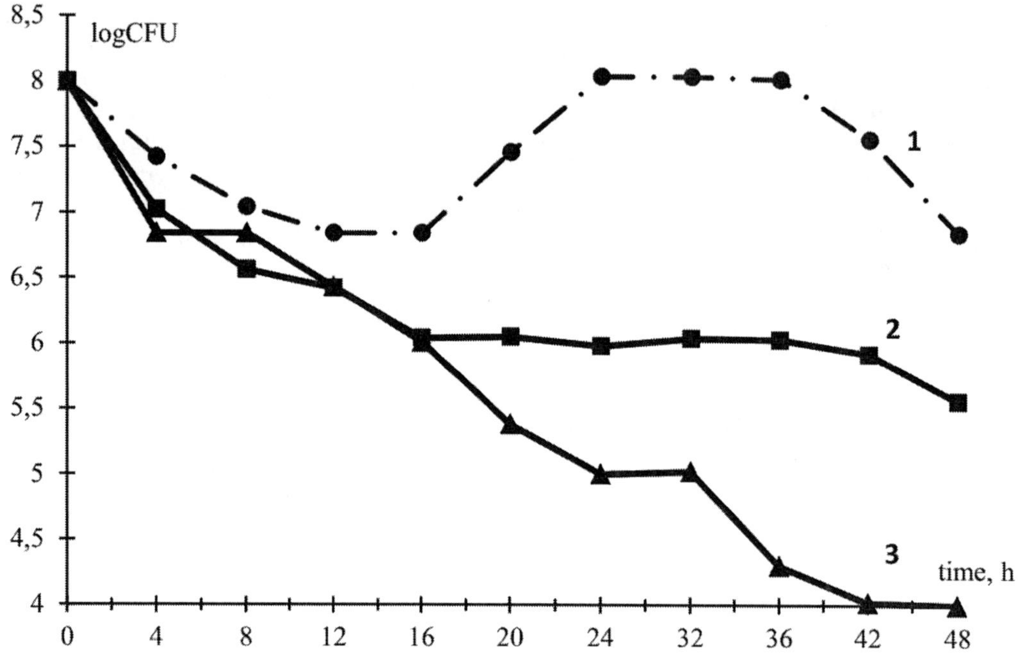

Figure 1. Changes in the number of bacteria of the genus Pseudomonas spp. on the nutrient medium with butanol-1of different concentration, %(vol.): 1- 1, 2-2, 3-3.

DOI: 10.1051/matecconf/20168501022

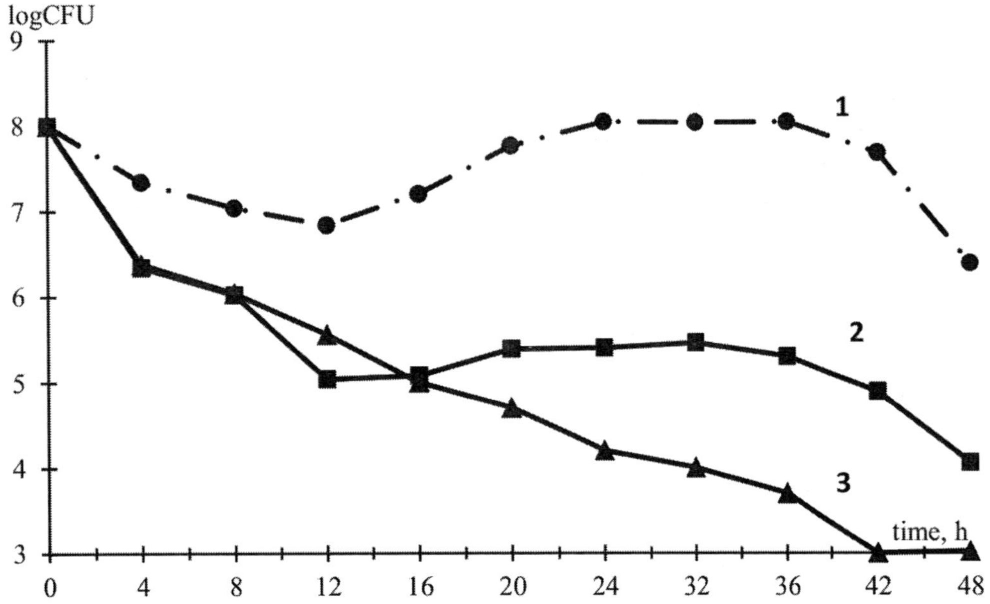

Figure 2. Changes in the number of bacteria of the genus Pseudomonas spp. on the nutrient medium with butanol-2 of different concentration, % (vol.): 1- 1, 2-2, 3-3.

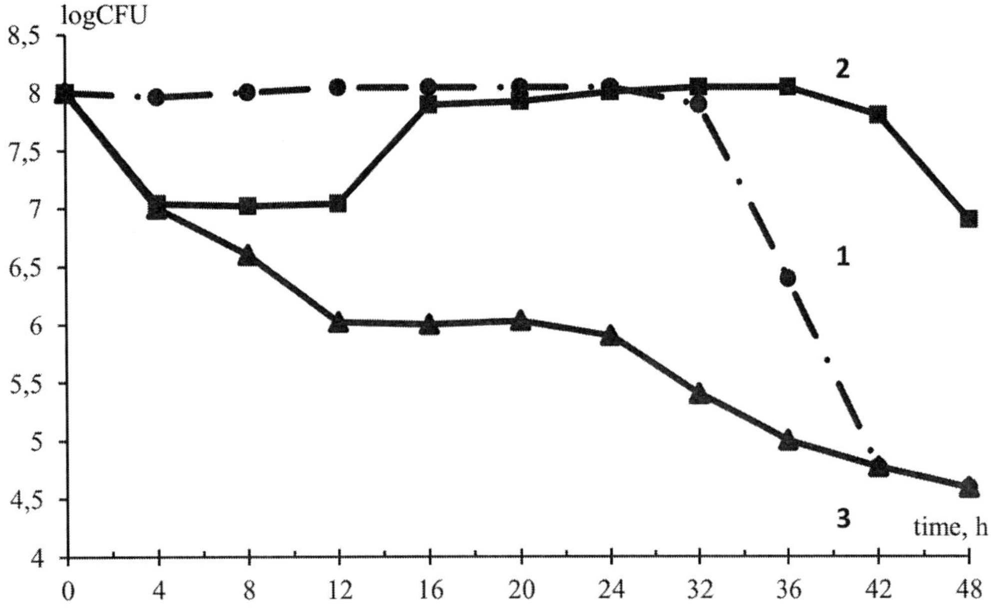

Figure 3. Changes in the number of bacteria of the genus Pseudomonas ssp. on the nutrients medium with 2-methylpropanol-2 of different concentration, %(vol.): 1- 1, 2-2, 3-3.

According to the obtained dependences, it can be concluded that the nature and amount of substrate influence the development of a culture. Butyl alcohols added into the medium in the amount of more than 1% (vol.) leads to its death. The growth curves of a culture (curves 2, 3) have complex character, which is similar to a typical growth curve with specific growth phases: induction, exponential, stationary, and death [2, 25-26]. According to figures 1-2 (curve 1), the phase of physiological adaptation of a culture to the medium continues 8-12 hours. Further, within the next 12 hours (12-24 hours), there is a maximum growth of a culture (exponential growth phase). The period after 24 hours corresponds to the stationary phase, in which the availability of essential nutrients substances becomes limited. The balance between the cell growth and division, as well as the process of cell death, is established. After 48 hours the phase of death begins as a result of the accumulation of toxic aromatic metabolites. It should be noted that in the case of 2-methylpropanol-2 (1% (vol.)) the induction and stationary phases are almost absent on the growth curve (curve 1). It shows that the studied culture was adapted to the given carbon source. However, the phase of death starts after 36 hours of cultivation due to the accumulation of toxic metabolites in all cases. On the basis of obtained data, the specific growth rate and generation time for a culture on the nutrients medium with butanol-1, butanol-2, 2-methylpropanol-2 were calculated according to the formula [2, 26]. The results are presented in table 1.

Table 1. Specific growth rate and generation time for bacteria of the genus Pseudomonas spp. on the nutrients medium with butyl alcohols

Substrate amount, % (vol.)	Butanol-1		Butanol-2		2-methylpropanol-2	
	K, h-1	g, h	K, h-1	g, h	K, h-1	g, h
1	0,15	4,40	0,16	4,20	-	-
2	0,01	1,40	0,05	3,20	0,10	4,20

According to data obtained, it can be said that the structure of isomer affects the specific growth rate and generation time of hydrocarbon bacteria by the following scheme:

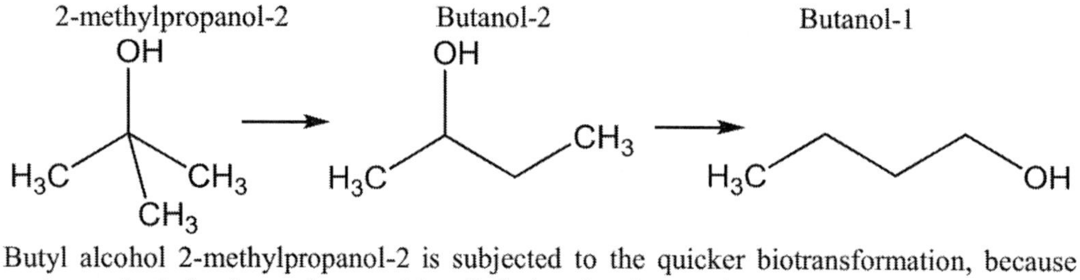

Butyl alcohol 2-methylpropanol-2 is subjected to the quicker biotransformation, because tertiary alcohols have less stable C-C bonds than primary and secondary ones. Consequently, the chain breaking occurs easier. Furthermore, the results of microbiological analysis showed that the bacteria Pseudomonas spr. are able to assimilate butyl alcohols and can be used as a power source and energy. Microbiological synthesis time should continue at least 48 hours.

For qualitative determination of methyl ethyl ketone in the reaction mixture, the gas chromatography method was used [27]. The results of the chromatographic investigations are presented in the form of chromatograms (figure 4-6).

DOI: 10.1051/matecconf/20168501022

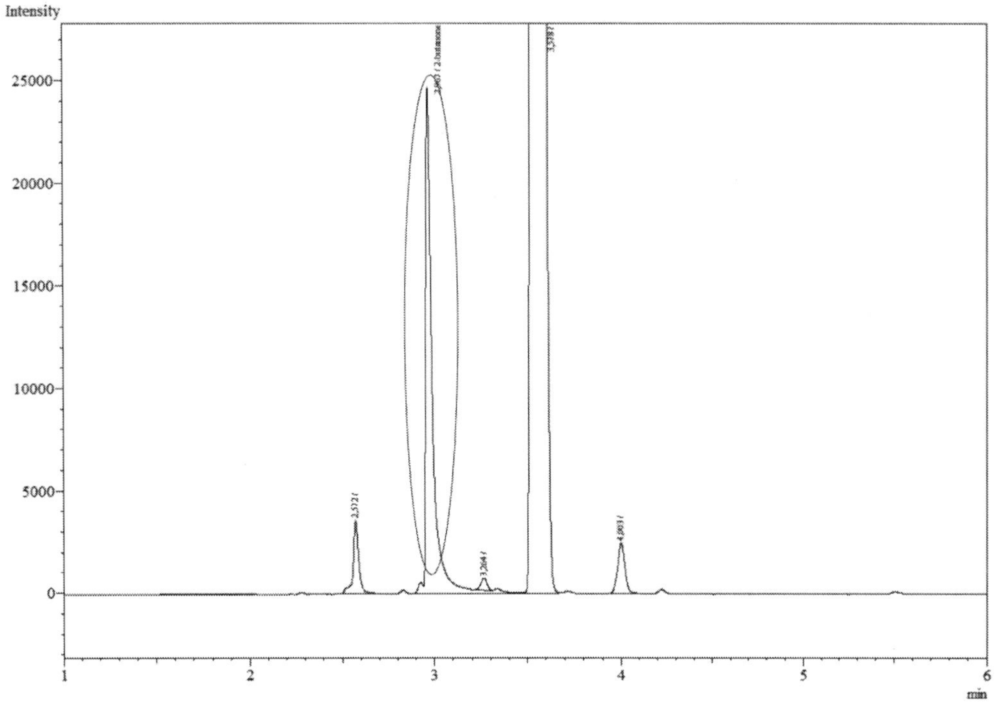

Figure 4. Chromatogram of MEK during the biotransformation of 2-methylpropanol-2 after 12 hours.

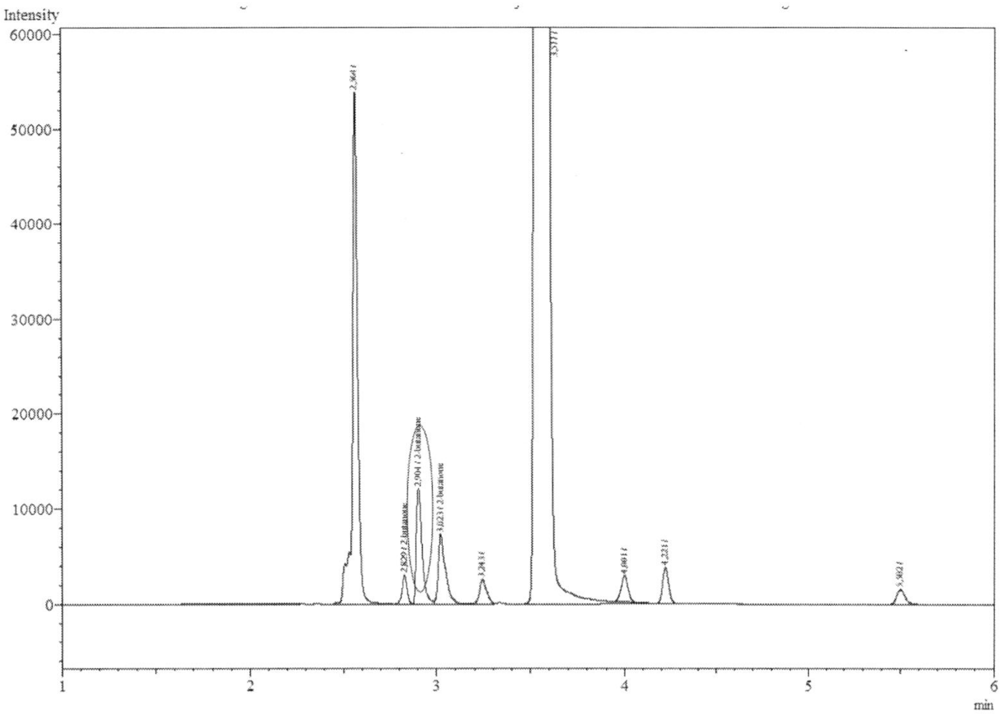

Figure 5. Chromatogram of MEK during the biotransformation of butane-2 after 48 hours.

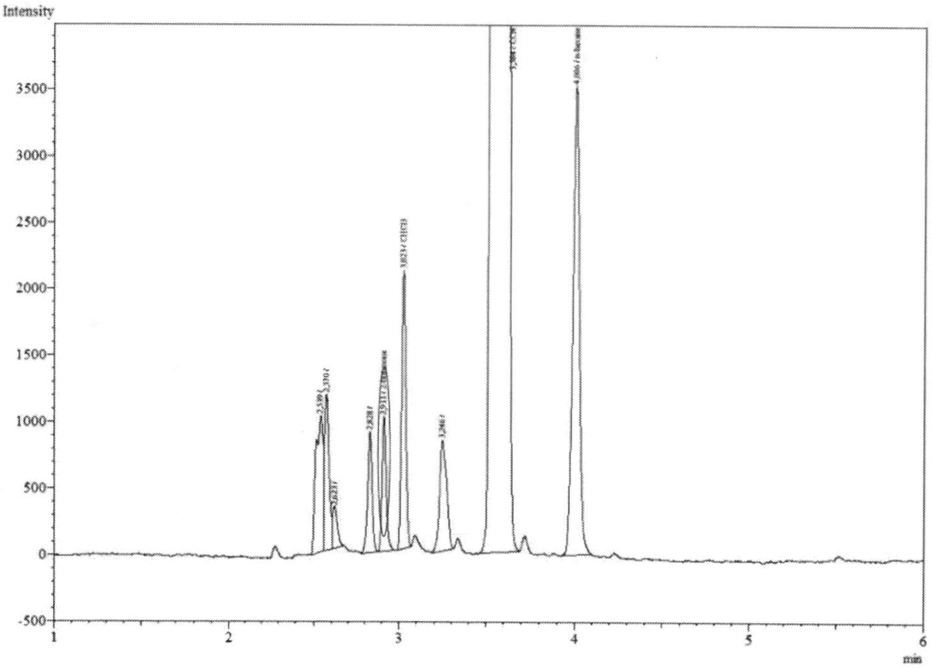

Figure 6. Chromatogram of MEK during the biotransformation butanol-1 after 72 hours.

It was determined that methyl ethyl ketone with a retention time of 3.025 was found during the cultivation of hydrocarbon bacteria of the genus Pseudomonas spp. on the nutrients medium with butyl alcohols.

Thus, the microbiological synthesis of MEK was carried out under the following conditions: mineral Adkins' medium containing 1% (vol.) of butyl alcohols, biosynthesis time – 72h, temperature – 28-30° C, stirring speed – 80 rev/min. Every day sampling from the reaction mixture for the GC analysis was carried out to identify MEK. To determine MEK assay in the reaction mixture, the calibration graph was made in the axes of the peak area and MEK concentration on model solutions. To prepare model solutions, MEK in the amount of 0.1-15% (vol.) was added to the mineral Adkins' medium, extracted with carbon tetrachloride, and gas chromatographic analysis in accordance with GOST R 52714-2007 [21] was carried out within 15 minutes. Calibration graph with correlation coefficient R^2 = 0.9931 was made, and the amount of MEK in the reaction mixture was calculated according to the equation $y=6247729x$. The results of the microbiological synthesis of MEK are shown in the diagram (figure 7).

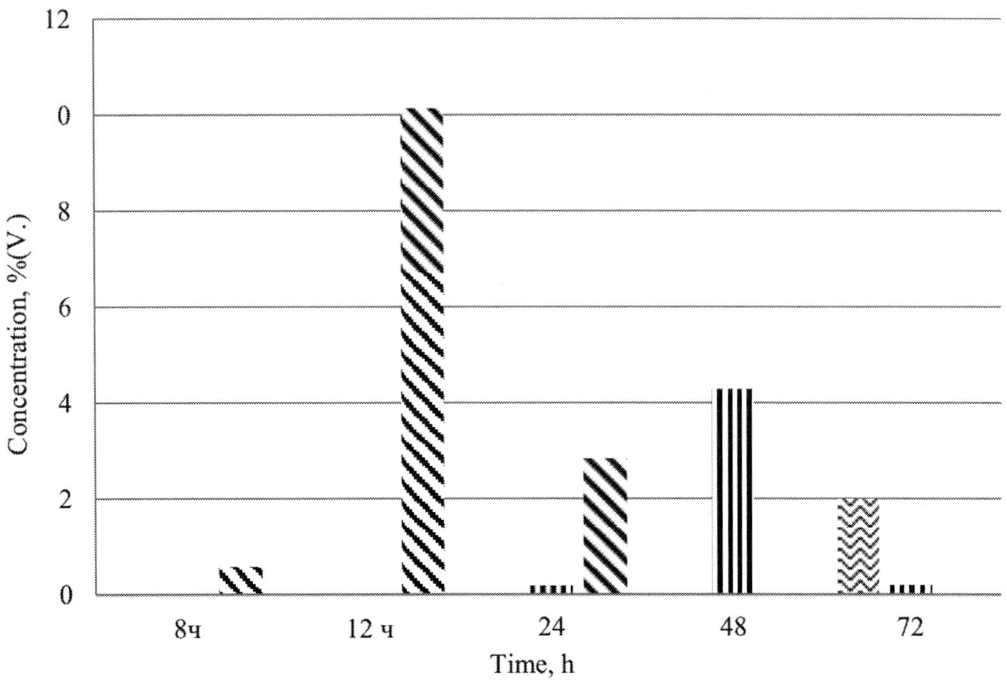

Figure 7. MEK content in the reaction mixture during the biotransformation by bacteria of the genus Pseudomonas spp. of butyl alcohols: butanol-1, butanol-2, 2-methylpropanol-2.

As it can be seen from the diagram (figure 7), MEK is observed in all cases, but the time of its formation is different. As expected, the isomer structure affects the rate of alcohol biotransformation. Therefore, MEK firstly forms in the reaction mixture in the case of using 2-methylpropanol-2 after 8 hours, and in the case of using n-butanol only after 72 hours. It should be assumed that the MEK formation occurs with the participation of the enzyme dehydrogenase complex [25], namely, alcohol dehydrogenase, which is responsible for catalyzing oxidation of alcohols and acetals to aldehydes and ketones in the presence of NAD (nicotinamide adenine dinucleotide). The highest MEK yield was observed when using 2-methylpropanol-2 after 12 hours of biosynthesis and accounted for 10%.

4 Conclusions

MEK was obtained on the basis of microbiological synthesis of butyl alcohols with using bacteria of the genus Pseudomonas spp. The optimal conditions for the synthesis are 1% (vol.) 2-methylpropanol-2, hydrocarbon-oxidizing bacteria of the genus Pseudomonas spp. with 108 CFU, temperature 30° C, time 12 hours, and stirring speed 80 rev/min. It should be considered that this method is an alternative method of obtaining MEK from the environmental and economic point of view.

References

1. T.A. Egorova, S.M. Klunova, E.A. Zhilukhina, Basics of Biotechnology, 208 (2008)
2. B.G. Sampat, Chemical Weekly July **5,** 198-201 (2011)
3. K. Foxall, Health Protection Agency Version **1,** 4 (2010)
4. D. Arora, M. Sharma, Chemical Weekly April **27,** 189-194 (2010)

5. S.A. Akimov, D.I. Perazich, A.N. Sokolov, A.I. Tsyvina, Patent USSR №583117 Method of MEK production (in Russian) (1977)
6. T.S. Zak, Patent US №4075128 Preparation of methyl ethyl ketone (1978)
7. S.I. Kryukov, A.V. Irodov, V.A. Smirnov, Patent USSR №825491 Process of MEK preparation (1981)
8. L.J. Velenyi, A.S. Krupa, Patent US №4329506 Isomerization of aldehydes to ketones (1982)
9. A.P. Khvorov, B.N. Gushchin, A.N. Petrov, D.M. Bayramgulov, S.M. Lakiza, V.I. Kotov, A.B. Guschevskiy, P.F. Tyugaev, Patent USSR №960160 Process of MEK preparation (1982)
10. B.L. Kozik, S.V. Pestrikov, Patent №165678 Process of MEK preparation (1984)
11. D.V. Mushenko, A.V. Grishchenko, G.M. Strashnova, V.S. Sobolevskiy, G.P. Cherkasov, A. P. Khvorov, V.I. Kotov, P. F. Tyugaev, V. N. Pavlychev, Patent USSR №2552716 Process of MEK preparation (1984)
12. A.B. Guschevskiy, M.L. Kolesov, et al. Modern problems of chemistry and chemical industry **8,** 1–46 (1987)
13. M.L. Cook, J.D. Fellmann, D.R. Hamm, S. Mahajan, M.T. Muraoka, T. Rostrup-Nielsen, A. R. Smith, J.H. Grate, R.J. Saxton, K.A. Klingman, Patent CA №2077549 Process for the production of methyl ethyl ketone (1991)
14. V.F. Odyakov, K.I. Matveev, T.A. Gorodetskaya, Z.P. Pai, L.N. Karandila, L.A. Stepanova, A.G. Nelin, D.B. Mushenko, F.A. Chernyshkova, G.N. Gvozdovskiy, Patent USSR №822417 Catalyst for the oxidation of ethylene to acetaldehyde or butylene to methyl ethyl ketone (1994)
15. K.I. Matveev, V.F. Odyakov, E.G. Zhizhina, Yu.A. Motornyi, S.A. Langue Patent №1669109 Catalyst for the oxidation of n-butelene to methyl ethyl ketone and process of its preparation (1995)
16. L. Berg, Patent US №5453166 Separation of ethanol from 2-butanone by extractive distillation (1995)
17. L. Berg Patent US №5876569 Separation of methyl ethyl ketone-from ethanol by extractive distillation (1999)
18. C.D. Murphy, Patent US №6121497 Process of recovering methyl ethyl ketone from an aqueous mixture of methyl ethyl ketone and ethanol (2000)
19. G. Marchionni, U.D. Patto, M. Avataneo, Patent US №7132051 B2 Liquid-liquid extraction of polar organic substances from their aqueous solutions with fluorinated extracting liquids(2003)
20. S.M. Aldoshin, P.E. Matkovsky, L.N. Russiyan, V.N. Troitsky, V.N. Parmon, Z.P. Pai, Patent №WO 2007055608 A2 Catalyst and process of methyl ethyl ketone preparation (2007)
21. GOST R 52714-2007 Motor gasolines. Determination of individual and group hydrocarbon composition by method of capillary gas chromatography //URL: http://allgosts.info/standarts/gost-r-52714-2007
22. T.H. Lee, J. Kim, M.J. Kim, H.W. Ryu, K.S. Cho, Chemosphere **63,** 315–322 (2006)
23. A.P. Astashkina, A.A. Bakibayev, E.V. Plotnikov, Yu.V. Kolbysheva, A.B Mukashev, Procedia chemistry **10,** 504–507 (2015)
24. V.V. Smirnov, E.A. Kiprianova, Bacteria of the genus Pseudomonas, 264 (1990)
25. A.M. Bezborodov, Biochemical basis of microbiological synthesis, 304 (1984)
26. K.A. Leonov, A.P. Astashkina, A.A. Bakibayev, Procedia chemistry **10**, 504–507 (2014)
27. G.K. Skryabin, L.A. Golovleva, Using micro-organisms in organic synthesis, 332 (1976)

DEVELOPMENT OF THE MATHEMATICAL MODEL OF DIESEL FUEL CATALYTIC DEWAXING PROCESS TAKING INTO ACCOUNT FACTORS OF NONSTATIONARITY

Evgeniya Frantsina [1,*], *Nataliya* Belinskaya[1], *Natalya* Popova[1]

[1]Tomsk Polytechnic University, Lenin av. 30, Tomsk, 634050, Russia

Abstract. The paper describes the results of mathematical modelling of diesel fuel catalytic dewaxing process, performed taking into account the factors of process nonstationarity driven by changes in process technological parameters, feedstock composition and catalyst deactivation. The error of hydrocarbon contents calculation via the use of the developed model does not exceed 1.6 wt.%. This makes it possible to apply the model for solution to optimization and forecasting problems occurred in catalytic systems under industrial conditions. It was shown through the model calculation that temperature in the dewaxing reactor without catalyst deactivation is lower by 19 °C than actual and catalyst deactivation degree accounts for 32 %.

1 Introduction

Production of various grades of diesel fuels according to the EURO-5 standard is a complex multistage chemical process scheme, which includes the stages of catalytic refining of wide diesel fraction. This fraction contents such hydrocarbon groups as paraffins, naphthenes, aromatic and olefins. One more stage of production implies the process of rectification and separation of the obtained product. These processes are operated in a system of adjacent and interconnected reactor apparatus. The feedstock of the process is a mixture of diesel fractions and atmospheric gasoil.

The process is aimed to produce a stable naphta, diesel fraction with improved low-temperature properties, and residue with a boiling point of above 340 degrees of Celsius [1].

The process occurs in three reactors. The first two reactors are hydrotreating reactors with a nickel-molibdenium catalyst, the third one is a dewaxing reactor with nickel catalyst. Process temperature is 350–400 °C. The reactions of hydrocracking, hydroisomerization, hydrogenation of unsaturated hydrocarbons, hydrogenation of mercaptanes followed by hydrogen sulfide formation, aromatization and coke formation take place [2].

* Corresponding author: evf@tpu.ru

After reactor section the products enters the stabilization column, where the light hydrocarbons, hydrogen sulfide and water are eliminated. The stable hydrogenate subsequently flows to the rectification section, where it is fractionated into products [3].

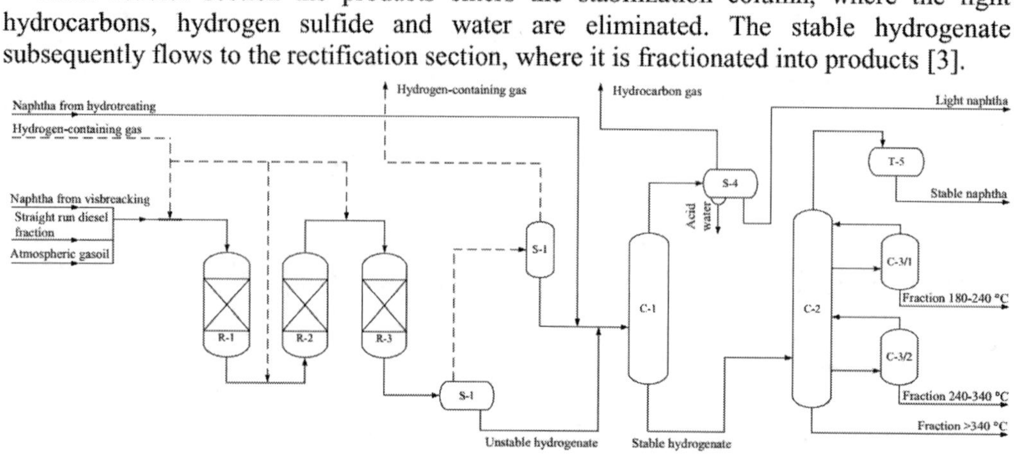

Fig. 1. Technological scheme of hydrodewaxing plant for production of diesel fuels
R-1, R-2 – hydrotreating reactors; R-3 – hydrodewaxing reactor; S-1 – high pressure separator; S-2 – low pressure separator; C-1 – stabilization column; C-2 – rectification column; S-3/1, S-3/2 – strippers; RT-5 – reflux tank.

2 Materials and methods

The method of mathematical modelling is widely used to study, improve, optimize and forecast the processes of petroleum refining industry [4-14].

The main task of the industrial process realization to be solved is sustaining high product yield with required quality. This is possible when developing the optimal technology of the catalyst operation which would allow using the catalyst resource as high as possible, following the required quality and quantity of the product.

To meet this challenge, the method of mathematical modelling was used. The model was developed on the basis of account for thermodynamic and kinetic laws of the process, factors of nonstationarity caused by variability of technological regime, feedstock composition and catalyst deactivation.

Accounting the thermodynamic and kinetic laws of the process allows developing the model sensitive to change in technological parameters and feedstock composition. This provides using the model for forecasting feedstock conversion degree, product composition and yield, as well as for process optimization within valid industrial technological conditions (temperature 335 – 365 °C, feedstock flow rate 295 – 325 m^3/h, hydrogen-containing gas flow rate 35000 – 65000 m^3/h).

The experimental data from the industrial unit has been used as the initial data for modelling. These data included operating conditions in a wide range of changing feed flow rate, temperature, HBG (hydrogen bearing gas) flow rate and its composition, and also the feed and the product content with a range of hydrocarbon group concentrations.

Mathematical model considering deactivation:

$$\begin{cases} G\cdot\dfrac{\partial C_i}{\partial z}+G\cdot\dfrac{\partial C_i}{\partial V}=\sum\limits_{j=1}^{m} a_j\cdot W_j \\ G\cdot\dfrac{\partial T}{\partial z}+G\cdot\dfrac{\partial T}{\partial V}=\dfrac{1}{\rho\cdot C_p^m}\sum\limits_{j=1}^{m} Q_j\cdot a_j\cdot W_j \end{cases}$$

Initial conditions:

$z=0$: $C_i=C_{i,0}$; $T=T_0$;

$V=0$: $C_i=C_{i,0}$; $T=T_0$,

where z is the volume of the refined feedstock from the moment of fresh catalyst loading, m^3; G is the feed flow rate, m^3/h; $z = G \cdot t$ (t is catalyst operating time from the moment of fresh catalyst loading, h); C_i is the concentration of i-component, mole/l; V is a catalyst bed volume, m^3; a_j is a catalyst activity in j^{th} reaction; ρ is mixture density, kg/m^3; C_p^m is specific heat capacity of the mixture, J/(kg*K); Q_j is heat effect of jth reaction, J/mole; T is a temperature, K; W_j is a jth reaction rate, mole/(l*s); m is number of reactions.

This model takes into account catalyst deactivation. Catalyst activity can be determined as a proportion of reaction rate constant at any time to the reaction rate constant at the initial moment of the time when the fresh catalyst had been charged:

$$a_j = \frac{k_{j,t}}{k_{j,0}}$$

where $k_{j,0}$ is j^{th} reaction rate constant at the initial moment of time (with the fresh catalyst); $k_{j,t}$ is j^{th} reaction rate constant at the current moment of time.

$$a_j = A_j \cdot e^{-\alpha_j \cdot C_c}$$

where A_j, α_j are the deactivation coefficients; C_c is coke concentration, % wt.

Deactivation coefficient includes the deactivation of metal and acid centers:

$$a_j = a_{a,j} + a_{m,j}$$

where a_a is the catalyst acid center activity; a_m is the catalyst metal center activity.

3 Results

Model adequacy examination was conducted by comparing the calculated data with experimental data from the industrial unit.

It is shown that the model is suitable for the process and can be applied for identification of the process laws, forecast calculations conduction and optimization.

Table 1. Calculated and experimental data comparison

Product (experimental), % wt.	Product (calculation), % wt.	Error (absolute), % wt.
	n-paraffins C_{10}–C_{27}	
10.17	11.18	1.01
10.08	10.19	0.11
9.32	9.48	0.16
	i-paraffins	
26.46	26.40	0.06
24.66	24.82	0.16
23.14	25.67	2.53
	Olefins	
1.74	1.99	0.25
2.63	2.58	0.05
2.25	2.35	0.10
	MNA	
20.53	20.06	0.47
19.90	20.09	0.19
19.30	19.73	0.43

The abbreviation in Table 1: MNA is monoaromatic compounds.

Computations via the use of the model showed that temperature inside the dewaxing reactor without catalyst deactivation is on the average lower than actual by 19 °C, considering the required product yield of the same quality concerning cold filter plugging point (CFPP) (Table 2).

Table 2. Computation of the deactivation compensating temperature

Date	Yield of fraction 240-340 (DF), %	CFPP, °C	T_{in} R-3 (actual), °C	T_{in} R-3 (comp.), °C	ΔT, °C
25.01.2016	41.4	-34	341	318	23
27.01.2016	41.8	-37	352	332	20
29.01.2016	41.5	-44	360	345	15
05.02.2016	43.0	-41	369	350	19

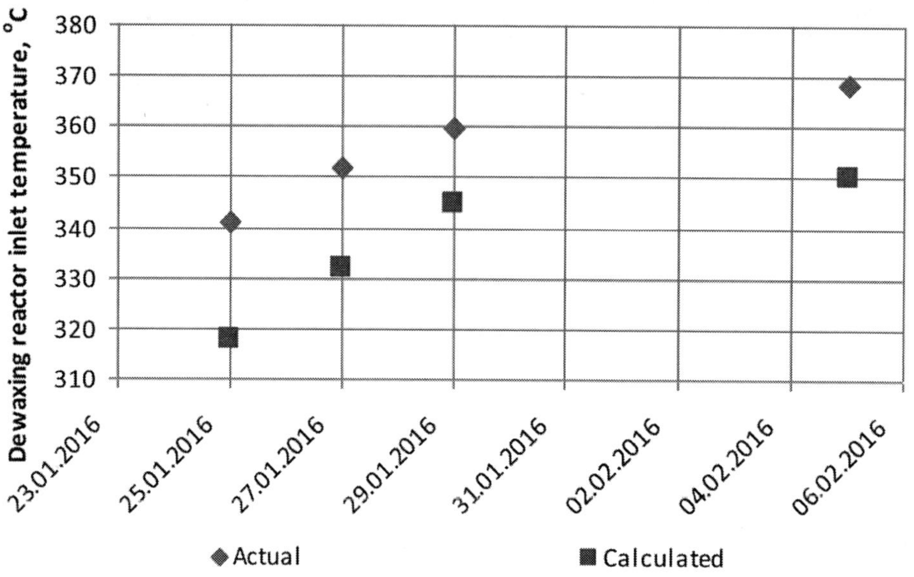

Fig. 2. Computation of the temperature which compensate catalyst deactivation

The level of relative catalyst activity at the cycle end (2016 year) equals to [15]:

$$Akt = \frac{\Delta - \Delta T}{\Delta} = \frac{60-19}{60} = 0.68$$

Where Δ is a largest possible temperature range in the process (330 – 390 oC); ΔT is a temperature that compensates deactivation. The activity of fresh catalyst equals to 1.

To sum up, deactivation of the catalyst in the period of operation accounts for 32%.

4 Conclusion

The developed mathematical model of catalytic dewaxing process is valid for description of the real process since the absolute calculation error of hydrocarbon contents in the product does not exceed 1.16 wt.%.

Accounting for factors of nonstationarity, caused by variability of technological regime, feedstock composition and catalyst deactivation, allows applying the developed model for solution to problems of industrial catalysts operation.

It was shown through the model calculation that temperature in the dewaxing reactor without catalyst deactivation is lower by 19 °C than actual and catalyst deactivation degree accounts for 32 %.

References

1. N. Belinskaya, E. Ivanchina, E. Ivashkina, E. Frantsina, G. Silko, IOP Conf. Series: Earth and Environmental Science **21**, Article number 012030, (2014)
2. N. Belinskaya, Petroleum and Coal **58 (1)**, 126-134, (2016)
3. N.S. Belinskaya, E.D. Ivanchina, E.N. Ivashkina, V.A. Chuzlov, S.A. Faleev, Procedia Engineering **113**, 68-72, (2015)
4. A.N. Zagoruiko, A.S. Belyi, M.D. Smolikov, A.S. Noskov, Catalysis Today **220–222**, 168-177, (2014)
5. F.A.N. Fernandes, U.M. Teles, Fuel Processing Technology **88(2)**, 207-214, (2007)
6. L. Pellegrini, S. Bonomi, S. Gamba, V. Calemma, D. Molinari, Chemical Engineering Science, **62(18-20)**, 5013-5020, (2007)
7. L.P. Oliveira, J.J. Verstraete, M. Colb, Catalysis Today **220–222**, 208-220, (2014)
8. A. Rufer, A. Werner, W. Reschetilowski, Chemical Engineering Science **87**, 160-172, (2013)
9. E. Frantsina, E. Ivashkina, E. Ivanchina, R. Romanovskii, Chemical Engineering Journal **282**, 224-232, (2015)
10. E.N. Ivashkina, E.V. Frantsina, R.V. Romanovsky, I.M. Dolganov, E.D. Ivanchina, A.V. Kravtsov, Catalysis in Industry **4 (2)**, 110-120, (2012)
11. E.D. Ivanchina, M.V. Kirgina, N.V. Chekantsev, B.V. Sakhnevich, E.V. Sviridova, R.V. Romanovskiy, Chemical Engineering Journal **282**, 194-205, (2015)
12. V. A. Chuzlov, K. V. Molotov, Petroleum and Coal **58(1)**, 47-55, (2016)
13. V.A. Chuzlov, E.D. Ivanchina, N.V. Chekantsev, K.V. Molotov, Procedia Engineering **113**, 131-137, (2015)
14. N.V. Usheva, O.E. Moizes, E.A. Kuzmenko, S.F. Kim, E.S. Khlebnikova, S.N. Dyalilova, T.V. Filippova, IOP Conference Series: Earth and Environmental Science **27(1)**, Article number 012047, (2015)
15. A.V. Kravtsov, E.D. Ivanchina, S.N. Averin, A.A. Fedorov, L.V. Krupenya, D.S. Polyboyartsev, Chemistry and Technology of Fuels and Oils **40(3)**, 176-180, (2004)

MATEC Web of Conferences 85, 01024 (2016) DOI: 10.1051/matecconf/20168501024
CCT 2016

Up-dated ore composition data (Central ore-field, Kuznetsk Alatau)

A. Bushmanov[1 a], V. Voroshilov[1], I. Reva[1], M. Rudmin[1], A. Chernyshov[2] and R. Abramova[3]

[1,3] Institute of Natural Resources, Tomsk Polytechnic University, Lenin av. 30, Tomsk, 634050, Russia
[2] Tomsk State University, Lenin av. 36, Tomsk, 634050, Russia

Abstract Applying scanning electron microscope (SEM), energy-dispersive microanalyzer and X-ray fluorescence microscope the ore mineral composition in Central gold-ore field (Kuznetsk Alatau) was investigated. Eleven new minerals were detected in this ore field. The differentiated behavior of mineral formation stages in veins and near-veined beresites was determined. The composition of native gold was studied, as well as the distribution of trace elements in pyrite.

1. Introduction

Based on the investigation of ore mineral composition and their zonation it is possible to identify the mineralization zoning genesis, to predict the ore bodies on the flanges and depth and to reveal new deposits. Previously, the ores in Central gold-ore field were investigated by optical microscopes. Such methods as scanning electron microscope (SEM) and X-ray fluorescence microscope revealed new mineral habits and their interrelationship behavior. Ore mineralization genesis and its distribution zoning were also considered [1,2,3].

2. Research methods

Ores and metasomatites in the Central ore field were studied via a detailed mapping of underground mining and analysis of numerous drill-hole cores. The selected ore samples were studied under optical, scanning electron microscope (SEM) and X-ray fluorescence microscope.

Scanning electron microscope Vega3 Tescan with energy-dispersive microanalyzer AZtec Energy X-max 50 were used to study the specific composition and structure of ore minerals and their behavior , as well as revealing new minerals. The polished section surfaces reflected the spectrum representing the chemical composition of every mineral habit. The petrochemical equivalent of each mineral formula was based on obtained data.

The distribution of the chemical elements within pyrite grains, a prevailing ore mineral, was studied by X-ray fluorescence microscope XGT-7200. The following modules were

[a] Corresponding author: bushmanovai@tpu.ru

 DOI: 10.1051/matecconf/20168501024

applied: spectrum analysis of the chemical composition (Spectrum) and areal mapping of the sample to identify the chemical element distribution in the sample (Mapping).

3. Results and discussion

Central ore-field is located near deep Kuznetsk Alatau fault in the northern part of Kuznetsk Alatau. The most ancient rocks in this ore-field are Prokopevsk-Vendian suite marmorized limestones with overlying Lower Cambrian Ust-Anzassk suite metamorphosed slates (fig. 1).

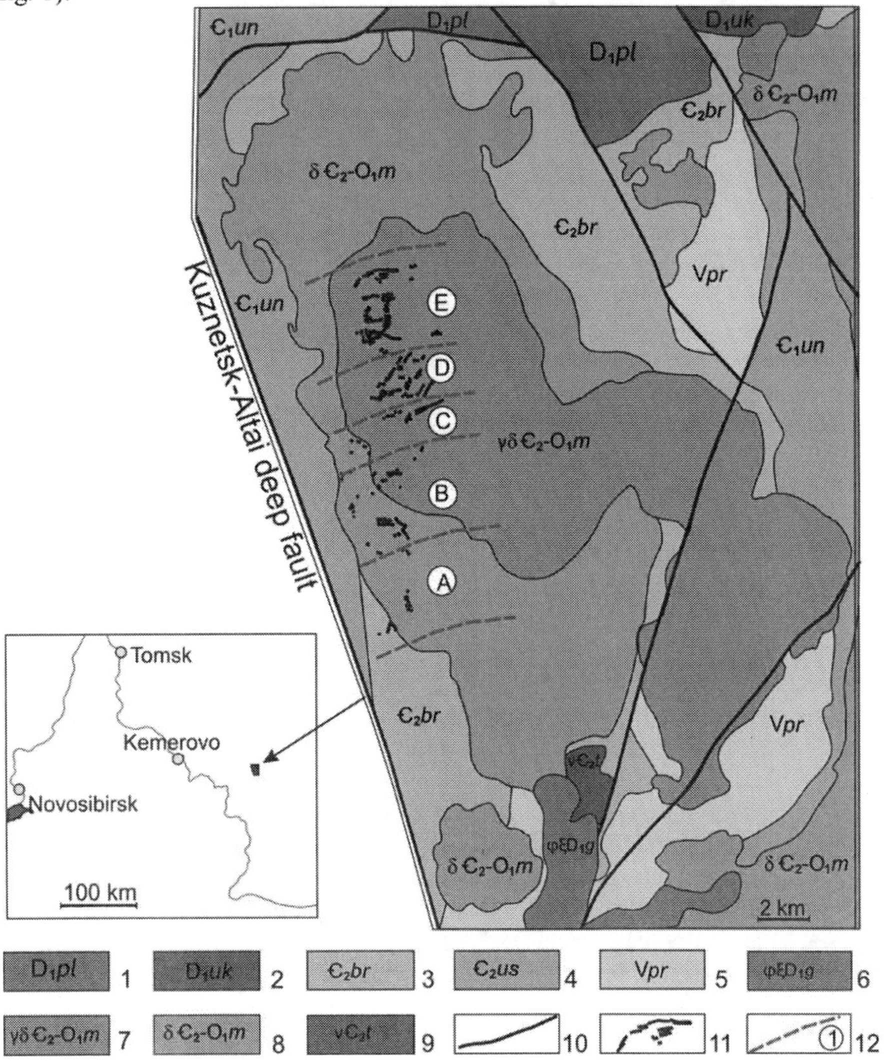

Figure 1. Geological map diagram of Central ore-field

1 – Palatine suite: basalt, andesite-basalt, rarely andesite, tuff sandstone; 2 – Ust-Kundusolskaya suite: conglomerates, sandstone, siltstone; 3 – Berikul suite: basalt, andesite basalt; 4 – Ust-Anzassk suite- chlorite- epidote-actinolite-albite orthoschist, carbon-siliceous schist; 5 – Prokopevsk suite: marmorized bitumen limestones; 6 – Goryachegorsk alkai-gabbroid complex: nepheline syenite; 7- Martaiginsk granitoid complex: granodiorite; 8 – diorite; 9 – Taskilski gabbro-pyroxenite complex: gabbro; 10 – basic ruptures; 11 –gold-sulphide-quartz veins; 12 – ore-field mineral zones: A- tourmaline; B- scheelite; C- molybdenite-chalcopyrite; D- sphalerite-galena; E- arsenopyrite

Mid-Cambrian island-arc basalt and Berikul suite andesite-basalt are overlyingVendian-Early Cambrian sediment cross-beddings. Centralninsk granitoid massif (Martaiginsk complex), resembling intruding south-eastward harpolith, is confined to the surface of the cross-bedding. This massif intrusion formed in the Mid-Cambrian to Lower Ordovician period during the collision stage of the territory development.

Devonian sediments in the northern area embrace two suites: Palatine volcanic suite and Ust-Kundusolyskaya terrigenous suite. Underlying Devonian sediment thicknesses revealed a tectonic contact [4].

Ore bodies include predominately east-west and north-east trending gold-sulphide-quartz veins which are within near N-S zone extending up to 15km. There are more than 200 known veins, some of which extend up to 2 km. and dipping downward to 1000m. Host rocks include Central massif granitoids. Areal and near-vein metasomatic processes preceded ore deposition, the earliest of which were quartz-feldspar metasomatites as mature thickness zone series from tens to hundreds of meters in extent. In the central parts of these zones intensive granodiorite K-feldspathization could be observed, while peripherally-postmagmatic biotitization. The reduced zones, where early potassium metasomatism was revealed, provided conditions for further propylitized solutions resulting in the intensive development of epidotization and chloritization within the central K-feldspathization zones.

Beresites with associated quartz-gold-sulphide veins extent towards the same increased permeability zones as propylites which are predominately located peripherally in echelon fractures. The beresite columns, forming along granodiorite (quartz-muscovite-ankerite facies) include the following 8 metasomatic zones:

1. plagioclase, K-feldspar, chlorite, quartz, magnetite;
2. plagioclase, K-feldspar, chlorite, quartz, pyrite;
3. plagioclase, K-feldspar, quartz, ankerite, sercite, pyrite;
4. ankerite, sercite, K-feldspar, quartz, pyrite;
5. ankerite, sercite, quartz, pyrite;
6. muscovite, quartz, pyrite;
7. muscovite, quartz;
8. quartz.

Progressively as the vein structures pinch out, the immature zones successively settle out of the beresite columns. Near- vein beresite column thickness decreases from 1-3m to 0.5-0.7m north south of the ore-field towards the deep fault. At the same time quartz-muscovite-ankerite facies grade to quartz- muscovite. In beresite-ankerite facies progressing along lamprophyre dykes the amount of sercite significantly increases (from 5–10 % to 30 %) from north to south [5, 6].

The gold-bearing vein composition involves 60 minerals the major ones of which are quartz, pyrite, arseno-pyrite, galena, sphalerite, calcite, chalcopyrite; as well as secondary minerals – tetrahedrite, molybdenite, scheelite, pyrrhotine, magnetite, tourmaline, telluride and sulphosalts, gold. Ore formation in the veins are subdivided into 4 successive steps:

- quartz-pyrite;
- arseno-pyrite;
- polysulphide;
- telluride-sulphosalt

Gold is associated with all sulphides and its maximum concentration being confined to the spatial overlapping areas of several mineral associations.

Mineral zonation throughout the ore-field is exceedingly contrasting. North south of the ore-field towards the deep fault these zones alternate (according to characterizing mineral): arseno-pyrite, galena-sphalerite, chalcopyrite-molybdenite, scheelite, tourmaline (fig.1). The sulphide content in the veins decreases from 50–80% to 5–8% in described direction. Horizontal zoning is distinctly observed and, even, along the vein strike, which is

DOI: 10.1051/matecconf/20168501024

conditioned by the distribution of late gold-bearing paragenesis in the central parts of the veins, as well as early prevailing quartz-pyrite mineralization on the periphery of the veins.

Contrasting mineralization zoning is vividly observed in the properties of pyrite itself (one of the most abundant ore minerals). Besides Fe and S, it embraces such element impurities as As, Co, Ni, Pb, As, Zn, Cu and rarely, Cs, Sm, Zr, Sn, Pr, Mn. In the north south of the ore-field pyrite composition varies significantly- from increasing Co and Ni content to decreasing concentrations of element impurities.

The evidence of mineralization zoning within ore-bodies is marked by relative intersecting vein areas embracing fluid leading faults. Such areas of ore mineralization involve only quartz and pyrite, which are significantly enriched in Ni and As. Progressively from the leading ore structures mineralization becomes more and more diverse, while pyrite becomes enriched in Pb, Zn, Cu. Microscopically investigated pyrite grains showed that enriched Pb, Zn, Cu pyrite is governed by such mechanical impurities as galena, sphalerite, chalcopyrite and fahlerz. Ni and As are evident as impurities in the pyrite crystal lattice and is distributed zonal-concentrically (fig. 2). Such an oscillating distribution pattern of element-impurities is generally typical of genetically different pyrite [7, 8, 9].

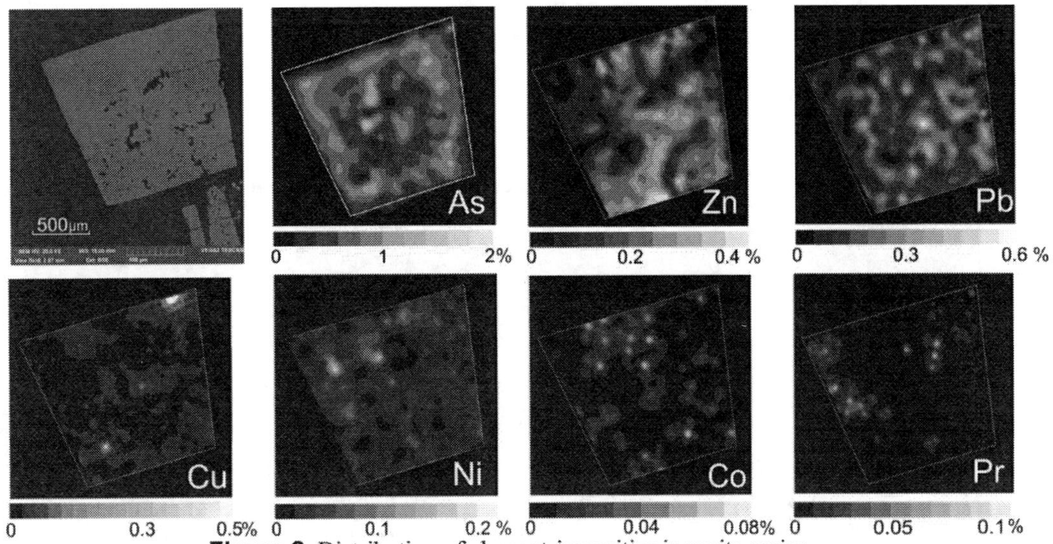

Figure 2. Distribution of element-impurities in pyrite grains

It was established that the mineral formation steps in veins and beresites demonstrate different behavior tendencies. Although arsenopyrite precipitates later than pyrite in the veins, both intersect the veinlets of late sphalerite-tetrahedrite- chalcopyrite -galena associations. However, within beresites arsenopyrite precipitates simultaneously with chalcopyrite, tetrahedrite, sphalerite, galena into carbonate-sulphide veinlets and intersecting pyrite, i.e. arsenopyrite mineralization step in beresites has not been identified (fig. 3).

DOI: 10.1051/matecconf/20168501024

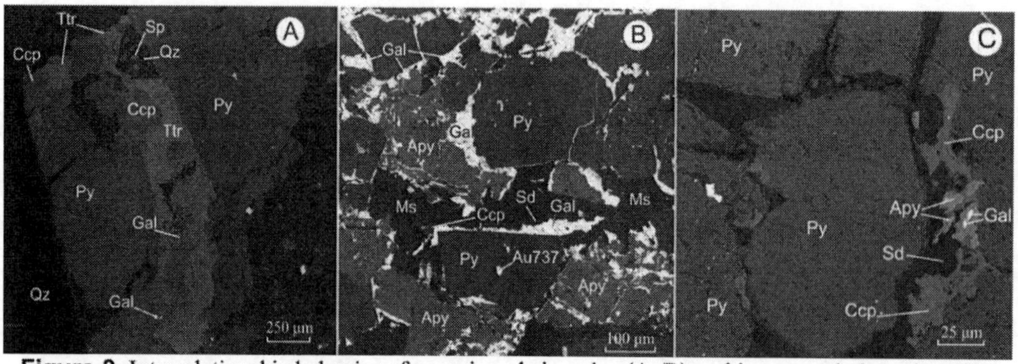

Figure 3. Interrelationship behavior of ore minerals in veins (A, B) and in near vein beresites (C): Py – pyrite; Apy – arsenopyrite; Gal – galena; Ttr – tetrahedrite; Ccp – chalcopyrite; Sp – sphalerite; Qz – quartz; Cal – calcite; Sd – siderite

Noble metal mineralization in veins and beresites is diverse. Native gold with rather high purity (730–788) is characteristic of these veins. Silver can be found in native gold and tetrahedrite as a concentration of 3.5 % to 13.4 %.

Only low-grade gold (500–560)and electrum (including silver content of 50-80%) can be found in beresites (fig. 4). Silver is also evident as argentite and as an impurity in tetraherite. Electrum and chalcopyrite often react resulting in the formation of a unique alloy composition: from $7(Ag,Au)*2CuFeS_2$ to $7(AgS)*2CuFeS_2$.

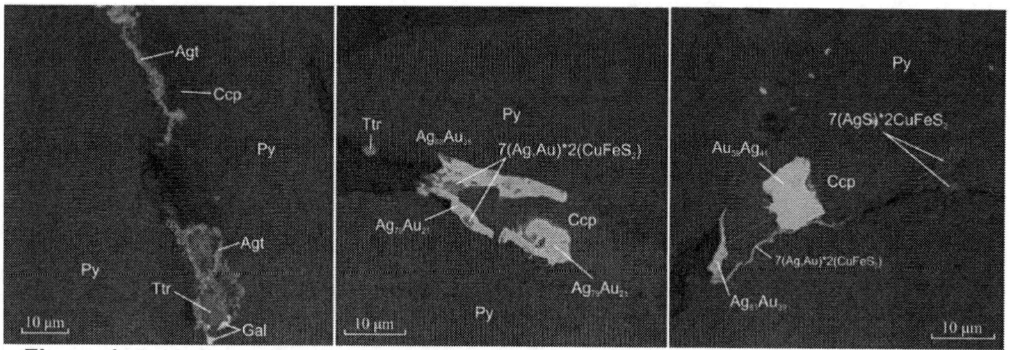

Figure 4. Gold-silver mineralization in beresites : Py – pyrite; Agt – argentite; Gal –galena; Ttr – tetrahedrite; Ccp – chalcopyrite

Late gold-galena-tellurium-bismuth mineralization was identified in two areas, i.e. Burlev and Khrebtov areas. Comparable to the total ore-field structure these two areas are centrally localized with respect to the developing mineralization zoning temperature. The presence of tellurium indicates the stabilizing convergent behavior of zoning in the late ore deposition stages. Tellurium-bismuth mineralization involves the following minerals: joseite (Bi_4Te_2S), pilzenite ($Bi_{11}Te_{10}$), wehrlite (Bi_3Te_2), tellurium-bismuth (Bi_2Te_3), gustavite (Bi_3S_6), volynskite ($AgBiTe_2$), matildite ($AgBiS_2$), stuetzite (Ag_5Te_3), hessite (Ag_2Te), tetradymite (Bi_2Te_2S), cosalite ($Pb_2Bi_2S_5$), tellurite (TeO2), native bismuth. Above-mentioned minerals, except for tellurium-bismuth and tetradymite, have been discovered in this ore-field for the first time. Gold in tellurites is often associated with hessite. Average purity of gold is more than 700, however, one can find electrum with a silver content of up to 50 % (fig. 5). There are no other impurities in gold, except for silver. One can find Fe and Cu rarely (less than 1 %).

 DOI: 10.1051/matecconf/20168501024

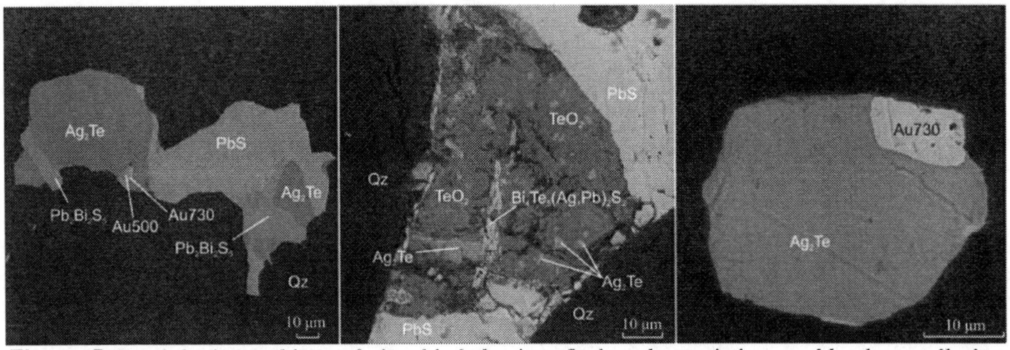

Figure 5. Precipitation and interrelationship behavior of mineral associations: gold-galena- tellurite-bismuth

Tellurite-bismuth minerals were not detected within other ore-field areas. Late galena is associated only with stibnite fahlerz (tetrahedrite) and sometimes accompanied by native bismuth. Zonal composition sequence of tellurite-bismuth mineralization was also revealed within other gold-ore deposits [10]. Spatial isolation of gold-galena-tellurite-bismuth and gold-galena-sulphosalt (fahlerz) associations, parallel in formation time, could possibly reflect the zonation of sediments in later gold-bearing mineralization and, in this case, could be used in predicting future mineralization.

4. Conclusion

Based on the investigation data results from Central ore-field, 11 previously undiscovered minerals were revealed. The distribution behavior and impurity deportment in pyrite were determined. Differential tendency of mineralization in veins and beresites was identified. The composition of native gold was studied. The zonation of sediments in gold-bearing tellurite-bismuth mineralization was established which could be used as a mineralogical prospecting criterion.

References

1. Gaskov I, Borisenko A, Babich V and Naumov E 2010 The stages and duration of formation of gold mineralization at copper-skarn deposits (Altai-Sayan area). *Russian Geology and Geophysics* **51(10)** 1091-1101
2. Kravtsova R 1998 Mineralogo-geohimicheskaja zonal'nost' i osobennosti formirovanija zoloto-serebrjanyh mestorozhdenij (north-eastern Russia) *Geology and geophysics* 39 763–777
3. Spiridonov E, Sokolova N and Gapeev A 1974 Mineral'nye associacii zoloto-telluridnogo mestorozhdenija Zhana-Tjube v Severnom Kazahstane *Geology of Ore Deposits* **1** 54–65
4. Kurtigesheva V, Rodchenko S, Mitrokhin D, Tumanova L and Babin G 2008 State geological map of Russian Federation, scale 1:200000. Edition 2 Kuzbass Series Sheet N-45-X (Central) *Exploratory record* St. Petersburg Lithographic Printing Office RGRI 171
5. Voroshilov V 1995 Formation conditions of Central ore-field in granodiorites (Kuznetsk Alatau) *Ore & Metals* **3** 68–80
6. Voroshilov V 2009 Anomalous structures of geochemical fields of hydrothermal gold deposits: formation mechanism. Methods of geometrization, typical models, and forecasting of ore mineralization. *Geology of Ore Deposits* **51** **(1)** 1-16
7. Duran C, Barnes S and Corkery J 2015 Chalcophile and platinum-group element distribution in pyrites from the sulfide-rich pods of the Lac des Iles Pd deposits, Western

DOI: 10.1051/matecconf/20168501024

Ontario, Canada: Implications for post-cumulus re-equilibration of the ore and the use of pyrite compositions in exploration *Journal of Geochemical Exploration* **158** 223–242
8. Craig J and Solberg T 1999 Compositional zoning in ore minerals at the Craig mine, Sudbury, Ontario, Canada *Can. Mineral* **37** 1163-1176
9. Franchini M, McFarlane C, Maydagán L, Reiche M, Lentz D, Meinert L and Bouhier V, 2015 Trace metals in pyrite and marcasite from the Agua Rica porphyry-high sulfidation epithermal deposit, Catamarca, Argentina: Textural features and metal zoning at the porphyry to epithermal transition. *Ore Geology Reviews* **66** 366–387
10. Sejkora J, Litochleb J, Plasil J and Buresh B 2009 Tsumoite and associated tellurides from the Au deposit Libshice near Novy Knín, Czech Republic: mineralogy and genetic significance *Journal of Geosciences* **54** pp 73–82

Features of chemical-mineralogical composition and processing properties for medium-ferrous bauxites

L.P. Govorova[1,*], *T.V.* Vakalova[1], *E.V.* Shvagrukova[1], *A.A.* Reshetova[1] and *A.P.* Potapova[1]

[1]Tomsk Polytechnic University, Tomsk, Russia

Abstract. The features of chemical-mineralogical composition and processing properties of bauxites from Iksinsk and Severoonezhsk deposits are investigated in the paper. It is found that the studied bauxites are coarsely dispersed and low-dispersive, high-alumina clay raw materials with a high content of colouring oxides. According to their mineralogical composition, Iksinsk and Severoonezhsk bauxites are boehmite-hydrargillite-kaolinite clay bauxites. Moreover, Iksinsk bauxite samples are characterized by the total absence of quartz impurities. The samples of plastic molding based on Iksinsk bauxites are fully sintered at 1500 °C with the formation of strong structures with a compressive strength of 70-120 MPa (depending on the variety of bauxites). The samples based on Severoonezhsk bauxites are fully sintered at 1450 °C, and they are characterized by a compressive strength to 70 MPa in the sintered state.

1 Introduction

Currently, oil and gas industry has faced a serious problem of keeping up the achieved level of oil production, which can be stabilized by intensification of old well operation. In this regard, Russian oil and gas industry pays much attention to the prospects of using hydraulic fracture for oil-and gas-bearing formations. These operations cannot be conducted without using a granular material – a proppant. It acts as a proppant agent and prevents a fracture closure in the process of fluid extraction [1].

Numerous studies and practical applications have allowed giving preference to the ceramic technology in proppant production, as far as ceramic proppants combine a high strength, low bulk density, chemical resistance, and high conductivity. Granular materials of the high and medium strength can be used as ceramic proppants. On a large scale, they have the alumina or silica-alumina composition of natural raw materials, mainly, bauxites or refractory clays and kaolins [2].

Production of ceramic proppants is connected with high energy costs, which can be cut by using inexpensive local aluminosilicate clay raw materials.

Therefore, the aim of this work is a comprehensive study of the bauxites from Severoonezhsk (mark GB-1) and Iksinsk deposits (marks GIO-2 and GIO-5) of

* Corresponding author: lgovorovatpu@yandex.ru

Arkhangelsk region in order to use them in the technology of aluminosilicate ceramic proppants.

2 Chemical and mineralogical features of studied bauxites

Severoonezhsk bauxite area is formed with the bauxites of a sedimentary origin. Three largest deposits in Severoonezhsk bauxite area are: Iksinsk, Plesetsk, and Denislavsk. Iksinsk deposit is situated in the north-west of the European part of Russia, Plesetsk district of Arkhangelsk region. It is represented by six deposits, and it is located on both banks of the Onega. The total area of the deposit is about 120 km^2 with a thickness of a layer from 0.8 to 16.0 m [3]. The distinctive feature of Iksinsk bauxite deposit is a high content of alumina along with a high content of silica and relatively low amounts of iron oxides. It is not typical for the other bauxite deposits in the world [4].

The analysis of granulometric composition for the tested samples of bauxites defined in accordance with GOST 21216.2-81 shows that the samples of Iksinsk clay bauxites are low-dispersive materials, and Severoonezhsk bauxites are coarsely dispersed raw materials (Table 1).

Table 1. Granulometric composition of studied bauxites

Sample code	Content,%, fraction size, mm				
	1-0,06	0,06-0,01	0,01-0,005	0,005-0,001	< 0,001
Bauxites of Iksinsk deposit					
GIO-2	9,97	27,99	7,76	25,28	29,00
GIO-5	14,79	37,65	7,52	19,08	20,96
Bauxites of Severoonezhsk deposit					
GB-1	28,45	68,65	0,96	-	1,94

Initial visual inspection of raw bauxite samples indicates that tested sample GIO-2 of Iksinsk bauxite contains the lowest amount of iron oxide, as far as it has a creamy colour. Tested sample GIO-5 has a red color, GB-1 is a light brown one, indicating the higher Fe_2O_3 content in their gross chemical composition. This fact is confirmed by the chemical analysis (Table 2).

Table 2. Chemical composition of studied bauxite in calcined state

Type of raw materials	Content of oxides, wt. %								
	SiO_2	Al_2O_3	TiO_2	Fe_2O_3	CaO	MgO	K_2O	Na_2O	MnO
Bauxites of Iksinsk deposit									
GIO-2	11,70	77,14	4,60	3,94	0,22	1,53	0,65	0,19	0,04
GIO-5	25,91	59,92	4,33	7,34	0,52	1,19	0,57	0,20	0,04
Bauxites of Severoonezhsk deposit									
GB-1	20,84	62,26	1,75	10,45	0,05	4,39	0,07	0,03	0,16

It is observed that both investigated samples of Iksinsk bauxite deposit are high alumina rocks (Al_2O_3 content – 60-77 wt.%) with a low content of alkaline and alkaline-earth oxides, and also with a high content of Fe_2O_3 (from 3.94 to 7.34 wt.%) and TiO_2 (from 4.33 to 4.60 wt.%) colouring oxides.

Severoonezhsk bauxite belongs to the ferrous type of bauxite with a high content of silica (20.84 wt.%), a high alkaline oxide content (the total content of 0.1 wt.%), and alkaline-earth oxides (the total content of 4.44 wt.%). In addition, this material belongs to high-alumina raw materials due to Al_2O_3 content of 62 wt. %.

In order to determine the mineralogical composition of the studied bauxite rocks, X-ray and integrated thermal analyses have been carried out.

The analysis of the results obtained while processing the diffraction patterns of Iksinsk bauxite samples indicates their qualitative identity. The difference is in the intensity of corresponding X-ray reflections, which allow making indirect conclusions about the quantitative content of minerals identified (Figure 1).

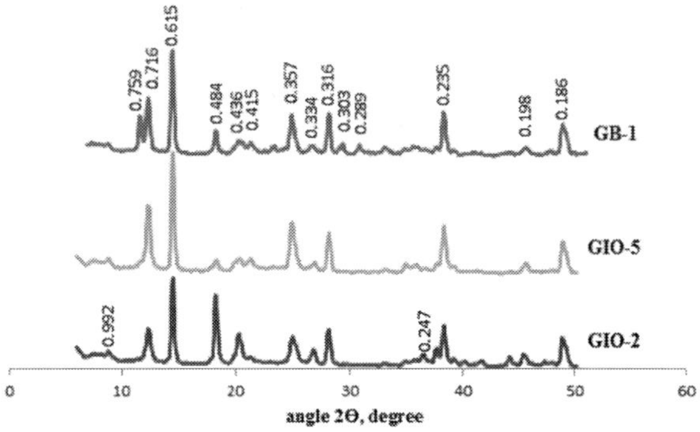

Figure 1. X-ray diffraction patterns of bauxites from Iksinsk (GIO-2, GIO-5) and Severoonezhsk (GB-1) deposits

The studied Iksinsk bauxites are polymineral clay-containing rocks, which are formed with a mixture of such clay minerals as kaolinite $Al_2O_3 \cdot 2SiO_2 \cdot 2H_2O$ (reflections with the interplanar distances of 0.717 - 0.720, 0.356, and 0.247 nm), with a low content of hydromica of an illite type $0.2 \cdot K_2O \cdot Al_2O_3 \cdot 3SiO_2 \cdot 1{,}5H_2O$ (0.992; 0.440 nm). X-ray kaolinite reflections are well-defined and rather intense, and this fact confirms its dominant content in the mixture with hydromica (illite) and its high crystallinity.

The mineralogical composition of Iksinsk bauxites has a special feature: it is the presence of aluminous minerals in the form of alumina hydrates of different water content – hydrargillite (gibbsite) $Al_2O_3 \cdot 3H_2O$ (differential peaks with the interplanar distances of 0.484 and 0.316 nm), and boehmite $Al_2O_3 \cdot H_2O$ (differential peaks with the interplanar distances of 0.615 and 0.185 nm), with a prevalence of the latter in both samples. In the case of GIO-5 sample, the prevalence of boehmite is particularly high (Figure 2). It means that the sample of bauxite will have less mass loss on calcination, and the fired samples based on GMO-5 will be characterized by lower shrinkage values.

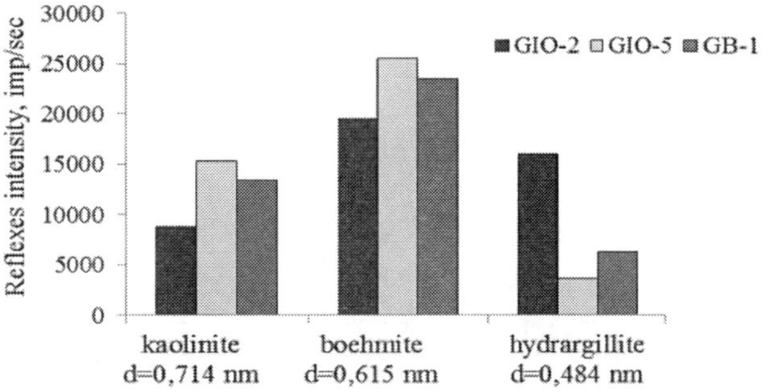

Figure 2. Intensity of X-ray reflections for basic minerals (kaolinite, d – 0.714 nm, boehmite, d – 0.615 nm, and hydrargillite, d – 0.484 nm) in X-ray diffraction patterns of Iksinsk and Severoonezhsk bauxites

The X-ray diffraction pattern of Severoonezhsk bauxite GB-1 is almost identical to the diffraction patterns of Iksinsk bauxites (Figure 1), and this confirms the closeness of their mineralogical composition.

The conclusions concerning the features of the mineralogical composition in Iksinsk and Severoonezhsk bauxites (according to the X-ray analysis) are consistent with the results of a complex thermal analysis for the investigated raw materials. The presence of kaolinite in the mineralogical composition of the studied clay bauxites is confirmed by the endothermic effect in DSC curves at 520 °C, and exothermic effect at 970 °C. The endothermic effect is associated with the removal of chemically bound water from a kaolinite lattice. The exothermic effect is caused by decomposition of the kaolinite lattice (Figure 3, 4). The high intensity of the endothermic effect with a minimum at 270-290 ° C caused by decomposition of hydrated aluminous minerals, DSC curve (Figure 3, 4) and GIO-2 samples confirm that its mineralogical composition has a large amount of watery alumina hydrate in the form of $Al_2O_3 \cdot 3H_2O$ hydrargillite, in comparison with sample GIO-5.

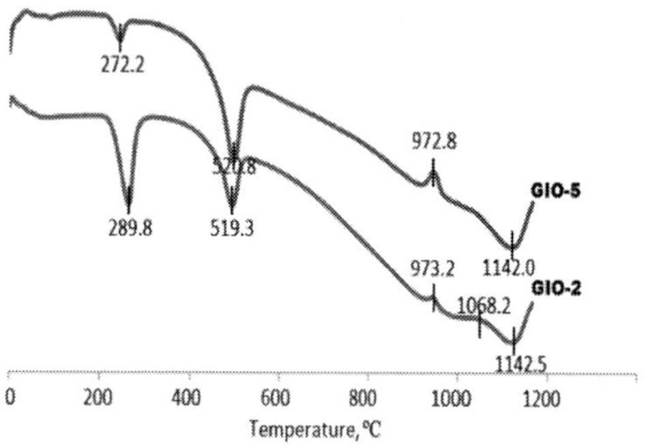

Figure 3. DSC curves of Iksinsk bauxite deposit

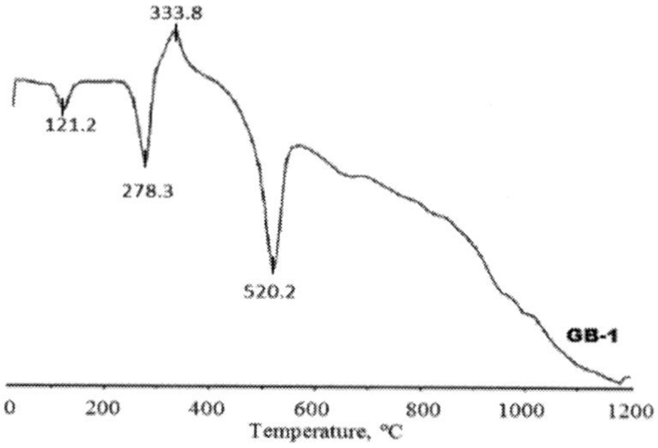

Figure 4. DSC curves of Severoonezhsk bauxite deposit

Thus, both Iksinsk and Severoonezhsk bauxite samples are varictics of a clay bauxite boehmite-kaolinite hydrargillite-type, and this agrees well with the data of their chemical composition.

3 Behaviour of studied bauxites at firing and sintering

Chemical and mineralogical characteristics of the investigated bauxites influence their behaviour at firing. So, sintering both samples of Iksinsk deposit is fully completed at the temperature of 1500 °C, which is proved by achieving water absorption values of less than 2% in the samples (Figure 5).

Moreover, the strength characteristics of the sintered samples of plastic molding from test sample GIO-5 (115-120 MPa) is almost 2 times higher than the strength characteristics of the samples from test sample GIO-2 (53-70 MPa). This feature may be associated with the difference in the ratio of minerals synthesized at firing.

Severoonezhsk clay bauxite (GB-1) is fully sintered at 1350 °C, providing the formation of sufficiently strong structures with mechanical properties (a compressive strength) at the level of 60-70 MPa (Figure 5).

Investigations of phase formation processes at Iksinsk bauxite heating from 1100 to 1400 ° C show that in the process of firing corundum is synthesized as a product of thermal decomposition of alumina hydrates, which are present in the initial samples in the form of boehmite and hydrargillite. In addition, the synthesis of mullite from kaolinite is observed, which is present in the mineralogical composition (Figure 6). A greater amount of mullite is found in the fired products of test sample GIO-5, and this mullite contributes into glass phase reinforcement. This fact could explain the higher strength characteristics of the fired samples based on the given bauxite.

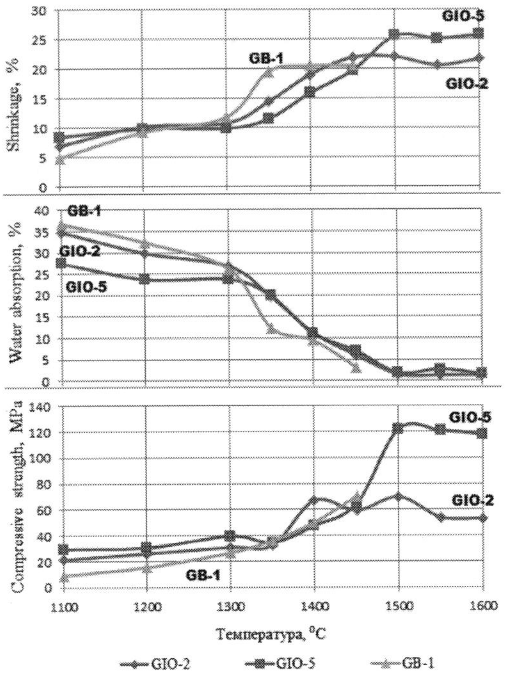

Figure 5. Sintering curves of Iksinsk and Severoonezhsk bauxites

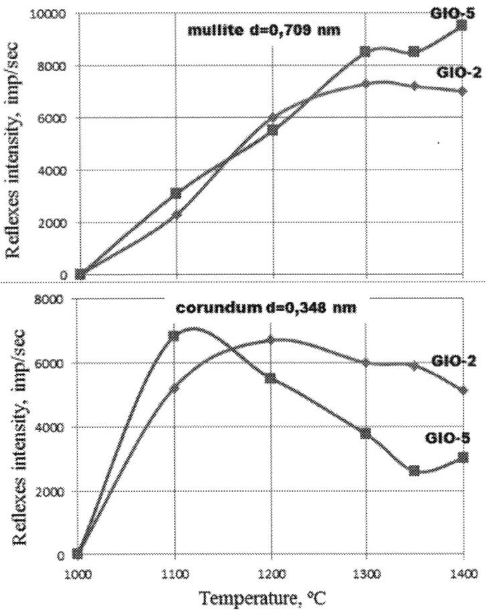

Figure 6. Intensity of X-ray reflections of main crystalline phases (mullite and corundum) in the sintered Iksinsk bauxite samples

It is necessary to take into consideration the fact that during the mullite synthesis from kaolinite amorphous silica is released, which can partially dissolve in the melt, and, partially, it is crystallized in the form of cristobalite. Consequently, the decrease of corundum reflection intensity at temperature increase up to 1400 °C (especially, with test sample GIO-5) is caused, probably, by its binding with this silica into the secondary mullite. A total absence of the fired Iksinsk bauxites with a silica component in the diffraction patterns can also be explained by the formation of the secondary mullite from the decomposition products of alumina hydrates and kaolinite.

4 Conclusion

Investigations of physical-chemical and structural-mineralogical characteristics of medium-ferrous bauxite rocks from Iksinsk and Severoonezhsk deposits (GIO-2, GIO-5, and GB-1), and high strength characteristics of plastic molding samples on their basis show that this raw material has a potential for its application in the technologies for aluminosilicate ceramic proppants of a high strength, which can withstand the pressure of 70 MPa.

References

1. H. Rischmuller. *Resource of Oil and Gas Ullmann's Encyclopedia of Industrial Chemistry,* **23**, 183 (2007).
2. A.A. Reshetova, A.I. Alferova *Proceedings of the XIV International scientific and practical conference of students and young scientists "Modern engineering and technologies",* 156 – 157 (2008). In Rus.
3. E.I. Malyutin, V.N. Shirobokov *Mineral Resources of Russia. Economics and Management,* **4,** 3-10 (2006). In Rus.
4. A.J. Menshenin *Author's abstract of the dissertation "Bauxites of Iksinskoye deposit as a complex raw material for production of alumina, synthetic zeolites and other products",* 90 (2001). In Rus.

The research of finely dispersed iron powder moistening applying the pH–metry method

Kseniya Ikonnikova[1,*], *Lyubov* Ikonnikova[1], *Ekaterina* Koltunova[1], and *Maxim* Murashkin[2]

[1]Tomsk Polytechnic University, Tomsk, Russia
[2]Institute of Physics of Advanced Materials, Ufa State Aviation Technical University, Ufa, Russian Federation

Abstract. This research is aimed at solving crucial problem to diagnose and prognosticate moistening of fine-grained (finely-dispersed) materials. The article considers the dependence of surface moistening from the nature of the primary acid-base center. It demonstrates the advantages of acid-base property study of the surface applying the hydrolytic adsorption method. The research of hydrolytic adsorption on the ferric oxide with a haematite structure was carried out. Ferric oxide is presented in various states (massive and in the form of oxide film on the iron core of nanometer size). The identification of the composition and structure of the ferric-oxide compound was carried out applying the method of transmission electron microscopy (TEM) and X-ray diffraction analysis (XRDA). The authors found out the identity of the behavior characteristics of hydrolytic adsorption on the haematite in various states. The paper also describes the inhibition mechanism of haematite moistening, which explains its hydrophobic behavior. The material of this article can be useful for its implementation in industrial, research and school laboratories.

1. Introduction

Finely dispersed metal powders are widely used in mechanical engineering, power engineering, medicine, construction, environmental protection and etc. If the process occurs in the system “solid body-liquid” then the efficiency of powder use depends on the compatibility of solid body and liquid. Therefore, it is very important to know such characteristic as surface moistening to solve the problem of obtaining products with desired functional properties.

Most of the methods used for quantitative assessment of moistening [1] are not appropriate for powders of nanometer size due to their high reactive capacity, small size and spherocity of particles. Therefore, the development of the methods used for quantitative assessment is very significant. These methods are based on the measurement of indirect values related to the moistening phenomenon. At qualitative level “moistening” is defined by opposite conceptions such as “hydrophilic behavior / hydrophobic behavior”.

* Corresponding author: ikonna@yandex.ru

Since the finely dispersed iron is pyrophoric, it is necessary to passivate it in advance in oxidizing medium. Whereby the oxide layer is formed on the surface of metal particles which phase compound depends on the passivating conditions [2-4]. The change of the oxidizing medium composition, temperature and curing period leads to the formation of nano-dimensional oxide layer which has a complex composition: FeO, Fe_2O_3 and Fe_3O_4 of different modifications. These compounds have different acid-base surface centers in massive crystalline state and adsorb water through different mechanisms. The research of hydrolytic adsorption conducted earlier on Fe_2O_3 and Fe_3O_4 samples in massive state shows considerable differences of pH change in aqueous suspension [5-7]. Acidification and alkalization of aqueous suspension were observed, which can be related with the transition of reaction products in the medium (H^+ or OH^- -ions). The pH value corresponds with the mass transfer unit [8].

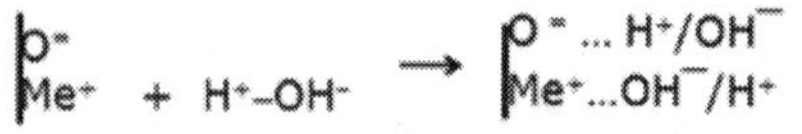

The analysis of the mass transfer speed ($\upsilon pH = \Delta pH_i / \Delta \tau, \text{min}$) with time allows us to establish the reason for inhibition process of moistening and assess the hydrophilic behavior/ hydrophobic behavior of the surface at qualitative level. The dimension of solid body particles significantly affects the absolute value of the indicators of the heterogeneous system. Therefore, the change of the mass transfer speed is only considered and does not require the consideration of the specific surface level.

All the above mentioned actualizes the study of moistening process of structures with nanometer dimension by means of hydrolytic adsorption method.

The aim of this work is to study the process of wetting in an aqueous medium of highly dispersed iron powder obtained by Electric Explosion of Wire method (EEW iron) and passivated in an oxidizing atmosphere.

2. Experimental setup and methods

The subject of inquiry is iron nanopowder («sample 1» - iron EEW) obtained applying the electric blasting method of a steel wire of the St 3 type (State Standard 380-2005) in the argon environment at gas pressure 0.2 megapascal [9]. The obtained powder was passivated by means of slow air inflow in the container with nanopowder over the period of 72 hours at 20° C. This process was carried out in accordance with the recommendations [10]. The sample was obtained in the Institute of Physics of Strength and Material Science (RAS, Tomsk).

The composition and structure of the oxide film on the iron core were studied applying the method of transmission electron microscopy (TEM). The «JEOL JEM - 2100F» equipment of Nano-Center of TPU was used for this purpose.

The identification of comparison-sample («comparison-sample 2» - haematite in massive condition (state) was carried out applying X-ray diffraction analysis (XRDA) on the "Diffray -401" diffractometer. To decrypt the x-ray used x ray diffraction of ASTM and JCPDS card files.

The moistening of sample surface was studied by variation of pH-metry methods of hydrolytic adsorption [11]. The initial water acidity was 6.7÷6.9 pH units. The suspension was prepared in the ratio water /solid body = 50. The change of pH of aqueous suspension over a period of time was registered by a special device of pH-150M type (measuring accuracy is ± 0.03 pH units). The parameters were measured every 5 seconds of the contact by continuous stirring of aqueous suspension using a magnetic stirrer within 90 minutes. The obtained kinetic mass transfer curve (pH = f (time)) was differentiated with various stepping time interval to get data about mass transfer ($\upsilon pH = \Delta pH_i / \Delta\tau, \min$) and moistening rate.

3. Results and discussion

It was necessary to solve a range of tasks to achieve the desired objective:

1. To identify the phase composition of the oxide film formed on the EEW iron core during passivation.
2. To investigate the process of hydrolytic adsorption on the nano-dimensional EEW iron
3. To give qualitative assessment of moistening (hydrophilic behaviour / hydrophobic behavior) applying mathematical treatment of kinetic curves of hydrolytic adsorption.

3.1. The identification of the phase composition of the oxide film formed on the iron core during passivation

The first stage of the given work was to carry out investigation on identification of the phase composition of the sample oxide layer. It is related to the fact that even minor deviation from the production procedures results in considerable changes of the sample dispersion, oxide film composition and structure [2].

The images of nanoparticles are presented in Figures 1-3. The analysis of TEM images shows (Fig.1) that the particles are polydisperse, they are not agglomerated with each other, they have spherical shape and different dimension (size). The diameter of the body of large particles together with oxide film is 1000 nm, whereas small particle diameter is only 50nm. The amount of large particles is small. The majority of particles are of different size from 60 till 200 nm.

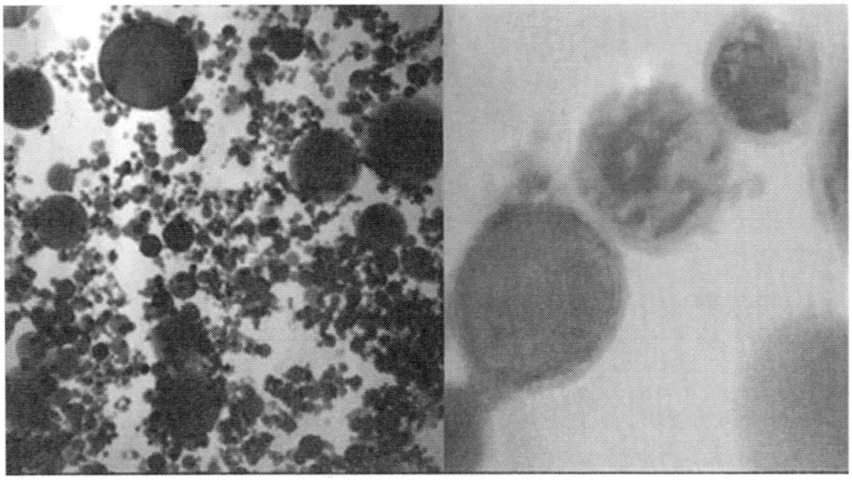

Fig.1. Microphotos with various expansion of electric blasting iron powder obtained applying the method of transmission electron microscopy (TEM).

Macrophoto (Fig. 3) allows defining the main characteristics of the particles of different size (Fig. 2). These characteristics are: oxide layer thickness, its continuity and crystallinity as well as the value of interplanar spacing.

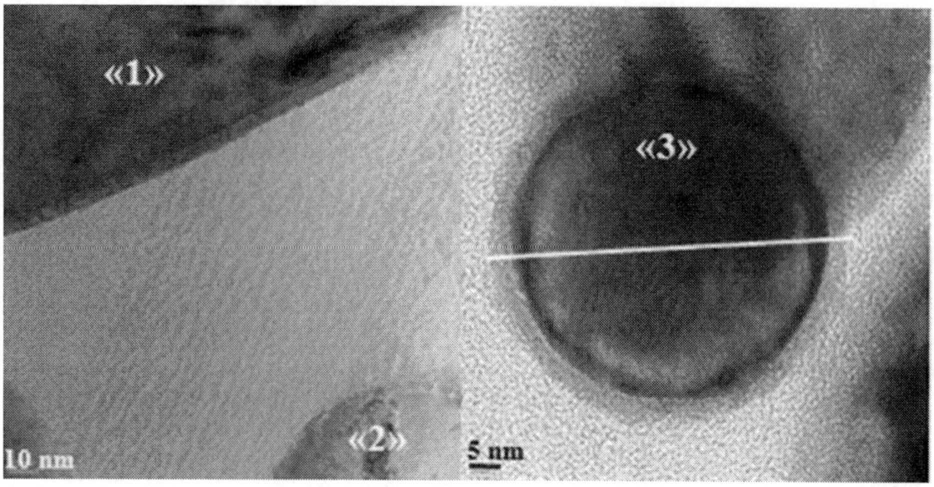

Fig. 2. Photo of TEM for particles of different size:
«1» - (d particles =1000 nm), «2» - (d particles = 100 nm), «3» - (d particles =70 nm).

The TEM analysis showed (Fig.3) that the parameters of oxide layer for the particles with different dimension are almost identical:

- the thickness of oxide layer is uniform and equals 4÷5 nm;
- the lines of atomic rows are uniform and have one direction. It testifies the single-layer coating and crystallinity of the oxide film;
- the distance between atomic rows is 0.269 nm. It corresponds to the reflex with d (104)=0.269 nm. This reflex has maximal intensity and is typical for ferric oxide with α-Fe_2O_3 structure (Fig.4). The absence of powder particles agglomeration is also a vivid evidence in the favor of α-Fe_2O_3 formation, which magnetic susceptibility is low.

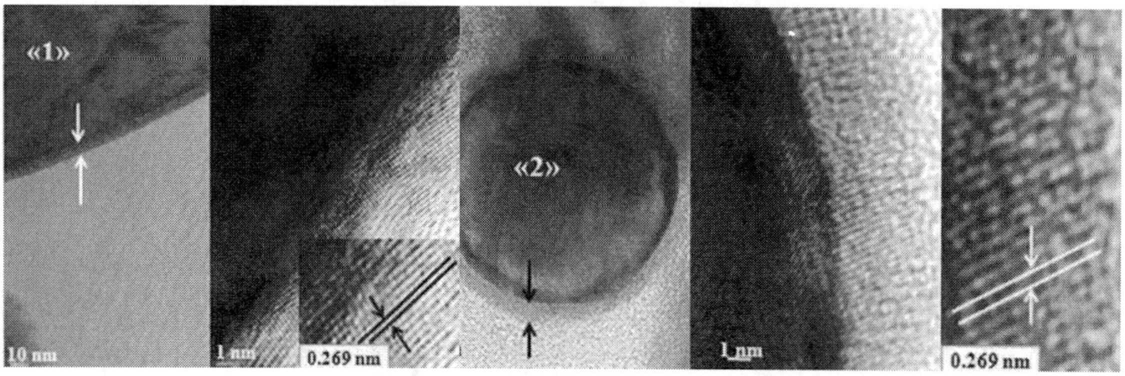

Fig. 3. The photo of TEM analysis of the oxide film of particles with different sizes (dimension):
«1» - (d particles =1000 nm), «2» - (d particles =70 nm).

3.2. The research applying the method of hydrolytic adsorption

The causes affecting the dimensional parameters and structure of low-dimensional particles are considered in monograph [2]. It is believed that one should be very attentive and careful with the idea of identifying phase composition of such structures by means of comparison between interatomic rows distances (defined by TEM method) and interplanar spacing of samples in massive state (defined applying X-ray diffraction analysis (XRDA)).

One should be very attentive when there is reflection of one plane only. Confirmatory findings are required that were obtained by applying other methods.

The next stage of our research was the comparison of findings concerning the hydrolytic adsorption obtained for electric blasting iron powder ("sample 1") and ferric oxide with α-Fe_2O_3 structure in massive state ("comparison- sample 2"). The identification of ferric oxide sample in massive state ("comparison- sample 2") was carried out applying X-ray diffraction analysis (XRDA). The findings of X-ray diffraction analysis show that the sample chosen for comparison is ferric oxide with α-Fe_2O_3 structure (Fig. 4).

The findings of the hydrolytic adsorption research showed (Fig. 5) that being in contact for 20 minutes the kinetic mass transfer curves (pH) have the same manner relative to the neutrality level and the kinetic curves of mass transfer speed (υ pH) display the same mechanism (regularity) by speed gain and drop. It is vivid evidence that the film on the surface of passivated metallic iron has ferrioxide composition with α- Fe_2O_3 structure. The rapid completion of speed instability (in two minutes after being in contact) is caused by the processes of external diffusion inhibition and characterizes the hydrophobic behavior of the surface. The explanation of external diffusion inhibition is as follows:

1. Primary active α-Fe_2O_3 surface centers are Fe^{3+}cations which have octahedral coordination (arrangement) on oxygen. The density of the negative charge is high;
2. The adsorption of water molecules from air forms the secondary Lewes basic centers of brandsted type (O^{2-} … H^+/OH^-) on the primary Lewes centers;

There is two-dimensional shielding film formed due to hydrogen bond between main brandsted centers. It offers resistance to the further moistening. The surface is hydrophobic [12].

Sample			(PDF-11053) Iron Oxide. Fe_2O_3 Hematite.		
2 teta	d (nm)	I %	2 teta	d (nm)	I %
26.40	0.424	3	-	-	-
30.31	0.370	22	30.50	0.368	18
42.21	0.269	100	42.18	0.269	100
45.26	0.252	69	45.37	0.251	75
51.97	0.221	31	52.21	0.220	18
63.67	0.184	39	63.48	0.184	63
69.58	0.170	53	69.89	0.169	63
74.37	0.160	11	74.46	0.160	13
81.27	0.149	28	81.03	0.149	50
83.56	0.145	36	83.76	0.145	50
91.31	0.135	2	91.62	0.135	3
94.60	0.132	12	95.28	0.131	18
100.33	0.126	6	100.40	0.126	13
104.15	0.123	1	103.81	0.123	3
108.70	0.119	5	108.87	0.119	8
112.40	0.117	5	113.13	0.116	5
115.98	0.114	8	116.24	0.114	13
122.46	0.110	10	123.29	0.110	10
132.50	0.106	12	131.91	0.106	8

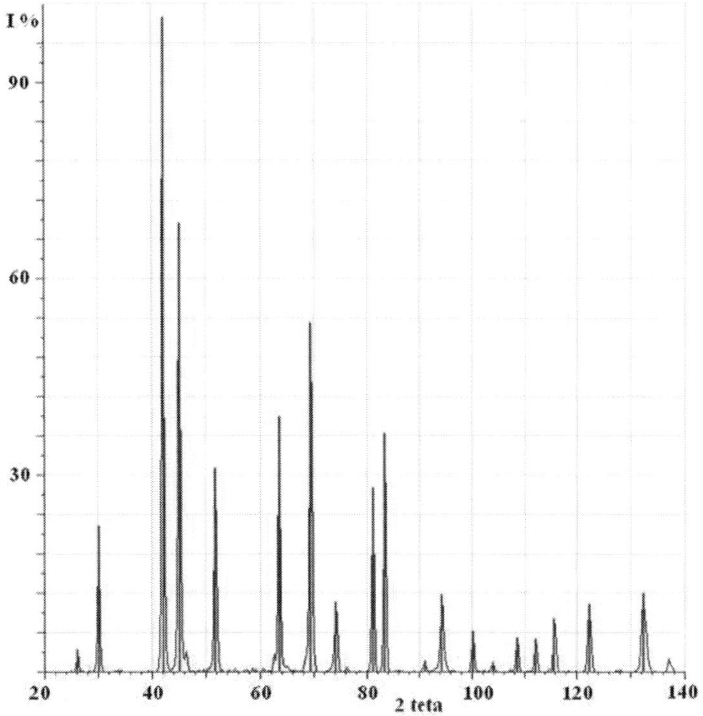

Fig. 4. Roentgendiffractogram «comparison-sample 2» (α-Fe_2O_3 in massive state) and comparison table of interplanar spacing (d, nm) of reflexes in the roentgenogram.

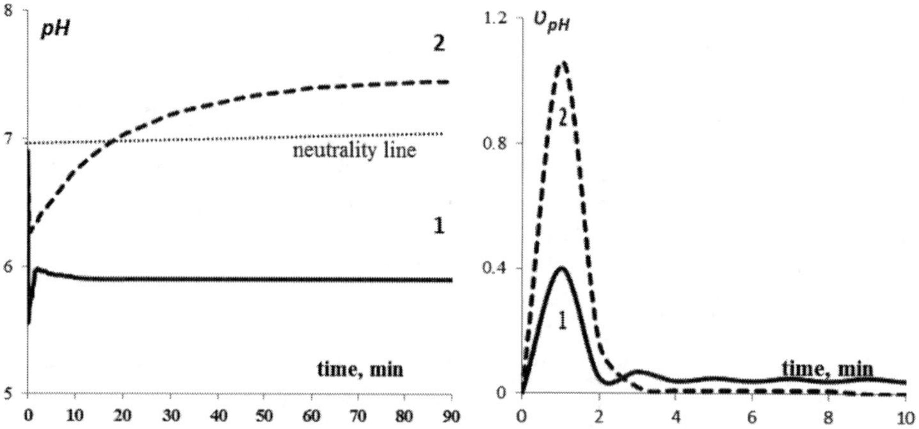

Fig.5. Kinetic curves of mass transfer variation (pH) and speed instability (υ pH) for the samples under study:
1) – «sample 1» (nanometric EEW iron)
2) – «comparison-sample 2» (ferric oxide with α-Fe_2O_3 structure)

The sample behavior in the process of hydrolytic adsorption is different when ferric oxide of another structure is formed. As an example, the variation in speed gain and drop of the mass transfer for Fe_3O_4 sample in massive state is presented (Fig. 6). Fast speed gain and slow speed drop of moistening is typical for this oxide. This oxide is characterized as hydrophilic. Moistening process occurs in accordance with external diffusion inhibition [13-19]. The causes of different behavior of haematite and magnetite during hydrolytic adsorption are thoroughly described in works [5-7].

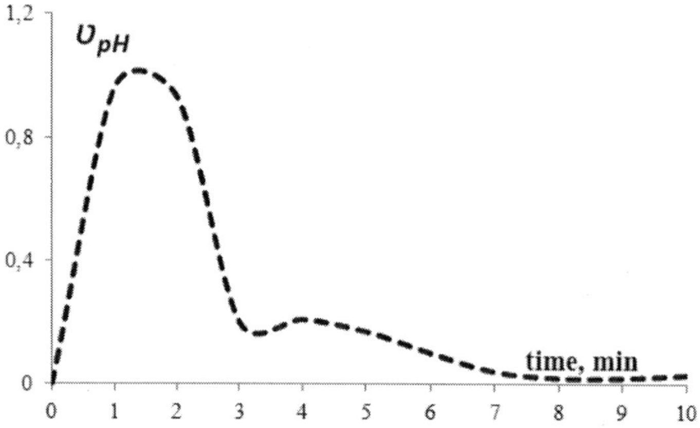

Fig.6. Kinetic curve of speed instability of mass transfer for ferric oxide with Fe_3O_4 structure.

Thus, the kinetic variation of the hydrolytic adsorption method allows us to give qualitative assessment of moistening of finely dispersed powders (hydrophilic behavior/ hydrophobic behavior). The diagnostic of moistening is based on the difference between speed gain and drop of moistening. The pH variations of aqueous suspension in a unit time is used as an analytic moistening speed signal. Speed drop of moistening for α-Fe_2O_3 haematite is fast which is caused by the processes of external diffusion inhibition and characterizes the hydrophobic behavior of surface. The speed gain of mass transfer for Fe_3O_4 magnetite is abrupt and its drop is slow and descrete which is caused by the processes of internal diffusion inhibition and characterizes the hydrophilic behavior of surface. The material of this article can be useful for its implementation in industrial, research and school laboratories.

4. Conclusions

It was shown, that using the findings of investigation of hudrolytic adsorption the moistening of low-dimension particles can be characterized qualitatively as «hydrophilic behavior/ hydrophobic behavior». Since the surface moistening is predetermined by the set of acid-base centers typical for each structure the hudrolytic adsorption process for solid body with the same structure is identical. The use of pH-metry results simplifies and accelerates the diagnostic process of surface moistening, which is economically viable and cost effective.

References

[1] P. A. Kouzov, L. Ya. Skryabina, *Methods of physicochemical property definition of industrial dusts* (L.: Chemistry, 1983).
[2] A. V. Korshunov, *Dimensional and structural effects in the oxidation processes of met*als (Tomsk: Publishing House TPU, 2013).
[3] A. I. Gusev, *Nanomaterials, nanostructures, nanotechnologies* (M.: FIZMATLIT, 2009)
[4] M. I. Lerner, V. V. Shamanskii, G. G. Savelyev, *Bulletin of Tomsk Polytechnic University, V. 310*. **2**. (Tomsk: Publishing House TPU, 2007) pp 132-136
[5] K. V. Ikonnikova, L. F. Ikonnikova, E. A. Koltunova, *Fundamental Research J.*, **10,** 2. pp 2134-2137 (2015)
[6] K. V. Ikonnikova, L. F. Ikonnikova, E. A. Koltunova, *Procedia Chemistry*, **15**. pp 167-173 (2015)
[7] K. V. Ikonnikova, L. F. Ikonnikova, E. A. Koltunova, *Key Engineering Materials: Scientific Journal* **683**. pp 301-305 (2016)
[8] A. A. Zhigal'skii, L. F. Ikonnikova, T. S. Minakova, V. A. Mukhachev, P. E. Troyan, *Russian Physics Journal*, **39**. *iss. 6*. pp 576-578 (1996)
[9] M. I. Lerner, N. V. Svarovskaya, S. G. Psakhe, O. V. Bakina, *Russian nanotechnologies*, **4**. 9, pp 6-18 (2009)
[10] M. Lerner, A. Vorozhtsov, Sh. Guseinov, P. Storozhenko, *Metal Nanopowders: Production, Characterization and Energetic Applications* p 440 (2014)
[11] K. V. Ikonnikova, L. F. Ikonnikova, Yu. S. Sarkisov, T. S. Minakova, *Theory and practice of pH-metry definition of acid-base properties of solid body surface.* (Tomsk: Publishing House TPU, 2011) pp 99
[12] L. F. Ikonnikova, K. V. Ikonnikova, E. A. Koltunova, *MATEC Web of Conferences* (The 2nd International Youth Forum "Smart Grids", 2014) **19**. 01017 p 3
[13] T. V. Kuznetsova, I. V. Kudryashov, V. V. Timashev, *Physical chemistry of cementitious materials: Textbook for Chem-primary process. specialist. universities.* (Moscow: high school, 1989) p 384
[14] S. C. McBain, H. H. Yiu, J. L. Dobson, *Journal Nanomedicine*. **3** *(2)* pp 169–180 (2008)
[15] R. M. Cornell, U. Schwertmann, *The Iron Oxides: Structure, properties, reactions, occurences and uses. Second edition. Weinheim, Willey-VCH Verlag GmbH & Co. KGaA* p 683 (2003)
[16] E. N. Naumova, A. Y. Kalinkov, A. I. Kostrzhitsky, *Refrigeration Tehnika i Tehnologiya* **5** *(74)* pp 46-51 (2001)
[17] S. P. Gubin, Yu. A. Koksharov, G. B. Chomutov, G. Y. Jurkov, *Chemistry successes Journal* **74** *(6)* pp 539-574 (2005)
[18] A. I. Kostrzhitsky, T. V. Cheban, R. A. Podolyan, *Electronic processing of materials,* **3** pp 50-55 (2007)
[19] I. S. Lyubutin, S. S. Starchikov, J. L. Dubinskaya, C.-R. Lin, *V-Russia Conference on Nanomaterials, Nano-2013 Zvenigorod* pp 314 (2013)

MATEC Web of Conferences **85**, 01027 (2016) DOI: 10.1051/matecconf/20168501027
CCT 2016

Applying paleomagnetic oriented coring in investigating the lithofacies heterogeneity of terrigenous sediments

Lyubov Krasnoshchekova [1,*], *Daria* Cherdantseva[2], and *Maria* Yurkova[1]

[1]TPU, INR, Department of Geology and Mineral Exploration, 634050, Tomsk, Lenin Avenue, 30, Russia
[2]TPU, INR, Department of Oil and Gas Industrial Facilities Design Construct, 634050, Tomsk, Lenin Avenue, 30, Russia

Abstract. This paper described the investigation of heterogeneous structure of oil field reservoirs via paleomagnetic oriented coring. As a rule, this heterogeneity develops during either sedimentation or secondary generation of terrigenous matter, which, in its turn, affects the lateral permeability in the layers. Applying oriented coring determines the vector behavior of anisotropic structure of oil fields which influence further production

1 Introduction

The heterogeneous reservoir structures in oil fields has been discussed in numerous geological research papers [1-4]. The concept that geological bodies have parallel-layered structures has been substituted by the fact that these bodies involve a more complex structure. These geological bodies are sediments of different facies that form the channels housing the further filtration of formation fluids.

The research target was the oil field located in S-E Western Siberia province within Tomsk Oblast. The investigated oil-bearing layer within this oil field was U_1^2 in Vasugan suite traced in all correlative well logs. The formation of layer sediments occurred within coastal-marine conditions. The upper and lower contacts are limited by extensive and stable interlayered coal thicknesses which can be revealed on gamma-neutron and induction (electromagnetic) logging. In this case it is possible to conduct reflection picking of investigated hydrocarbon reservoirs. Layer sediment thickness is up to 10-15 m composed of aleurolites (siltstones) and sandstones which is intersected by carbonated interlayers (0.5-1.0 m) grading laterally to argillaceous varieties. In singled-out field areas the upper sand bodies of U_1^2 could be replaced by clay matter often with glauconite. Glauconite presence on the boundary layers confirms the sediment genesis. Detailed study of glauconite features in the sedimentary rocks in this region is given in [5].

Currently, a sharply increasing layer water cutting has been observed within this oil field which indicates the existence of litho-petrophysical heterogeneity in the reservoirs. The present paper describes the investigation results of the lithofacies zonation of the layers via paleomagnetic oriented coring.

2 Research methods

Paleomagnetic orientation of rock samples was conducted in several stages. The diagram of oriented coring is depicted in fig.1. Algorithm analysis of the specific features of oriented rocks and thin sections, including their structural, lithological and petrophysical features, are presented in fig. 2 and in [6, 7]. Methods investigating the morphological and micro structural characteristics of rocks are described in details in [8].

3 Results and discussion

According to petrophysical data, the predominate N-E direction of investigated magnetic rock properties and subnormally N-W orientation of elastic sandstone behavior was determined. Dominating development of elongated quartz grains was observed in N-E and N-W directions. Based on stated orientations, not only the microstructure patterns of optical axes of quartz individuals, but also the orientation of cracks and void space in thin sections were correlated.

* Corresponding author: krasnl@yandex.ru

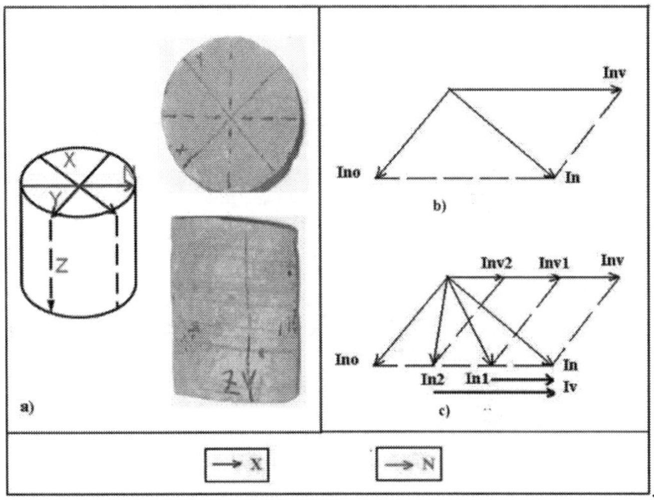

Fig. 1. Diagram of spatial oriented coring via paleomagnetic method: a) core sample with arbitrary coordinates (X, Y, Z) and calculated magnetic meridian plane (to viscous magnetization -N); b) vector components of natural remanent magnetization (NRM): In – primary remanent magnetism Ino; viscous remanent magnetization – Inv; c) diagram of step-by-step demagnetization of natural remanent magnetization (In-In1–In2) with successive destruction of viscous magnetization (Inv–Inv1–Inv2) and separation of differential viscous element (Inv).

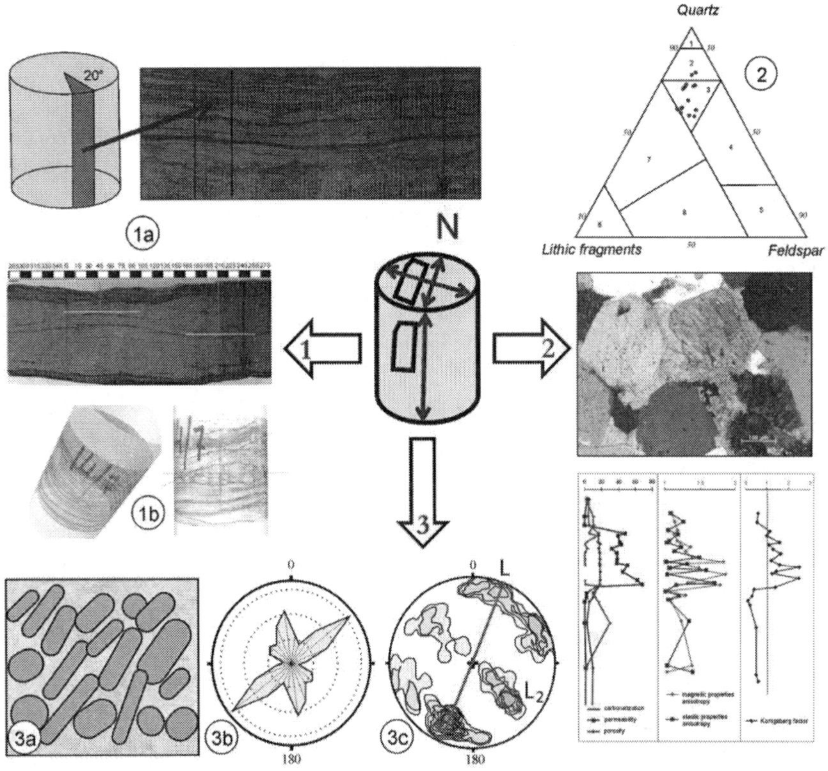

Fig. 2. Diagram of investigated oriented cores and thin sections: 1a – 3-D scanning (photo-scan) 1b – 3D core imaging; 2 – lithological and petrophysical investigations; 3 – morphological and micro structural analysis: 3a – morphology and grain packing, 3b – diagram of quartz grain elongation, 3c – diagram of optical quartz axes emergence

DOI: 10.1051/matecconf/20168501027

Investigating quartz grain fracturing in rock thin sections showed random distribution of crack occurrence frequency in the cross-section. Development of cataclasis cracks, pores and channels, where fluids pass through, indicate the previously defined N-E and N-W orientation of magnetic parameters and elongated mineral skeleton grains. In mutually perpendicular thin sections of one sample, one can observe intensive fracturing anisotropy, sometimes up to 2-5 times. Rigid quartz grain skeleton deforms or cracks during stadial and/or epigenetic alterations, which, in its turn, increase, rock permeability.

To determine the facies conditions and formation mechanisms of sedimentary rock layers in investigated wells 50m coring was plotted and described. The structure features of investigated sandstones and aleurolites (siltstones) were specified as photo scans according to oriented cores. The plotting method of panoramic core view is illustrated in fig. 3.

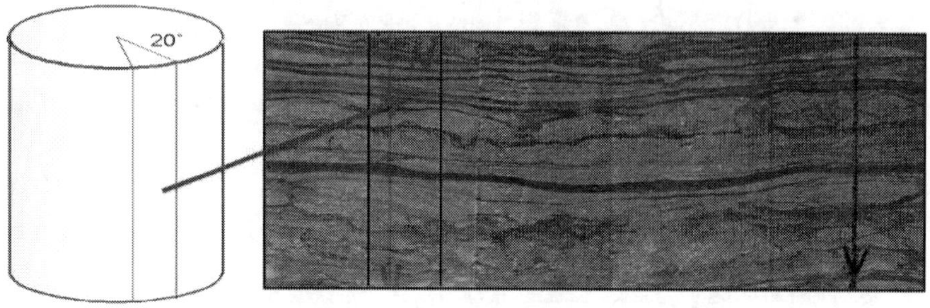

Fig. 3. Example of plotting panoramic core view

Plotting panoramic view of a core sample correlated with well depth makes it possible to determine the structure not by the flat rock cross-section, but by tracing the characteristic alterations in bulk. In this case, the orientation and layer dip angle, their quantitative relation, distribution pattern and interrelations are also determined. Panoramic coring is spaced so that northerly oriented samples coincide with zero degree. All studied rock reservoir structure features in combination with sediment grain-size analysis furthers the reconstruction and identification of sedimentogenesis facies. Besides above-mentioned method modeling 3D imaging of oriented coring in graphic editor Adobe PhotoShop CS 5 was applied.

Investigating the core was based on a sophisticated tool- plotting 3D transparently structured sample model. Input data included dimensional core features with constant diameter (9.8cm), which is conditioned by the application of conventional drilling tools. The processing chain was divided into several sub-tasks: 1) designing graphic view of rock core sample; 2) plotting 3D model as a 9.8 cm cylinder surface which would be the further computer prototype.

The modeling itself involved a rather non-standard approach. The software product in investigating the model was the most common graphic editor Adobe Photoshop (version CS5 and higher) [9]. Preliminary investigation of the core surface involved designing standard 3D object based on a primitive cylinder with the diameter corresponding to the diameter of the circular image of the original sample. Property alterations of this object resulted in semi-transparent dimensional sample. Core model was rotated within the range of 0-3600 in all three coordinating plane axes. The angles were measured by adding two straight angles and reflecting angle value (fig. 4). The following illustration shows an example of measuring layer incident angle of 13.40.

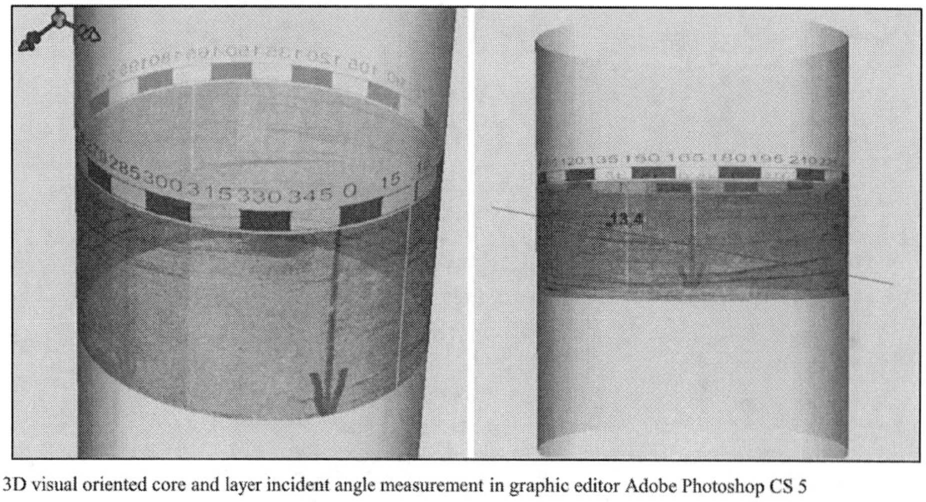

Fig. 4. 3D visual oriented core and layer incident angle measurement in graphic editor Adobe Photoshop CS 5

4 Conclusion

Applying paleomagnetic oriented coring opens up new opportunities in reconstructing and identifying the facies conditions of sedimentation in oil-bearing fields. Furthermore, oriented samples make it possible to investigate the specific features of reservoirs in details, taking into consideration all heterogeneous layers and as a result, plot real-time models of different oil fields.

References

1. N. J.C Farrell., D. Healy , C. Taylor, *Journal of Structural Geology*, **63** 50-67 (2014)
2. D. Healy et al, Geological Society special publication, **367** 247-253 (2012)
3. L. Louis, T.-M. N. Chen, C. David, P. Robion, T.-f. Wong and S.-R. Song, *Journal of Structural Geology*, 30 948-962 (2008)
4. V. Merkulov, D. Aleksandrov, L. Krasnoshchekova et al, *UKOS Scientific Vestnik*, 10 33-36 (2004)
5. M.A. Rudmin, I.V. Reva, A.P. Gunko, A.K. Mazurov, R.N. Abramova. *IOP Conf. Ser. Earth Environ. Sci.* 27. 012026 (2015)
6. M. Korovin and V Merkulov, *Petroleum Engineering*, 1 24-30 (2015)
7. V. Merkulov, L. Krasnoshchekova, *Gazovaya promyishlennost*, 3/703 22-27 (2014)
8. D. Cherdantseva and L Krasnoshchekova. *IOP Conf. Ser. Earth Environ. Sci.* 21 012007 (2014)
9. L.A Krasnoschekova, V.P Merkulov, D.A. Cherdantseva, I.V. Ryashentsev. Geomodel 2015 17th Scientific-Practical Conference on Oil and Gas Geological Exploration and Development, Pages 255-259 (2015)

MATEC Web of Conferences **85**, 01028 (2016) DOI: 10.1051/matecconf/20168501028
CCT 2016

MODELLING OF THE KINETICS OF SULFURE COMPOUNDS IN DESULFURISATION PROCESSES BASED ON INDUSTRY DATA OF PLANT

Nadezhda Krivtcova [1,*], *Anton* Tataurshikov[1], *Elena* Kotkova[1]

[1]Tomsk Polytechnic University, Lenin av. 30, Tomsk, 634050, Russia

Abstract. Modelling of sulfur compounds kinetics was performed, including kinetics of benzothiophene and dibenzothiophene homologues. Modelling is based on experimental data obtained from monitoring of industrial hydrotreating set. Obtained results include kinetic parameters of reactions.

1 Introduction

Desulfurisation - the process of catalytic conversion of hydrocarbons under the influence of substances at high pressure and temperature. Petroleum fractions subjected to desulfurisation to reduce the content of sulfur compounds in gasoline and diesel fuels. Concomitant effect of this process is the saturation of unsaturated hydrocarbons, reduced resin content of oxygenates, and hydrocracking of light hydrocarbon molecules. Desulfurisation - large-refining process and is used in almost all refineries [1].

Due to the increasing role of the hydrotreating processes in the refining industry matures the need to develop software systems that implement mathematical models hydrotreating [2].

The main objective of this study is to model the kinetics of the process of desulfurisation of diesel fuel, with a focus on the dynamics of sulfur-containing compounds (SC), such as sulfides, benzothiophene and dibenzothiophene, the choice of the reactor model.

2 Object of research

Typical schematic diagram of a hydrodesulfurization unit is shown in Figure 1. Preheated raw material moves to the hydrotreating reactor R, a major unit of the whole process, where the catalytic hydrotreating of diesel fraction is performed. There are the reactions of sulfur compounds hydrogenation with hydrogen to form hydrogen sulfide performed on the catalyst surface. The hydrogen-containing gas (HCG) is separated from

* Corresponding author: Krivtcovani@mail.ru

the vapor-liquid mixture in the separators S-1 and S-2 and enters the column C-2 for pure hydrogen to recover. The column C-1 is used for separation of vapor-liquid mixture into gas fraction and the basic product - hydrotreated diesel fuel [3].

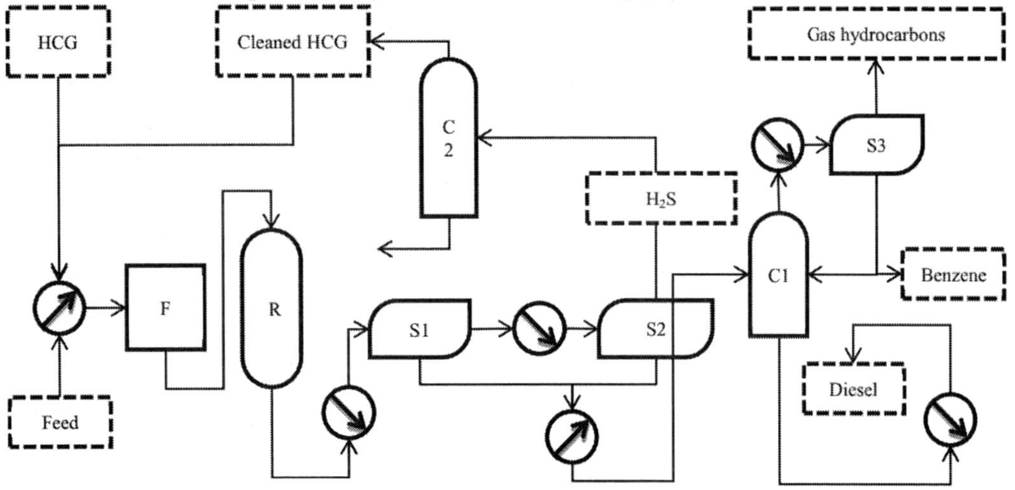

Fig. 1. Typical schematic diagram of hydrodesulfurization unit

The resolution №118 About approval of the Technical regulation about requirements to automobile and aviation gasoline, diesel and ship fuel, jet propulsion fuel and fuel oil dated February 27, 2008 approving technical requirements to the motor fuel was signed by the Russian Federation government. The requirements of GOST R 52368-2005 are brought into compliance with the legal requirements of Euro-3 standard and higher. Prior to that, the fuels were produced in accordance with GOST 305-82. The main difference from the EN 590 standard in the GOST R 52368-2005 is that diesel fuel is divided into three categories depending on the amount of sulfur compounds. Diesel fuel of Type 3 may also be called completely purified.

The plant used French catalyst HR-538, which properties are given in Table 1. The catalyst activity in respect of SC is numerically considered according to the reaction rate constants.

Table 1. HR-538 catalyst properties

Diameter of catalyst particles, mm	1.2; 1.6; 2.5
Composition: Nickel (NiO), % wt. Molybdenum (MoO3), % wt.	 3.5 17.0
Surface area, m2/g	210

3 Results

When modeling multicomponent processes can not account each individual component which participates in the chemical process - the number of components 300-400 in the case of the desulfurisation [4]. But, at the same time, the mathematical model must be sensitive to the changing composition of the raw material. Only in this case, the model will be adequate to the real process and will have predictive ability [2,4].

The change of Gibbs free energy provided by the reactions calculated using software packages HyperChem and Gaussian 09 [5]. In previous work, studies were conducted to determine the kinetic parameters of the hydrodesulfurization reactions [6,7,8] (Table 2).

Table 2. Changing the Gibbs energy of reaction hydrogenation SC

№	ΔG, kJ/mol	Reaction
1	-95,3	$C_9H_9S + 3H_2 \rightarrow C_9H_{13} + H_2S$
2	-83,5	$C_{10}H_{11}S + 3H_2 \rightarrow C_{10}H_{15} + H_2S$
3	-71,1	$C_{11}H_{13}S + 3H_2 \rightarrow C_{11}H_{17} + H_2S$
4	-65,1	$C_{12}H_8S + 2H_2 \rightarrow C_{12}H_{10} + H_2S$
5	-54,3	$C_{13}H_{11}S + 2H_2 \rightarrow C_{13}H_{13} + H_2S$
6	-53,7	$C_{14}H_{13}S + 2H_2 \rightarrow C_{14}H_{15} + H_2S$
7	-49,4	$C_{15}H_{15}S + 2H_2 \rightarrow C_{15}H_{17} + H_2S$

On the basis of calculations of thermodynamic parameters of the reactions taking place in the process of desulfurization of diesel fuel is made up of substances mathematical model. The developed mathematical model based on the law of mass action and is a system of differential equations, which reflect changes in the concentrations of the reactants:

benzothiophene

$$\begin{cases} W_1 = k_{C_1BT}C_{C_1BT}C_{H_2}^{v} \\ W_2 = k_{C_2BT}C_{C_2BT}C_{H_2}^{v} \\ W_3 = k_{C_3BT}C_{C_3BT}C_{H_2}^{v} \end{cases}$$

$$\begin{cases} \frac{dC_{C_1BT}}{dt} = W_1 \\ \frac{dC_{C_2BT}}{dt} = W_2 \\ \frac{dC_{C_3BT}}{dt} = W_3 \end{cases}$$

dibenzothiophene

$$\begin{cases} W_4 = k_{DBT}C_{DBT}C_{H_2}^{v} \\ W_5 = k_{C_1DBT}C_{C_1DBT}C_{H_2}^{v} \\ W_6 = k_{C_2DBT}C_{C_2DBT}C_{H_2}^{v} \\ W_7 = k_{C_3DBT}C_{C_3DBT}C_{H_2}^{v} \end{cases}$$

$$\begin{cases} \frac{dC_{DBT}}{dt} = W_4 \\ \frac{dC_{C_1DBT}}{dt} = W_5 \\ \frac{dC_{C_2DBT}}{dt} = W_6 \\ \frac{dC_{C_3DBT}}{dt} = W_7 \end{cases}$$

Initial conditions: $dC_i = C_{i,0};\ t = 0$

where k_i - rate constant of the component i; C_i - the current concentration of the substance i; v - the stoichiometric ratio.

To study the data taken from the monitoring installation LG-24/7, reflecting the state of technological parameters are shown in Table 3.

Table 3. Technological parameters of plant LG-24/7

Day	23.12.14	24.12.14	25.12.14
Rate, m^3/h	80,15	79,97	80,24
Volume rate of feed, m^3/m^3	1,336	1,333	1,337
Temperature, °C	333,12	333,30	333,12
Presure, kgf/sm^2	33,709	33,774	33,734

Raw sulfur content, % mas.	0,8200	0,9000	0,8800
Product sulfur content, % mas.	0,0030	0,0040	0,0030
The length of the reaction zone, m	8		
Volume of catalyst, m^3	26		

As the basis of hydrodynamic models taken quasi-homogeneous stationary model of a plug flow reactor according to equation:

$$\frac{dC_r}{dl} = -U * \sum_{i=1}^{p} W_i$$

where l - the length of the reactor; U - the linear velocity of feed; p - quantity of reagents and products by route r; W_i - the rate of the chemical reaction.

Calculated data obtained as a result of the mathematical model are shown in Table 4. The pressure in the reactor vapor-liquid mixture is from 30 to 34 kgf/cm2. The temperature in the reactor for the entire installation work cycle was in the range from 320 to 350 ° C.

Table 4. The calculated rate constant transformation of components h^{-1}

Day	23.02.15	24.02.15	25.02.15
$C_9H_9S + 3H_2 \rightarrow C_9H_{13} + H_2S$	1,58	1,51	1,60
$C_{10}H_{11}S + 3H_2 \rightarrow C_{10}H_{15} + H_2S$	1,51	1,45	1,54
$C_{11}H_{13}S + 3H_2 \rightarrow C_{11}H_{17} + H_2S$	1,51	1,44	1,53
$C_{12}H_8S + 2H_2 \rightarrow C_{12}H_{10} + H_2S$	1,35	1,30	1,37
$C_{13}H_{11}S + 2H_2 \rightarrow C_{13}H_{13} + H_2S$	1,37	1,32	1,39
$C_{14}H_{13}S + 2H_2 \rightarrow C_{14}H_{15} + H_2S$	1,27	1,22	1,29
$C_{15}H_{15}S + 2H_2 \rightarrow C_{15}H_{17} + H_2S$	0,70	0,67	0,71
Calculated product sulphur content, % mass.	0,0031	0,0042	0,0030
Experimental product sulfur content, % mas.	0,0030	0,0040	0,0030
Relative sulphur calculation error, %	3,333	5,000	0,000

4 Conclusion

The result of the study is successful in building a mathematical model, including the analysis of experimental data, the aggregation kinetic scheme, the development of mathematical models and software, performing calculations.

The relative error of calculation that was conducted on the data for the entire installation of the monitoring period, amounted to no more than 5% on the total sulfur content in the product desulfurisation. It confirmed a direct correlation conversion rate constant calculated from the sulfur-containing component of the Gibbs energy of the hydrogenolysis reaction.

According to the latest Euro-5 standards governing the quality of diesel fuel, the maximum sulfur content in it should not exceed 10 ppm. According to the results of calculations a residual sulfur content exceeds the threshold of 10 ppm even at 356 °C and raw material consumption from 50 to 60 m^3/h.

References

1. Anne Belinda Bjerre, Emil Sorensen. Ind. Eng. Chem. Res. 1992, V. 5, 31, 1577–1580.
2. Ed Palmer, Stan Polcar and Anne Wong, PTQ, 2009. 91-100.
3. Emiel J.M. Hensen. Hydrodesulphurization catalysis and mechanism of supported transition metal sulfides. Eindhoven: Technische Universiteit Eindhoven. 2000. 254 p.
4. Krivtsova N. I., Ivanchina E. D. , Tataurshchikov A. A. XXI International Conference on Chemical Reactors (CHEMREACTOR 21). Delft, September 22-25, 2014. 392-393.
5. Belinskaya N.S., Ivanchina E.D., Ivashkina E.N., Silko G.Yu., Frantsyna E.V. Izvestiya vuzov: Chemistry and Chemical Technology, 2014. V. 57. №. 11, 90-92.
6. N.I. Krivtcova, A.A. Tataurshikov, I.D. Ivanchina, E.B. Krivtsov, A.K. Golovko. Procedia Engineering / Volume 113, 2015. 73–78
7. Tataurshikov A., Ivanchina E., Krivtcova N. , Krivtsov E., Syskina A. IOP Conference Series: Earth and Environmental Science. Volume 27, Issue 1, 10 November 2015.
8. Krivtcova N.I. , Tataurshikov A.A., Ivanchina I.D., Krivtsov E.B., Golovko A.K. Procedia Engineering. Volume 113, 2015, Pages 73-78.

Development and exploitation features of methane-coal wells using downhole pumping equipment

Elena Kurganova[1], *Alexandra* Makarovskikh[1*], *Tatyana* Chernykh[1]

[1]Tomsk Polytechnic University, 30 Lenin Ave., Tomsk, 634050, Russia

Abstract. In this article extraction of methane from coal beds is presented. The article presents results of the study of downhole pumping equipment used in the exploitation of coal bed methane wells. The features of various types of equipment such as piston sucker-rod pumping units, electrical centrifugal pumping units and sucker-rod screw pumping units have been considered. The results show benefits of sucker-rod screw pumping units because of their versatility and resistance to higher content of mechanical impurities.

1. Introduction

Nowadays natural resources use tends to be effective. As a result there is a necessity of natural methane gas use for industrial and domestic needs (fuelling, lighting, etc).
A coalfield in Kuznetsk (Kuzbass) is one of the most perspective and prepared regions for methane extraction from gas-bearing coal beds in Russia.

Actually Kuzbass has no equals among the famous coal basins of the world. It has unique coal-gas-related characteristics:

- coal-bearing formation presented by the most preferable coal rank in terms of methane-bearing capacity with a stack height up to 90-120 meters with a methane content of 25-30 m^3/t;
- high concentration of methane resources in coal beds - up to 3 bcm/km^2;
- cumulative methane influx in coal beds of the basin (up to the depth of 1800 m) is about 13 tcm over the area of 21 thousand square kilometers. Moreover, in the most promising regions of the south (Erunakovskij, Tersinskij, Tom-Usinskij, Mrasskij) estimated resources are ranged from 0.9 to 3.2 tcm, and in some promising areas - 100 - 480 bcm. Gas is substantially pure and does not require additional purification [3].

Industrial infrastructure which is established in Kuzbass creates exceptionally favorable conditions for the organization of methane production from coal deposits. Methane which is extracted from coal beds via drilled wells is environmentally friendly hydrocarbon feedstock used for the region gasification as well as in industry, gas chemistry as a motor fuel, etc.

The optimization of gas-related equipment at all stages of production and preparation of gas for transportation plays an important part in solving the above-listed problems. Moreover, the stage of gas production largely determines the efficiency of the methane-coal complex operation as a whole. Therefore, in this situation, the optimal use of downhole equipment and the extension of service life get their applicability [4].

2. Materials and Methods

The aim of this work is to analyze the efficiency of wells mechanized with gas-oil equipment on Naryksko-Ostashkinskaya field, as well as a competent evaluation and selection of advanced technologies for removing operational defects that lead to equipment failure in definite wells to ensure their safe performance and reduce the extraction cost of formation fluid.

The development of methane-coal wells is fulfilled in two stages. The first stage is characterized by intense carrying out of fluid after fracturing with a high content of mechanical impurities: proppant, clay mud, coal slime. When the formation water is pumped out, a dynamic fluid level of the well decreases and it is necessary to control a start of the gas desorption, which is accompanied by changing of gas pressure in the annular space. It is particularly important to monitor condition and properties of the pumped liquid (color, contamination, etc). For example, dark water with coal dust indicates a start of the coal bed destruction, which leads to the mudding of the formation zone bottomhole. The

* Corresponding author: alexandravm1@rambler.ru

results obtained in a specialized laboratory have shown that the average content of solids in pumping fluid at the wells of Taldinskaya area is 12 mg/l. At the wells of Narykskо-Ostashkinskaya area the same value is 406 mg/l. Consequently, at the initial stage of the well development there is a considerable cuttings lifting and while an operating process the content of suspended matters is multiply reducing. Therefore, the objectives of the first stage of the development are:

- mechanical impurities removal;
- minimization of the failures of downhole equipment which will be used during the trial operation;
- indirect identification of potential flowrates of water and gas.

At the initial stage of the development pumping equipment arrangement for pumping formation fluid was lowered into the well to a depth of 300-450 m. 50 m above the upper bed depth is the best depth for downhole pumping equipment arrangement at the first stage of the development [4]. If a well is not subjected to hydraulic fracturing, the first stage of the development is excluded. The stage makes it more accurately determine the potential of the well flowrate of gas and water, as well as choose the most appropriate type and size of pumping equipment for the second stage which consists of the output to the steady-state exploitation and maximum elimination of the risk of clogging up of gas-oil equipment.

At the second stage of the development different types of downhole equipment units are used: piston sucker-rod pumping units, sucker-rod screw pumping units and electrical centrifugal pumping units (figure 1).

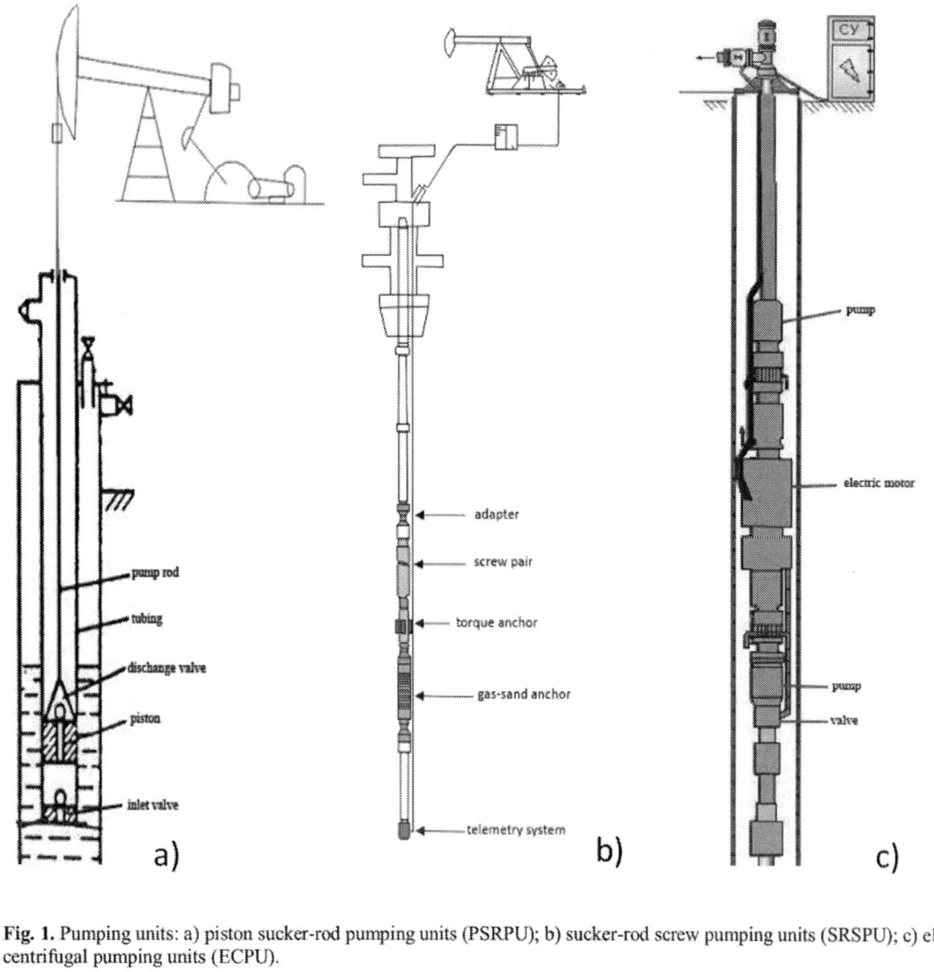

Fig. 1. Pumping units: a) piston sucker-rod pumping units (PSRPU); b) sucker-rod screw pumping units (SRSPU); c) electrical centrifugal pumping units (ECPU).

In 2002-2003 in Kuzbass during the gas-dynamic research at the exploration well the use of piston sucker-rod pumping units to pump formation fluid was first experienced [5].

The application of piston sucker-rod pumping units has not received extensive use due to some reasons: clogging of valves with mechanical impurities, inability to backwash and absorption of flush fluid by productive formations. Besides, at the second stage of the development there are electrical centrifugal pump units, which are used, if water influx permits and it is a steady-state exploitation. This type of pumping equipment has a high overhaul period at methane-coal wells. The main reasons of failure of this type of equipment:

- isolation reduction in cable link;
- work tool wear;
- pump jamming.

About 84% of the reasons for failure are connected with clogging with mechanical impurities. The mean time between failures of electrical centrifugal pumps is 162 days in the fund (for the period from 2009 to 2012) [1]. Sucker-rod screw pumping units receive widespread application because they are the most versatile for methane-coal field conditions [3]. The main reasons of failure of this type of equipment:

- leakage and drop of tubing;
- drop and breakage of rods;
- pump jamming.

The mean time between failures of sucker-rod screw pumps is 99 days in the fund (for the period from 2009 to 2012) [1].

3. Results and Discussion

Thus, for the exploitation of methane-coal wells the most optimal types of downhole equipment are sucker-rod screw units. Therefore, it is reasonable to use sucker-rod screw pumping units because of their resistance to higher content of mechanical impurities in the pumped formation fluid and versatility (at all stages of the development). It is recommended to meet the following conditions:

1. Exclusion of a gas-sand anchor from the packaging arrangement to minimize the risk of further failure of sucker-rod screw pumping units at the initial stage of wells development;
2. Optimal centralizers application and allocation to reduce mechanical wear upon contact with the sucker rod of tubing and extend its service life;
3. Application of the dynamic torque anchor A-168 as part of the sucker-rod screw pumping units to avoid threaded connections drops when screw pair jamming;
4. Use of a downhole telemetry system with a minimum grade value of 0.01 Bar.
5. If possible, elimination of gas-oil equipment operation in periodic mode [2].

Currently, 17 downhole pumping units are exploitated at the LLC "Gazprom dobycha Kuznetsk" at the wells of Taldinskaya and Naryksko-Ostashkinskaya areas. There are sucker-rod screw pumping units (SRSPU) and electrical centrifugal pumping units (ECPU) of various types and sizes. Figure 2 shows distribution of the current fund by the number of pumping units at the wells of Taldinskaya and Naryksko-Ostashkinskaya areas. Preference is given to using sucker-rod screw pumping units (71%).

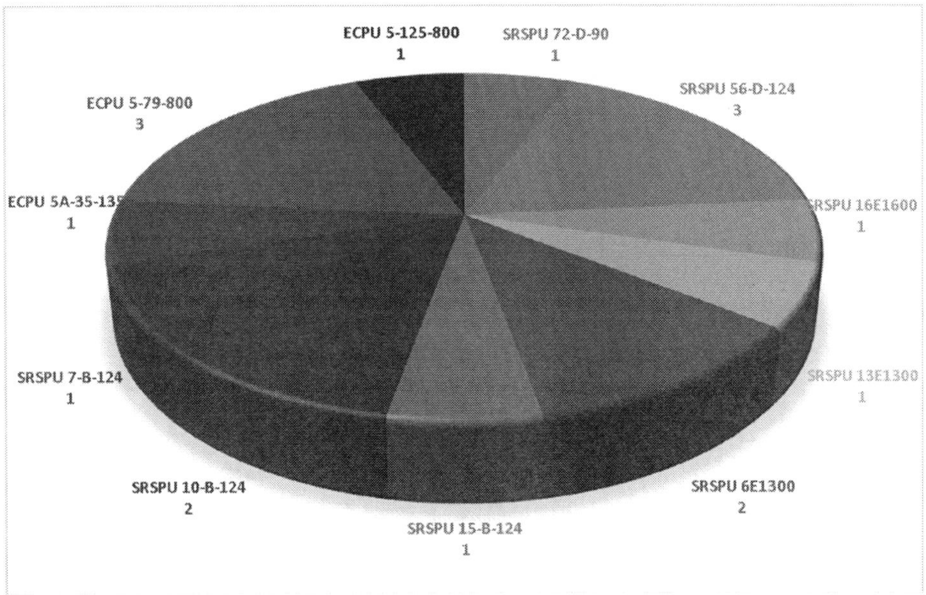

Fig. 2. Current fund of pumping units used for pumping formation fluid at the wells of Taldinskaya and Naryksko-Ostashkinskaya areas.

4. Conclusion

As a result a sucker-rod screw pumping unit has been offered due to the duration of the equipment, revenue position and operating efficiency. This type of setting is primarily used for pumping out formation water. The use of methane-coal beds is economically advantageous [5], first of all due to the comparatively inconsiderable transportation leg from the field to the consumer. In addition, the methane content in the coal gas reaches 98-99%, which means no additional costs for purifying and processing gas before shipment [6].

References

1. *Gas-dynamic research at the well № 16320 at the Taldinskaya area in Ermakovskij region of Kuzbass* (by contract report №4 dated 02.10.2002)
2. A.N. Drozdov, The use of downhole fluid-jet pumping units for operation in abnormal wells *Oil and gas vertical* **12**, 70–73 (2009)
3. A.M. Karasevich, Methodology and management of the regional gas-supply systems development *The Abstract of doctoral thesis in Engineering Science* (Irkutsk, 2003)
4. R.R. Lopatin, Models and algorithms of the frequency-regulated process of jamming the electrical centrifugal pump in oil production in abnormal operating conditions *The Abstract of PhD thesis in Engineering Science* (Tomsk, 2011)
5. P.N. Parmuzin, Estimate the economic effect in the project for methane recovery from coal seams *Problems of economy and management of oil and gas complex* **3**, 15–17 (2013)
6. G.S. Spitsin *Gas from coal beds* Electronic resource, URL: http://twidler.ru/referat/geografiya/gaz-iz-ugolnyh-plastov-39611 (Access date 02.07.2016)

Modification of natural sorbent for providing it with bactericidal and bacteriostatic properties

Irina Martemianova [1,*], Luiza Nadeina[1], *Evgenii* Plotnikov[1] , *Dmitrii* Martemianov[1]

[1]Tomsk Polytechnic University, Lenin av. 30, Tomsk, 634050, Russia

Abstract. Currently, the problem of water purification from chemical and microbiological contaminations is very important. Sorption technologies are one of the effective and easy-to-use water purification techniques. To prevent the growth of microbiological contaminations on the surface of investigated sorbents, the surface of the investigated sorbents was treated by aqueous solution of copper sulphate with the further study of bactericidal and bacteriostatic activity. The results of sorption efficiency of original and modified zeolite samples of Kholinsk deposit are presented in this paper. These results were obtained, when Zn^{2+} and Pb^{2+} ions were removed from the simulated solutions. It was concluded that it is possible to provide natural zeolites with bactericidal and bacteriostatic properties.

1 Introduction

One of the current problems is to provide people with clean drinking water. Natural mineral sorbents are often used in water purification due to their low cost and higher sorption capacity. The review of periodic printed publications has shown that sorption processes have not been studied well yet. They need to be studied in more detail. Zeolite is one of well-known natural minerals possessing good sorption properties. According to these properties, zeolites are used to purify industrial wastewater from heavy metals such as lead, zinc, copper, nickel, cadmium, chromium, and others. Microbiological pollution is known to appear on the surface of sorbent in the water purification module. It is necessary to provide sorption materials (zeolites) with bacteriostatic properties in order to prevent the growth of microbiological pollution on their surface. In order not to reproduce microbes in the filtered water, sorption material needs to be provided with bactericidal properties.

Copper is known to possess a number of useful properties. It prevents bacteria, harmful fungi and other organisms to produce on the different surfaces. It has been proved that copper is required for eubolism. It has anti-inflammatory, antiulcer, anticonvulsive, and anticarcinogenic effect. Copper oxygenates cells, and it participates in energy metabolism. It has a good effect on skin. Copper benefits when vascular malformations and diabetes

* Corresponding author: martemiv@yandex.ru

mellitus. But there is a basic requirement. The thing is, the copper content in water should not exceed maximum permissible concentration (MPC).

The aim of this investigation is to immobilize copper salts on the surface and inside natural zeolite pores of Kholinsk deposit (the Republic of Buryatia, Russia) in order to provide them with bactericidal and bacteriostatic properties. Physico-chemical investigations should be carried out according to original and modified samples of natural sorbents (zeolites). Sorption properties of investigated materials should be identified in removing Zn^{2+} and Pb^{2+} ions from the simulated solutions.

2 Materials and methods

The natural zeolites samples of Kholinsk deposit were modified with copper sulphate water solution ($CuSO_4*5H_2O$). These samples were the subject of investigation in the paper. To obtain required fractions, zeolite of Kholinsk deposit was crushed in the agata mortar. After that it was sifted using the sieves with 0.1mm, 0.5 mm and 1 mm mesh sizes. As a result of sifting the fractions of zeolites with the sizes less than 0.1 mm and 0.5-1mm were selected.

The modification of zeolite samples was carried out under 3 different conditions:

Modification 1. Three grams of zeolite sample were kept in 100 cm^3 of the copper sulphate water solution with 5 g/dm^3concentration for 24 hours. It was mixed intermittently. Then the solution poured out, and the sample was dried for 6 hours at the temperature of 120 ^0C;

Modification 2. Three grams of zeolite sample were kept in 100 cm^3 of the copper sulphate water solution with 10 g/dm^3concentration for 24 hours. It was mixed intermittently. Then the solution poured out, and the sample was dried for 6 hours at the temperature of 120 ^0C;

Modification 3. Three grams of zeolite sample were kept in 200 cm^3 of the copper sulphate water solution with 10 g/dm^3concentration for 24 hours. It was intermittently mixed. Then the solution poured out, and the sample was dried for 6 hours at the temperature of 120 ^0C.

Eight different zeolite samples of Kholinsk deposit were used to carry out the investigations: 1). # 1. The fraction size of sample was less than 0,1 mm; 2). # 2. The fraction size of sample was less than 0,1 mm. Modification 1; 3). # 3. The fraction size of sample was less than 0,1 mm. Modification 2; 4). # 4. The fraction size of sample was less than 0,1 mm. Modification 3; 5). # 5. The fraction size of sample was less than 0,5-1 mm; 6). # 6. The fraction size of sample was less than 0,5-1 mm. Modification 1; 7). # 7. The fraction size of sample was less than 0,5-1 mm. Modification 2; 8). # 8. The fraction size of sample was less than 0,5-1 mm. Modification 3.

The method of thermal nitrogen desorption was used to determine specific surface and specific pore volume of the original and modified samples. The analyzer "SORBTOMETR M" (LLC "Katakon", Russia) was used.

The sorption of Zn^{2+}, Pb^{2+} ions from the simulated solutions was carried out under static condition with the use of investigated zeolite samples. These samples contained different quantity of copper on their surface and inside pores. The weight of zeolite samples was 0.2 g. They were poured with 20 cm^3 of the simulated solution containing Zn^{2+} or Pb^{2+} ions. The initial concentration of Zn^{2+} ions in the solution was 5.16 mg/dm^3. The initial concentration of Pb^{2+} ions in the simulated solution was equal to 5.03 mg/dm^3. The simulated solutions were prepared with the use of bidistilled water as well as state standard samples of the composition of zinc and lead ions. The sorption carried out in stirring with a magnetic stirrer. The contact time was different: 1; 5; 15; 30 and 60 minutes. After stirring, the solutions were separated from the sorbent by filtration with a help of "Blue ribbon" paper filter. Then the equilibrium concentrations of Zn^{2+}, Pb^{2+} ions were defined by the method of inversion voltamperometry with a help of the analyzer TA-07 (LLC "Technoanalit", Russia).

The determination of the copper washing out from the investigated zeolite samples was carried out under static condition, with the use of bidistilled water, and under the conditions described above (0.2 g of sample per 100 cm^3 of bidistilled water).

Bacteriostatic properties of original and modified zeolite samples of Kholinsk deposit were defined in the following way. The culture of Escherichia coli (concentration: 2×10^7 CFU/cm^3 and volume - 0.1cm^3) was placed on Petri dish containing a standard medium (meat infusion agar). Then, Escherichia coli was distributed with a glass spreading rod on the surface of Petri dish. Medium on Petri dish was divided into halves. 0.1 g of the investigated zeolite sample was put on one half. After that, Petri dish was shut, and placed in the thermostat at a temperature of 37°C. Petri dish was placed in the thermostat for a day. After that, the culture growth in the zone of application of the sample, and in the checking half were compared.

3 Results

Physical-chemical parameters of Kholinsk's zeolite samples such as granulometric composition, specific surface and specific pore volume are presented in Table 1.

Table 1. Physical-chemical parameters of zeolites of Kholinsk deposit and their modified analogues.

Zeolite sample	Particle size, mm	Specific surface, m^2/g	Specific pore volume, cm^3/g
# 1	less than 0.1	22.4	0.009
# 2		19.7	0.008
# 3		19.1	0.008
# 4		18.7	0.008
# 5	0.5-1	27.65	0.012
# 6		25.2	0.011
# 7		25.1	0.011
# 8		21.5	0.009

As Table 1 shows, zeolite samples settled in the strong concentration solutions have a decrease of specific surface and specific pore volume. This can be explained in a certain way: Cu^{2+} ions penetrate into the branched porous structure of zeolites, and they are kept there (mechanical adsorption), and thereby their presence decreases the internal pore volume. The decrease of specific surface is carried out by the same principle. The internal surface is covered with Cu^{2+} ions, and they reduce the internal surface.

Degree of Cu^{2+} ions washing out from the modified zeolite samples according to different contact time of the sample with bidistilled water is presented in Fig.1. The study of Cu^{2+} ion washing out in water was carried out with the use of only modified zeolite samples. This is due to the fact that there is practically no Cu^{2+} ions washing out in water in the initial samples (non-modified samples).

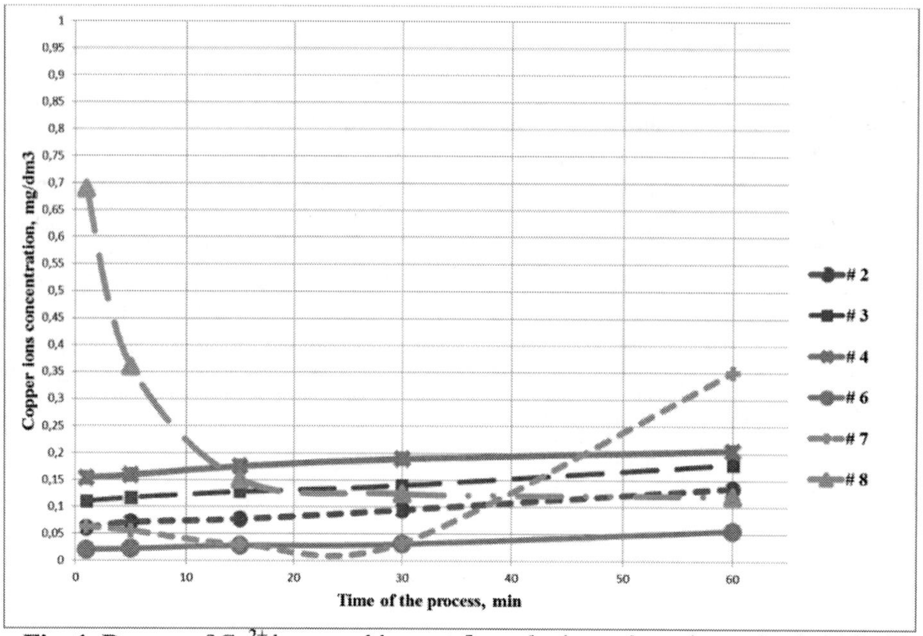

Fig. 1. Degree of Cu^{2+} ions washing out from the investigated samples in bidistilled water.

Figure 1 shows that the degree of Cu^{2+} ion washing out in water in all modified zeolite samples is within the range of maximum permissible concentration in drinking water (MPC in drinking water (Russia) – 1 mg/dm^3).

Sorption characteristics of the investigated zeolite samples of Kholinsk deposit in removing of Zn^{2+} ions from the simulated solutions are presented in Figure 2. Sorption was carried out under static condition in stirring.

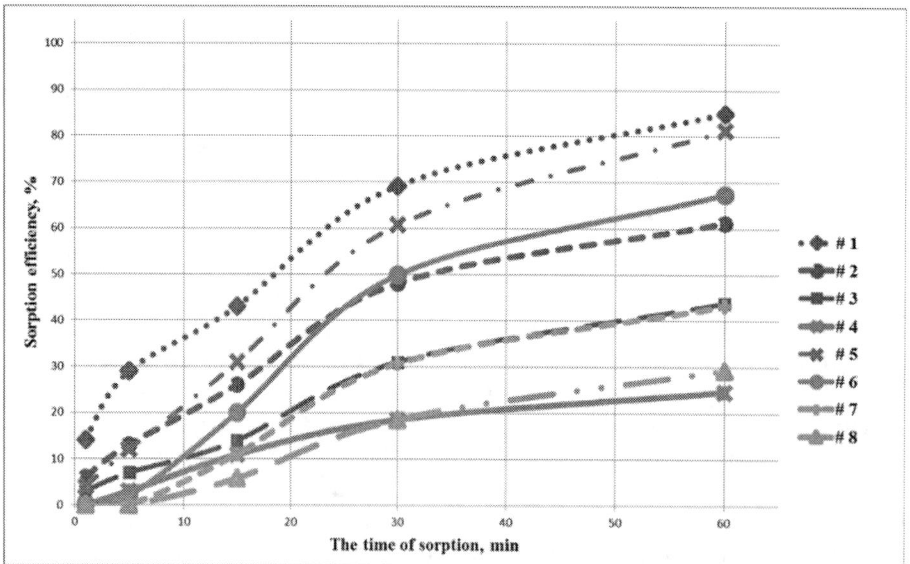

Fig. 2. Sorption capacity of investigated zeolite samples in removing of Zn^{2+} ions from the simulated solution.

As shown (Figure 2), the best sorption properties revealed for the intact (not modified) zeolite samples. Opposite, the modified zeolite samples (with high content in the structure of Cu^{2+} ions) showed the lowest sorption characteristics.

Sorption properties of investigated zeolite samples while extracting of Pb^{2+} ions from the simulated solution under static condition are given in Figure 3.

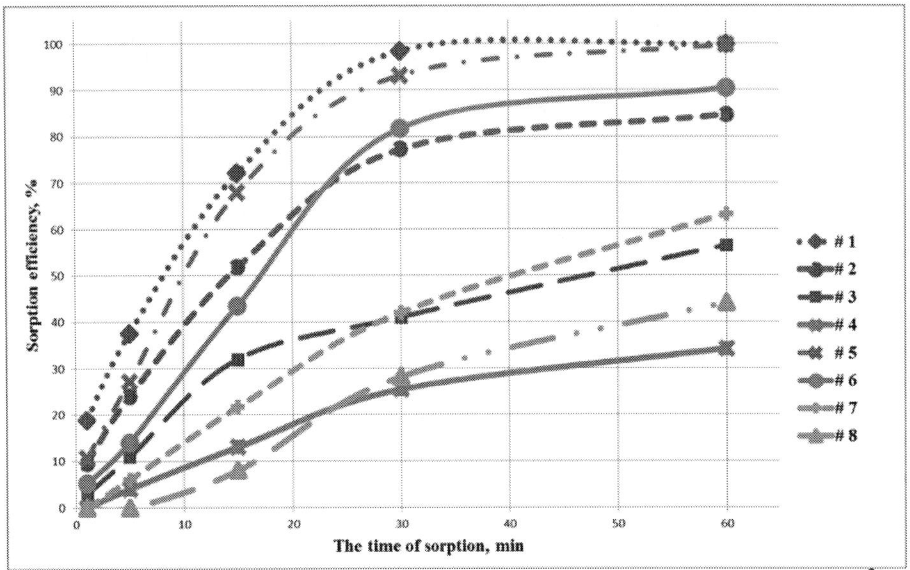

Fig. 3. Sorption capacity of investigated zeolite samples while extracting of Pb^{2+} ions from the simulated solution.

Figure 3 shows that intact (not modified) zeolite sample revealed the best sorption properties for removing Pb2 + ions from the model solution. The modified zeolite samples possessed lower sorption properties which gradually decrease with increasing content of Cu^{2+} ions in the structure of the zeolite. Modified samples with a particle size less than 0.1 mm showed better sorption properties at short contact time, but worse ones during prolonged sorption process. Bactericidal and bacteriostatic characteristics of investigated zeolite samples in the process of influece on the culture of Escherichia coli are presented in Table 2.

Table 2. Determination of bactericidal and bacteriostatic properties of original and modified zeolite samples of Kholinsk deposit.

Zeolite sample	Particl e size, mm	Concentration of Escherichia coli, CFU/ml	Inhibition of Escherichia coli growth (radius), mm
Petri dish with Escherichia coli (check)			Confluent growth
# 1			Confluent growth
# 2	less than 0.1		Confluent growth
# 3			5
# 4		$2*10^7$	8

# 5	0.5-1.0	Confluent growth
# 6		Confluent growth
# 7		2
# 8		4

As Table 2 shows, confluent growth of Escherichia coli is observed on Petri dish (check sample) in twenty-four hours of settling in the thermostat. The different fractions of the original zeolite samples (# 1 and # 5), and the samples # 2 and # 6 modified by the solution, containing the least content of copper sulphate, have the same result. Some bactericidal and bacteriostatic effects can be observed in samples # 3 and # 7. The growth inhibition of Escherichia coli is observed in a radius of 5 mm (fraction less than 0.1 mm) and 2 mm (fraction 0.5-1 mm) appropriately round the zeolite sample. Zeolite samples # 4 and # 8 show better bacteriostatic and bactericidal properties.

This is due to the fact that more Cu^{2+} ions wash out from the porous structure of zeolites. Modified zeolite samples with lesser fractional composition have better bactericidal and bacteriostatic properties.

As the investigation indicates, in providing zeolite samples with bactericidal and bacteriostatic properties, the successful result can be observed only in samples # 3, # 7 (modification 2), and samples # 4, # 8 (modification 3).

The modified zeolite samples # 2, # 6 (modification 1) do not produce bactericidal and bacteriostatic effect, and that is why they also don't correspond with the aim of the study. Zeolite samples # 3 and # 7 (modification 2) are optimal, because they have the capacity to remove efficiently Zn^{2+} and Pb^{2+} ions from the simulated solutions, and they have bactericidal and bacteriostatic properties.

4 Conclusion

1. The natural zeolite samples of Kholinsk deposit modified by Cu^{2+} ions were obtained.
2. Specific surface and specific pore volume of investigated samples were identified.
3. The degrees of Cu^{2+} ions washing out from presented natural zeolite samples of Kholinsk deposit were investigated before and after modification.
4. Sorption were carried out with the use of investigated samples under static conditions in removing Zn^{2+} and Pb^{2+} ions from the simulated solutions.
5. Bacteriostatic characteristics of the investigated zeolite samples in the process of Escherichia coli growth inhibition were identified.
6. The most optimal modified zeolite samples with bactericidal and bacteriostatic properties, and sorption activity to Zn^{2+} and Pb^{2+} ions have been identified (samples #3 and #7).

This study was supported with the grant of the President of the Russian Federation to support of young Russian scientists MK-5939.2016.8

References

1. Frog B, Levchenko A 1996 *Water treatment* (MSU, Moscow)
2. Telitchenko M, Ostroumov S. 1990 *Introduction to Problems of Biochemical Ecology: Biotechnology, agriculture, protection of the environment* (Moscow: Science).
3. Kosoy V, Bazhenov A 2001 *Water and Ecology: Problems and Solutions (Rus)* **1**: 40-45

4. Plotnikov E, Martemianova I, Martemianov D, et al 2015 *Procedia Chemistry* **15**: 219 – 224
5. Baidina N 1994 *Soil science*. **9**:121 – 125
6. Martemianov D, Galanov A, Yurmazova T et al 2014 *Proceedings of the higher educational institutions. Chemistry and chemical technology*. **57**:30-33
7. Martemianov D, Galanov A, Yurmazova T 2013 *Basic Research*. **8**: 666-670
8. Martemianov D, Korotkova E, Galanov A 2012 *Bulletin of the Karaganda University*. **3:** 61-65.

Chemical Composition of Bottom Sediments in Small Mountain Rivers (Altai, the Russian Federation)

Oleg Savichev [1,2], *Irina* SMatveenko[1] , and *Vladimir* Paromov[2]

[1]Tomsk Polytechnic University, Institute of Natural Resources, 634050, Lenin Avenue, 30,Tomsk, Russia
[2] Tomsk State University, 634050, Lenin Avenue, 36,Tomsk, Russia

Abstract. The article deals with the chemical composition of bottom sediments in the small mountain rivers Aktru, Tyute, and Uznezya (Altai, Russia). It is established that as compared to small and middle plain rivers in Taiga zone, Siberia bottom sediments of the Aktru River contain less amount of K, Na, V, Cr, Mn, Ni, Al, Si, Ti, Sr, approximately the same – Fe, Co, Cu, Zn, Pb and more – As, Ca, Mg. It is suggested that there is a sharp increase of relative Al, Na, K and Si shares from the upland to lowland due to both weathering products accumulation of primary alumosilicates in bottom sediments and export of Ca and Mg from them when dissolving carbonate minerals. Composition transformations of bottom sediments is non-linear and considerably depends on composition of underlying and water-bearing rocks and water exchange intensity that manages time, conditions, and stages of interactions in "water – organic matter – rock" system.

1 Introduction

Humic Rivers consisting of surface water and lands covered with it within the shore line with bottom sediments (solid particles) formed as a result of different physical, chemical, and biological processes in a water body and its watershed are a boundary layer between rivers and subsurface, their chemical composition being one of the most essential factors for evaluating general state of water ecosystems and indicator of natural-anthropogenic changes in climate and environment [1, 2]. Therefore, the adequate study in ecological and geochemical conditions of water bodies should include investigation of chemical composition of not only water, but also bottom sediments. In this connection, in the recent decades the study in bottom sediments (hereinafter referred to as river sediments) in vicinity of residential areas and industrial centers is being given focused attention in many parts of the world, particularly in the high anthropogenic-risk areas [2, 3]. Research of rivers not exposed to direct impact is of considerable interest. It is explained by the fact that the results of such a research would contribute to determining or specifying the conditions of natural geochemical anomalies development as an important indicator of hydrogenic mineral deposits. Of particular interest is the study in chemical composition of mountain river bottom sediments with abrupt height and slopes, which the elevated element concentrations are often associated with. Besides, the data on river conditions with minimal anthropogenic impact may and should be used in evaluating background geochemical indicators.

The arguments presented above serve as a basis for research in chemical composition of bottom sediments of small rivers Aktru, Tyute, and Uznezya in Gorny Altai (Russia). The Uznezya River is a right tributary of the Katun River, its length – 25 km, watershed line head – 1459 m, there is no ice in the basin. The Aktru River (can also termed as "Ak-

Turu") is an element of "Aktru – Chuya – Katun – Ob" drainage network, a left tributary of the Chuya River. The river length is 25 km, catchment area – 161.4 km^2, source altitude – 2750 m, the maximum elevation – 4044 m. Within the catchment area there are seven glaciers of total area 16.7 km^2. The largest glaciers among those of the catchment involved are mountain-valley glaciers of the Left, Right, and Small Aktru. The major part of water flow of the Aktru River forms in the upstream within 20–25 km from the mouth. The Tyute River is also the left tributary of the Chuya River, its length – 33.8 km, catchment area – 163.5 km^2, the maximum elevation – 3556 m; in catchment area of the Tyute River there are 4 glaciers of total area 3.0 km^2 [4, 5].

The choice of the water bodies mentioned above is conditioned by the following reasons: 1) the Aktru, Uznezya, Tyute Rivers are typical water streams of Gorny Altai with minimal economic impacts on the river ecosystems; 2) the data on the Aktru and Tyute Rivers can be used to study other mountain-glacier basins of the Southern Siberia, the data on the Uznezya River – small mountain water streams without glacier inflow; 3) in 1962–1994 hydrological monitoring was performed at the Aktru River, but at present the researchers of Tomsk State University (TSU) and Tomsk Polytechnic University (TPU) are performing observation of glacier dynamics and sporadic hydrometeorological and hydrochemical analyses [4–7]. The main goal of the research is to estimate the average chemical composition of bottom sediment water in small mountain rivers with minimal anthropogenic impact and to reveal the trends in its length changes as well as relations with water chemical composition and mineral composition of bottom sediments. These facts define the research structure.

2 Materials and Methods

The research suggests field and laboratory work, analysis of field and laboratory data.

1. The fieldwork is carried out in term of the requirements [1, 3] and includes sampling of river water and bottom sediments: the Aktru River – in July of 2013; the Tyute River – in July of 2014; the Uznezya River – in August of 2013. Bottom sediments are sampled in low flow zones 0.2 – 1.0 m from the water edge from the layer 0.2 m using plastic sampler. The river waters are sampled at the same point (before bottom sediment sampling) from the layer 0.3 – 0.5 m from the surface. Water discharge is measured in the Aktru and Tyute Rivers additionally (using current meter GR-21M). The layout of sampling points is given in [7, 8].

2. The laboratory work is performed in TPU and TSU, including analysis of: 2.1) composition of river water – in TPU under supervision of N.V. Guseva, A.A. Khvashchevskaya and V.A. Tarbokov using the following methods: pH – potentiometric; SO_4^{2-}, – turbidimetric; Ca^{2+}, Mg^{2+}, HCO_3^-, Cl^-, dichromate oxidizability (DO) – titrimetric; nitrogen compounds, phosphates, Si – photometric; Al – fluorometric; Na^+, K^+, Fe – atomic absorption; Zn, Pb, Cu – stripping voltammetry; Li, Ti, V, Cr, Mn, Co, Ni, As, Sr, Cd, Sn, Ba – atomic emission spectroscopy with inductively coupled plasma; 2.2) composition of bottom sediments with the particle diameter of up to 1 mm – in TPU as well as using the following methods: pH of water extract – potentiometric; concentrations of Ca^{2+}, Mg^{2+}, HCO_3^-, Cl^-, carbon of water-soluble organic substances $C_{org.}$ in water extract – photometric; SO_4^{2-} in water extract – turbidimetric; Na^+, K^+ in water extract – ion chromatography; NO_3^-, NO_2^-, NH_4^+ in water extract – photometric; Li, Na, Mg, Al, Si, K, Ca, Ti, V, Cr, Mn, Fe, Co, Ni, Cu, Zn, As, Sr, Cd, Sn, Ba, Pb – atomic emission spectroscopy with inductively coupled plasma with acid-soluble element extraction (the fractions of up to 1 mm are analysed); 2.3) mineral composition of bottom sediments – in TSU (X-ray phase analysis; the analyst is T.S. Nebera). The details of the analysis are presented in [6, 7].

3. Statistical, landscape, and geochemical analyses of the data obtained are carried out by the authors taking into consideration the results of the previous research [5–7] and requirements [1] in the following sequence: 3.1) analysis of river water composition; 3.2) analysis of bottom sediment composition and its relation to hydrochemical and other conditions.

3 Results and discussion

3.1 River water. The analysis of hydrochemical data obtained in 2013 and earlier [5–7] showed that river water is fresh with mineralization from very low (up to 0.1 g/dm^3) to medium (up to 0.5 g/dm^3), hydrocarbonate calcium, from neutral to alkaline (table 1).

Table 1. Water chemical composition of the Aktru and Uznezya Rivers in July – August, 2013 and the Tyute River in July, 2014 (distance from the mouth, km)

Para-meter	Aktru glacier [5]	Small rivers *	The Aktru River [7]						The Uznezya River		The Tyute River
			24.5	20	16.9	12.4	7.4	0.8	4.0	1.4	22.0
Date	–	–	19.07	18.07	18.07	18.07	18.07	20.07	06.08	06.08	08.07
Q, m^3/s	–	–	1.3	5.2	5.3	5.3	5.4	5.8	–	–	1.9
pH	5.43	8.08	8.23	8.90	9.00	8.40	8.40	8.90	7.90	7.75	8.01
mg/dm^3											
TDS	21	141	63	114	126	153	150	142	188	181	128
NO_3^-	1.24	0.77	0.72	0.58	0.51	0.59	0.24	0.39	1.74	1.40	0.78
PO_4^{3-}	0.044	0.093	0.15	0.07	0.14	0.03	0.04	0.04	0.08	0.06	<0.01
Si	0.46	3.78	1.78	1.18	1.57	1.06	0.77	1.48	5.77	5.15	1.02
Fe	0.329	2.23	12.1	6.22	5.64	1.22	1.28	1.31	0.69	0.31	0.03
DO	4.72	1.60	<1.0	<1.0	<1.0	<1.0	<1.0	<1.0	<1.0	<1.0	1.20
mkg/dm^3											
Al	135	1055	5298	2501	1178	553	153	102	426	395	102
Ti	–	–	263	107	40.0	17.0	3.0	0.0	11.0	10.0	–
Mn	–	–	185	75.0	32.0	21.0	13.0	11.0	22.0	10.8	–
Cu	1.7	4.1	14.0	4.6	6.5	2.1	2.2	2.7	3.0	2.3	1.3
Zn	44.7	14.9	37.0	16.0	18.0	15.0	5.6	12.0	8.0	8.9	15.0
Cd	0.33	0.06	0.20	0.10	0.10	0.10	*	*	*	*	*
Hg	0.248	0.230	*	*	0.36	0.28	*	0.450	0.09	0.02	–
Pb	0.9	2.5	11.2	2.8	2.8	0.9	0.7	1.4	1.2	0.4	0.4

Note: Q – water discharge; DO – dichromate oxidizability; «–» – absence of data or N<3; * materials [5–7] and the data obtained by the authors on the Ursul, Sema, Uznezya, Aktru, Tyute Rivers in 2013–2014; * <0.01 mkg/dm^3

As compared to glacier water (average content of principle ion sum TDS 21.3 mg/dm^3; TDS is calculated as a sum of concentrations of Ca^{2+}, Mg^{2+}, Na^+, K^+, HCO_3^-, CO_3^{2-}, SO_4^{2-},

Cl^-), water of the Aktru and Tyute Rivers contains high amount of dissolved salts and is characterised by rather a sharp increase in concentrations of several macro- and microelements in glacier zone as a result of melt water interaction with rocks. Downstream in different hydrological conditions one can observe different tendencies in the changes of hydrochemical parameters along the river stream [5–7]. In some cases at the sites of midstream there is a decrease in concentrations (for example, Al, Ti, Mn, Zn, Cu, Pb in water of the Aktru River, table 1) that can be explained by sedimentation of suspended particles, on whose surface the sorption of some elements and low-soluble compounds takes place (carbonates, metal compounds with humic acid) [9]. The effect of surface and ground water inflow with different content of soluble matter is also possible. For example, in the study area of the Katun River basin in Quaternary water-bearing deposits the TDS value amounts 155–297 mg/dm^3; in the fracture zone of Lower Cambrian deposits the average value of TDS – 255 mg/dm^3, in the fracture zone of Upper- and Middle Cambrian deposits – 329 mg/dm^3 [8].

Comparison of the obtained data with the values of maximum permissible concentrations (MPC) adopted in Russia has shown that nearly in all points of study rivers there is an excess of water body quality standards: in fishery and household use – in concentrations of Fe and Al; fishery – in concentrations of Mn, Cu. In addition, the violation of water quality standards is found in the Aktru River (mostly in glacier zone) in concentrations of Ti, V, Zn, Hg, Pb. In authors' opinion, the main reason for MPC excess is a natural factor, i.e. intensive interaction of melt water with rocks and dust. The chemical composition of glaciers is expected to be connected with transport of atmospheric pollutants to some extent, particularly in case of radioactive elements and pesticides. But the key features of river water and sediment chemical composition of study water streams and the whole basin of the Ob River develop, first of all, under the impact of natural factors [10].

3.2 Bottom sediments. The field survey and study in bottom sediment composition have shown that the largest fractions (>1 mm) are associated with moraine complexes and river course and floodplain narrowing sites. The increase in the particle diameter within the fraction from 0.01 to 1 mm, is, by contrast, observed along the river length from the source to mouth. The sufficient changes of particle share with the diameter of up to 0.01 mm are not stated. The composition analysis of water extracts from bottom sediments showed that they are referred to fresh water in terms of mineralization and to hydrocarbonate calcium one in terms of chemical composition. According to pH values the water extracts are characterized as neutral and weakly alkaline. Total dissolved salts in water extracts of bottom sediments in the Aktru River changes along the river length without profound trends, while in the Uznezya River there is a definite increase in TDS as approaching to the mouth. The consistent DO growth is also recorded in the water extracts of bottom sediments in the Uznezya River as the water surface slopes and current velocity increase. As for the Aktru River, DO value grows rather sharply at the boundary of glacier zone and remains high enough down to the mouth. Probably, it is explained by both soil erosion and intensity of biogeochemical processes at interaction of melt water and mineral particles as well as simultaneous increase in soil erosion and concentration of humic substances in soil (table 2).

The analysis of data on acid-soluble elements has shown that, firstly, in comparison with small and medium valley rivers in Taiga zone of Siberia without anthropogenic pollution [11], in bottom sediments of the Aktru and Uznezya Rivers there is less amount of K, Na, V, Cr, Mn, Al, Ti, Si (in Taiga zone of Siberia – the average content of Si – nearly 36 % of sample weight), approximately the same – Fe, Co, Cu, Zn, Ni, Pb and more – As, Ca, Mg (table 2). Therefore, one might suggest that there is not always a strong link of rather high concentrations of some elements with economic activity impact in the bottom sediments of Western Siberia and Altai.

Table 2. Chemical composition of bottom sediments of the Aktru (18-20.07.2013), Uznezya (06.08.2013) and Tyute Rivers (08.07.2014; distance from the mouth, km)

Para-meter	Average for Taiga [11]	the Aktru River						the Uznezya River	
		24.5	20	16.9	12.4	7.4	0.8	4.0	1.4
pH(w.)	–	7.76	7.68	7.76	7.68	7.68	7.66	7.35	7.72
mg/kg									
TDS(w.)	–	72.4	66.2	80.3	61.1	60.6	83.4	69.1	111.9
NO_3^- (w.)	–	0.70	0.22	0.45	0.26	0.28	0.12	4.78	6.11
DO(w.)	–	2.5	65.75	64.35	66.15	64.8	72.8	2.5	25.9
Na	8131	313	315	294	238	290	252	281	422
Mg	2768	17840	8318	11200	11060	10770	12380	11030	9868
Al	42271	13890	9256	9543	10150	11430	11360	16270	26190
Si	–	112	188	105	77	119	74	133	92
K	12200	2009	3598	2335	2562	3513	2097	1502	2764
Ca	6563	36000	27150	29800	27780	24960	20380	8575	10400
Ti	4232.3	1413	667.3	1005	907.6	688	1067	679.9	390.4
Mn	839.7	664.9	546.0	589.0	526.2	514.8	691.0	593.3	706.5
Fe	22548	25490	15590	18970	17680	17390	19250	21910	20400
Cu	22.3	29.3	15.4	18.9	10.8	12.4	16.6	14.0	15.4
Zn	46.9	67.3	33.2	44.5	39.7	37.7	74.6	57.8	55.7
Cd	0.35	0.59	0.33	0.40	0.37	0.37	0.49	0.47	0.45
Pb	9.9	9.8	4.3	5.7	3.8	4.2	12.3	5.9	7.1

Note: w.e. – water extract; for Tyute river: pH(w.)=7.6; TDS(w.)=130.4 mg/kg; NO_3^- (w.)=0.05 mg/kg

Secondly, in the Aktru River the maximum concentrations of most study elements are found in the bottom sediments of the glacier zone (24.5 and/or 20 km from the mouth); approximately the same content or definite decrease from source to mouth in the Aktru River, Ca, Fe, Mn, Na are revealed; in the Uznezya River the increase in Si, Fe, Co, As is observed from the sources to the mouth (table 2).

Thirdly, the statistically significant relationships of river water concentrations and bottom sediments (at correlation coefficient r more than double error of its determination) are found for Cu (r = 0.97±0.10), Fe (r = 0.71±0.22), Ti (r = 0.63±0.27), NO_3^- (r = 0.59±0.29), but taking into account a few number of samples the conclusion on relationship of water geochemical parameters and bottom sediments of small mountain rivers are still indicative. In terms of pH, TDS, DO, Pb, Cd, Mn, Al, Si the statistically significant relationships were not stated.

Fourthly, there is a weak link of geochemical parameters with mineralogical composition of bottom sediments and it is observed only in the fraction with the particle diameter of up to 0.01 mm. Particularly, a notable reduction of albite is established in the sample taken 0.8 km up the mouth of the Aktru River, as compared to the rest five samples (in the middle part of the fraction <0.01 of approximately 10–15 % of its weight, the following minerals are revealed in the mouth section, km: 24.5 km – clinochlore, albite, calcite; 20 km – chlorite, albite; 16.9 km – chlorite, albite; 12.4 km – mica, kaolinite, albite, calcite; 7.4 km – mica, kaolinite, albite, calcite; 0.8 km – chlorite, mica. To some extent, this fact is correlated with the trend towards reduction of sodium concentration in the bottom sediments from the source to the mouth (table 2). In other cases any relationship between mineral and chemical composition of bottom sediments is not revealed in the Aktru River (in all fractions of the particle diameter from 0.01 to 1 mm quartz and plagioclase are revealed and, almost in all samples – potassic feldspar, mica, calcite, chlorite, and kaolinite).

Fifthly, the most significant inverse dependencies of geochemical parameters of the bottom sediments in the Aktru and Tyute Rivers are established for acid-soluble Mg, Al, Ca, Ti, Fe, Cu, Cd on water discharge at the time of study. However, these links are generally weak (one can observe an excessively wide spread of points with their small amount). A more stable link is the relationship of some element concentrations in bottom sediments and the part of glaciers in the general catchment area. This parameter is also an indirect characteristic of water exchange that reflects, first of all, the area of water interaction with rocks. Thus, there is the relationship $S_{Ca}=11907.94\cdot(100\cdot A_g/A_b)^{0.26}$ of calcium concentration S_{Ca} in the bottom sediments of the Aktru River, where A_b and A_g – catchment and glacier areas (at the cross-section); the square of correlation ratio $R^2=0.76$.

Quality standards of bottom sediments have not been stated in Russia. However, if one suggests that bottom sediments are responsible for soil cover development of the river flood plains, it is possible to compare the geochemical parameters of bottom sediments with soil standard values. As for the Aktru and Uznezya Rivers such a comparison has shown that floodplain soil with similar chemical composition do not correlate with standards of Cr, Mn, Co, Ni, Cu, Zn, As, Cd, Pb concentrations. As is the case of river water, this fact is mostly explained by natural factors, first of all, composition of underlying and water-bearing rocks and water exchange rate defining, based on [9, 12], time and interaction area in "water – rock" system.

4 Conclusion

Chemical composition of bottom sediments in the Aktru, Tyute and Uznezya Rivers is characterized by concentrations of Fe, Co, Cu, Zn, Pb typical for valley rivers of Siberian Taiga zone. At the same time, the bottom sediments of study rivers contain less amount of K, Na, V, Cr, Mn, Al, Ti, Si and more – As, Ca, Mg. As a first approximation, one could suggest that there is a sharp increase in relative parts of Al, Na, K and, particularly, Si from mountain to valley regions due to both accumulation of weathering products of primary aluminosilicates in the sediments and washout of Ca and Mg from them at solution of carbonate minerals.

Transformation of chemical composition of bottom sediments is a non-linear process and depends on both composition of underlying and water-bearing rocks and water exchange rate that regulates time, conditions, and stages of interactions in "water – organic matter – rock" system (features of water exchange are not only water discharge or water flow module, but also relationship of some hydromorphological parameters). As for the Aktru River it is expressed in maximum concentrations of most study elements in the glacier zone of catchment area, where the main part of water flow develops and the initial contact of melting ice with rocks occurs.

Along the water flow one can observe decrease in their concentrations and subsequent fluctuations that are weakly connected with mineral composition of bottom sediments.

Presumably, it is due to prevalence of longitudinal advective transport of suspended and bed load of a mountain river as compared to the other parts of the turbulent diffusion equation. Definite relationship between chemical and mineral compositions is established only for albite, whose decrease in the fraction of diameter less than 0.01 mm is correlated with the general sodium reduction from the source to the mouth.

Acknowledgements

The research has been supported by the grant issued in accordance with VIU_VAF_144_2014.

References

1. A.A. Golovin, N.N.Moskalenko et al. *Requirements for production and results of multipurpose geochemical mapping of 1: 200 000 scale* . Moscow. IMGRE Publishing House, 92 (2002) [in Russian]
2. D.P.Loucks and E. Van Beek. *Water resources systems planning and management. An Introduction to Methods, Models and Applications*. Turin, UNESCO PH, 680 (2005)
3. *Water Quality Assessments - A Guide to Use of Biota, Sediments and Water in Environmental Monitoring*. London, Chapman & Hall, 348 (1996)
4. Yu.K. Narozhny, S.A. Nikitin, A.A. Lukianov and A.B. Osipov. Mountain-glacier basin of the Aktru River: new morphometric and resource characteristics. *Voprosy Geograpgii Sibiri.* Volume 26, pp 67–74 (2006) [in Russian]
5. O.G. Savichev and V.V. Paromov. Chemical composition of glacial meltwaters and river waters within the Aktru river basin (Gornyi Altai) *Geography and Natural Resources.* Volume 4, pp 364-370 (2013)
6. O.G. Savichev, V.V. Paromov, Yu.G. Kopylova, A.A. Khvashchevskaya and N.V. Guseva. Ecogeochemical condition of surface waters in Katun River basin (Gorny Altai). *TSU J.*, Volume 366, pp 157-161 (2013) [in Russian]
7. V.V. Paromov, O.G. Savichev, L.N. Shantikova and Ks.A. Torgasheva. Ionic runoff and chemical composition of glacial waters of the Aktru river (Gorny Altai). *TSU J,* Volume 383, pp 226-231 (2014) [in Russian]
8. S.L. Shvartsev, V.S. Kuskovsky, O.G. Savichev, A.A. Lukin and Ye.V. Domrocheva. On criteria of distinguishing the earlier stages of groundwater anthropogenic pollution (by the example of the Katun River basin). *Proceedings of Tolstikhin's readings.* pp 90-95 (1999) [in Russian]
9. O.G. Savichev. Influence of the interactions within the water - earth materials system on the formation of river water composition in the Ob basin *Geography and Nat Res.* Volume 30. Issue 2, pp 161-166 (2009)
10. O.G. Savichev, A.K. Mazurov, I.I. Pipko, V.I. Sergienko and I.P. Semiletov. Spatial Patterns of the Evolution of the Chemical Composition and Discharge of River Water in the Ob River Basin *Doklady Earth Sciences* . Volume 466, pp 47–51. (2016). DOI: 10.1134/S1028334X16010141
11. O.G. Savichev and P.T. Duong. Zone laws of change of chemical composition of river sediments of Siberia and condition of its formation *Bulletin of the Tomsk Polytec Univ.* Volume 323, Issue 1, pp 157-161 (2013) [in Russian]
12. M. Meybeck. Global chemical weathering of surficial rocks estimated from river dissolved loads *Am. J. of Sci.* Volume 287, Issue 5, pp 401–428 (1987).

Typomorphic characteristic features of accessory ilmenite in granitoids of the polyphase Aleisk-Zmeinogorsk complex (N-W Rudny Altai area)

Novoselov K.L.[1], Gavrilova K.A.[1]
[1]National Research Tomsk Polytechnic University, 30 Lenina Ave., Tomsk, 634050, Russia

Abstract. The paper material is devoted to the research of crystallomorphology, accessory ilmenite expansion, characteristic features of its chemical composition, paragenetic association and conditions for mineral formation in Devonian granitoids of the polyphase Aleisk-Zmeinogorsk complex. Two ilmenite generations formed at various crystallization stages of the magmatic melt have been identified by the morphological features of the ilmenite crystals, chemical composition and paragenesis. Early-magmatic Ilmenite1 singled out at the protocrystallization stage during reduction and weakly-oxidizing environment conditions. It is characterized by the crystalline-faced shapes, low content of the pyrophanite minal and V admixture. Ilmenite 2 singled out at the major crystallization stage of the melt in the conditions of higher oxygen fugacity as the forms of the flattened, xenomorphic grains with the enhanced pyrophanite concentration and Nb admixture.

1 Introduction

Generally speaking, Rudny Altai is considered to be a unique metallogenic province with the precious, rare and iron-ore deposits. Though, the problems on the geochemical trend and potential granitoids ore-bearing in Aleisk-Zmeinogorsk complex (D_{2-3}), which occupies the larger part of the north-west of Rudny Altai, are obscure and nowadays relevant.

Target of the research is to analyze accessory ilmenite distribution relationships as the essential mineral and Ti concentrator, characteristic features of its crystallomorphology and chemical composition in granitoids of the successive intrusive phases.

Research methods besides traditional (mineralogo-petrochemical) included accessory mineral one. It analyzed granitoid crushed samples according to the routine scheme [1], sampling monomineral ilmenite fractions to calculate the content and provide analytical research. The ilmenite chemical composition was identified by the electronic scanning microscope JSM–6510LV (Jeol Ltd) (SNIIGGIMS laboratory, Novosibirsk, M.V Khlestov, analyst-researcher) and X-ray fluorescence microscope (Department of Geology and Minerals Prospecting, Institute of Natural Resources, (analyst M.A. Rudmin, candidate in science (Geology & Mineralogy)).

2 Petrologo-geochemical description of granitoids from Aleisk-Zmeinogorsk complex

Aleisk-Zmeinogorsk complex (D_{2-3}) includes three big granitoid massifs (Novonikolayevskiy, Aleiskiy, Ustyanskiy) and some smaller bodies localized in the north structures of Aleisk uplift in Rudny Altai. These are between Devonian volcanogenic sedimentary depressions, controlled by the deep-seated faults of the north-western and sublatitudinal strike, characterized by comagmatic connection with Devonian vulcanites of Kamenevskiy complex [2]. The biggest territory (more than 1000 km^2) is occupied by Novonikolayevskiy batholith, located in the north end of Aleisk anticlinorium. It has a comlex polyphase texture, comprising the whole range of homodromous succession – from early gabbroid to the fourth final leucogranitic phase with biotite and biotite- muscovite differences. Gabbroids – derivatives from the first phase intrusion – are intensively metamorphosed and form small xenolith bodies in the granitoids of the later phases. The derivatives of the second, major phase are widespread. These are plagiogranits, tonalites; of the fourth final phase – the leucogranits. The basic petrochemical characteristics of granitoids are the following: (from plagiogranits and tonalities of the early phase to leucogranits of the final) 1) agpaitic and total alkalinity ratios increase in the rocks due to Na_2O domination over K_2O; 2) rock alumina *al'* notably increases, so does the ferruginosity ratio *f* (from 0.77 to 3.75), when the sustained iron oxidation ratio is F (0.81 – 0.84); the rock calcareousness and titanium rates decrease (9.89 – 5.44).

The characteristic features of granitoids are the following: sulfer scarcity and absence of admixture elements referring to the chalcophile group, limited set and low content of siderophiles like Co, Ni, Cr, Sc, V which mainly concentrate in the earliest granitoid melt differentiates. Lithophile elements represent the most abundant group, where Ti content in the major phase plagiogranits exeeds the bulk earth value (concentration criterion Ti (KкTi) is 1.22 – 2.2. KкTi decreases to 0.95 – 0.43 in the later rock phases. Moreover, granitoids are substantially enriched by rare earth elements, including Y, that forms excess concentrations (KкY = 1.25 – 1.95).

Titanomagnetite-ilmenite-zircon-apatite-sphene-orthite association is considered to be the leading accessory minerals association in plagiogranits and tonalities relating to the major phase. Hematite, rutile, martite, rare earth elements significance becomes substantially greater when the ilmenite and sphene contents decrease. This occurs in granitoids of the second, third and the final (the fourth) intrusion phases among the accessory minerals.

By the formation conditions the granitoid massifs of the complex are thought to be the individualized parts of the large pluton formed from the unique magmatic center of the activated continental margin in the Middle-Late Devonian. These are associated with the granites of the I-type [2]; hypabyssal facies is responsible for the formation depth. Both, the calculated rocks crystallization temperature and oxygen fugacity according to mineralogical and petrochemical geothermometres [3] are at 900 – 850°C when $\lg f_{O2}$ is from 17.5 to 13.5.

3 Results and discussion

Ilmenite is a low-magnetic and electromagnetic fraction in the association with the titanomagnetite, hematite, ferrorutile, and forms low concentrations in granitoids; is irregularly distributed both along the plagiogranits massif area in the major phase, and in the derivatives of the successive phases. It is widespread in plagiogranite, tonalite and in the fourth final leucogranitic phase. Ilmenite occurs in the third leucoplagiogranite phase, but only in the solid solution of titanomagnetite decomposition texture in the form of the parallel-oriented thin lamella. Ilmenite crystallizes at the magmatic stage in the wide temperature range of reduction and weakly oxidizing conditions [4, 5, 6, 7]. The absence of accessory ilmenite grains in the third leucoplagiogranite intrusion phase is due to the oxygen activity increase during their formation. This resulted in ferrorutile occurrence, characterized by Mn admixture as a substitute for ilmenite [3].

Ilmenite constantly occurs in the major phase plagiogranite, but its content is not constant and doesn't increase 30 – 90 g/t. The lowest ilmenite concentrations are in the final leucogranitic phase where it is identified in the individual grains; its sample content is not more than 8 g/t.

By the crystallomorphological characteristics two types of ilmenite emanation have been determined– crystalline-faced and flattened or platy-angular grains.

The crystalline-faced form is definitely rare one and mostly occurs in the major-phase plagiogranites. Its habit is tabular with the size of 0.3 – 0.5 mm, well-defined rhombohedron faces {1011} and pinacoid {0001}. Cross striation often occurs on the faces (0001) as well as numerous micron -size pores and blebs of quartz, and plagioclase grains. But the widespread ilmenites in rocks are the ones which don't have simple forms faces. These are xenomorphic grains with the size of 0.1 – 0.3 mm, angular or a little elongated, having indented, sinuous edges; the pinacoid faces are shown on the individual grains.

Ilmenite is usually ferrum-black; steel-grey or brownish shade appears in the crystals. The ilmenite streak on the porcelain plate is red.

Ilmenite grain forms research of petrographic thin sections has also identified its two morphological types: intergranular spaces are characterized by the ilmenite idiomorphic grains and usually accumulate early plagioclase grains (An_{30-50}) as well as quartz grains. Ilmenite xenomorphic grains occupy the intergranular spaces and form the hornblende intergrowths. So, the ilmenite crystalline morphology, grains interrelation with major rock-forming minerals show the ilmenite isolation during the whole magmatic stage of granitoid melt crystallization. This process provides the identification of two generations formed at different rock crystallization phases. Ilmenite1, crystalline-faced, early-magmatic had singled out from the melt at the protocrystallization stage before major rock-forming minerals occurrence (plagioclase, potash feldspas). Ilmenite 2, xenomorphic, singled out at the mass melt crystallization stage, time-associated with dark-colored minerals (potash feldspas, quartz).

Ilmenite1is widespread in the major-phase plagiogranites, Ilmenite 2 grains generally occur at the final-stage leucogranites; early ilmenite crystals are rare.

The chemical composition of the identified ilmenite generations is presented in the Table.

By the obtained chemical composition data, ilmenite is characterized by the limited set of admixtures. The increased concentration of Mn is constantly present, Al admixture is occasionally determined, V is in Ilmenite1, and Nb is in Ilmenite 2. These elements presence is due to their occurrence in the ilmenite texture by the scheme of isovalent or heterovalent isomorphism $Fe^{2+} \leftarrow Mn^{2+}$; $Ti^{4+} \leftarrow Al^{3+}, Nb^{5+}, V^{5+}$ [8]. In most Ilmenite1 and Ilmenite 2 crystallochemical formulae the calculated ratios are approximate to stoichiometric values; their individual deviations for the ferrous iron might be related to Fe^{3+} presence that appeared to have no identification. Moreover, while calculating it was considered to be Fe^{2+}.

As it has been already mentioned, Mn has the greatest concentrations in ilmenite, the abundance of which sharply increases in Ilmenite 2 (up to 15.71 mas. %) as its manganese variety – mangan-ilmenite [9]. Ilmenite manganese is in the form of minal – pyrophanite $MnTiO_3$, isostructural with $FeTiO_3$ forming two-component isomorphic mixture. The calculated pyrophanite content in Ilmenite1 changes from 2.9 to 19.2 mol.%; in Ilmenite 2 the concentration of pyrophanite molecule sharply increases to 21 – 33 mol. %. Al admixture is not identified in all studied crystals of the early generation, though the increased concentrations (up to 5.95 mas. %) characterize Ilmenite1 as more aluminia-saturated of all plagiogranites. The Al admixture is present in Ilmenite 2 (leucogranites), but in smaller amount (up to 1.85 mas. %); only separate grains have higher content (up to 5.54 mas. %). V admixture in the amount of 0.31 – 0.6 % is occasionally identified in early magmatic Ilmenite1(plagiogranites) of the major phase. In early magmatic Ilmenite1 (leucogranites) of the final phase the V admixture has not been identified. This absence relates to V occurrence in the crystal lattice of the widespread accessory titanomagnitite [3]. Moreover, the V admixture absence is explained by the increased alkalinity of the melt, significant potassium trend and volatile elements activity, particularly F, which in complex compound with V [VF_5] is capable of migration or accumulation in the residual melt [10]. Niobium is present in Ilmenite 2 (leucogranites) of the final phase, its concentration is not more than 0.3 %.

Table. Ilmenite chemical composition in plagiogranites (major phase) and leucogranites (final phase) mas. %

№	TiO_2	FeO	MnO	Al_2O_3	V_2O_5	Nb_2O_5	total
Ilmenite1							
1.	48.29	42.6	5.41	3.1	–	–	99.4
2.	44.62	47.54	3.13	4.11	0.6	–	100
3.	46.68	46.33	2.51	4.48	–	–	100
4.	48.09	48.08	3.35	–	0.48	–	100
5.	43.13	44.55	6.18	5.95	–	–	99.81
6.	49.67	47.73	3.06	–	–	–	100.46
7.	50.38	45.75	4.74	–	–	–	100.87
8.	49.94	41.23	8.95	–	–	–	100.12
9.	54.06	42.39	4.56	–	–	–	101.01
10.	52.06	38.7	4.3	–	0.5	–	95.56
11.	52.38	41.26	5.27	–	0.31	–	99.22
12.	52.44	38.69	7.53	–	0.35	–	99.01
13.	49.62	47.65	3.54	–	–	–	100.81
14.	49.56	44.46	3.78	2.17	–	–	99.97
15.	59.81	35.34	2.05	2.81	–	–	100.01
16.	59.17	35.09	1.41	4.33	–	–	100
Ilmenite2							
17.	51.63	37.67	12.19	–	–	–	101.49
18.	51.34	32.79	13.96	–	–	0.23	98.32
19.	48.92	37.05	12.31	1.57	–	0.15	100
20.	48.87	40	9.17	1.74	–	0.22	100
21.	48.7	39.73	10.18	–	–	0.18	98.79
22.	46.13	36.47	15.19	1.85	–	0.2	99.84
23.	47.29	33.39	14.86	4.17	–	0.29	100
24.	46.12	32.08	15.71	5.54	–	0.3	99.75

Note: 1) FeO as FeO + Fe_2O_3; 2) №№1 – 12 Ilmenite 1 (plagiogranite); 3) №№13 – 16 Ilmenite 1 (leucogranite); 4) №№ 17 – 24 Ilmenite 2 (leucogranite); 5) dash – element is not identified

Crystallochemical formulae (correspond to the numbers in the table):

Ilmenite1

1. $(Fe_{0.903}Mn_{0.116})_{1.019}(Ti_{0.922}Al_{0.092})_{1.014}O_3$
2. $(Fe_{1.017}Mn_{0.068})_{1.085}(Ti_{0.858}Al_{0.123}V_{0.006})_{0.987}O_3$
3. $(Fe_{0.976}Mn_{0.053})_{1.029}(Ti_{0.885}Al_{0.133})_{1.018}O_3$
4. $(Fe_{1.043}Mn_{0.073})_{1.116}(Ti_{0.937}V_{0.004})_{0.941}O_3$
5. $(Fe_{0.947}Mn_{0.133})_{1.08}(Ti_{0.825}Al_{0.181})_{1.006}O_3$
6. $(Fe_{1.021}Mn_{0.066})_{1.087}Ti_{0.956}O_3$

14. $(Fe_{0.939}Mn_{0.063})_{1.002}(Ti_{0.998}Al_{0.067})_{1.065}O_3$
15. $(Fe_{0.699}Mn_{0.04})_{0.739}(Ti_{1.073}Al_{0.077})_{1.15}O_3$
16. $(Fe_{0.689}Mn_{0.028})_{0.717}(Ti_{1.052}Al_{0.119})_{1.171}O_3$

Ilmenite 2

17. $(Fe_{0.804}Mn_{0.252})_{1.056}Ti_{0.971}O_3$

7. $(Fe_{0.973}Mn_{0.102})_{1.075}Ti_{0.962}O_3$
8. $(Fe_{0.883}Mn_{0.193})_{1.076}Ti_{0.961}O_3$
9. $(Fe_{0.881}Mn_{0.096})_{0.977}Ti_{1.011}O_3$
10. $(Fe_{0.832}Mn_{0.094})_{0.926}(Ti_{1.01}V_{0.01})_{1.02}O_3$
11. $(Fe_{0.884}Mn_{0.113})_{0.997}(Ti_{0.995}V_{0.007})_{1.002}O_3$
12. $(Fe_{0.812}Mn_{0.164})_{0.976}(Ti_{0.997}V_{0.012})_{1.009}O_3$
13. $(Fe_{1.023}Mn_{0.059})_{1.082}Ti_{0.958}O_3$
18. $(Fe_{1.123}Mn_{0.245})_{1.368}(Ti_{0.815}Nb_{0.002})_{0.817}O_3$
19. $(Fe_{0.78}Mn_{0.26})_{1.04}(Ti_{0.93}Al_{0.04}Nb_{0.001})_{0.971}O_3$
20. $(Fe_{0.844}Mn_{0.21})_{1.054}(Ti_{0.927}Al_{0.05}Nb_{0.002})_{0.979}O_3$
21. $(Fe_{0.85}Mn_{0.23})_{1.08}(Ti_{0.94}Nb_{0.008})_{0.948}O_3$
22. $(Fe_{0.78}Mn_{0.33})_{1.11}(Ti_{0.89}Al_{0.05}Nb_{0.002})_{0.942}O_3$
23. $(Fe_{0.70}Mn_{0.316})_{1.018}(Ti_{0.894}Al_{0.124}Nb_{0.003})_{1.02}O_3$
24. $(Fe_{0.67}Mn_{0.333})_{1.01}(Ti_{0.868}Al_{0.163}Nb_{0.003})_{1.034}O_3$

In comparison with V, Nb accumulation is favoured by the alkalinity of the melt, and the concentration increase is fully matched to these rocks alkalinity increase, particularly of potassium trend.

Thus, ilmenite presence in two generations in granatoids indicates its crystallization to magmatic stage within the wide range of temperature. According to the data [4] ilmenite can also be formed in the moderate and intensive oxidizing conditions, where a certain amount of Fe^{3+} is present in the ilmenite composition. If there is Fe^{3+} in Ilmenite 2 composition, both the increased amount of manganese and the increased alkalinity of leucogranites as the final stage derivatives are identified.

4 Conclusion

The research analysis of the accessory ilmenite for Devonian granatoids in polyphase Aleisk-Zmeinogorsk complex allows to conclude the following:

1. Ilmenite in granitoids of various phases is distributed irregularly and forms low concentrations; plagiogranits of the major phase are significally enriched by ilmenite.
2. The accessory ilmenite is present in granitoids in two generations singled out at different magmatic stages of alumosilicate melt crystallization.
3. Typomorphic features for ilmenite generations identification have been established: Ilmenite1, mostly crystalline-faced form, presence of V admixture, low aluminia content; Ilmenite 2, with xenomorphic or platy grains, Nb presence, low aluminia content, the increased amount of manganese.
4. The V admixture presence in Ilmenite1, Nb admixture presence in Ilmenite 2. This is the early-phase plagiogranites typomorphic feature – the geochemical trend for V, leucogranites final phase – for Nb.

References

1 Lyakhovitch V. V. *Accessory rock minerals*. Course book. *Moscow. Nedra*. 296, (1979)
2 Turkin Y. A., Novoselov K. L. *Petrologo-gechemical features of Devonian granitoids in Rudny Altai N-W*, 2012, Izvestiya TPU, Volume 321, Issue 1, pp 5-15
3 Novoselov K. L., Turkin Y. A. *Typomorphism of accessory titanomagnetite for Devonian granitoids in Rudny Altai N-W*, 2014, Izvestiya TPU, Volume 324, Issue 1, pp 5-16
4 Nikolskiy N. S. Fluid regime of endogenous mineral formation. *Moscow. Nauka*. 199, (1987)
5 Vakhrushev V. A. Ore minerals of igneous metamorphic rocks. Handbook. *Moscow. Nedra*. 199, (1988)
6 Kryvdik S. G. *Geochemical features of ilmenites from the alkaline complexes of the Ukrainian shield: LA-ICP MS data*, 2014, Geochemistry International, Volume 52, Issue 4, pp 287-295
7 Alekseev V. I., Marin Yu. B. *Accessory mineralization of rocks from late cretaceous intrusive series with Li – F granites in the Far East*, 2015, Geology of Ore Deposits, Volume 57, Issue 7, pp 537-551
8 Makarov E. C. Isomorphism of atoms in crystals. *Moscow. Atomizdat*, 288, (1973)
9 Minerals. Reference Book Volume 2, Issue 3. *Moscow. Nauka*, 675, (1967)
10 Borisenko L. F. Vanadium (mineralogy, geochemistry and types of endogenous deposits). *Moscow. Nedra*. 192 (1973)

Voltammetric estimation of the content of antibiotics in veterinary preparations

Galina Slepchenko[1,*], *Tatyana* Gindullina[1], *Valentine* Deryabina[2], *Tatyana* Shchukina[1], *Aydin* Okanov[3] and *Saule* Kabieva[3]

[1] National Research Tomsk Polytechnic University, Tomsk, Russia
[2] National Research University Belgorod State University, Belgorod, Russia
[3] Karaganda State Technical University, Karaganda, Kazakhstan

Abstract. The voltammetric method for determination of tylosin tartrate, gentamicin sulfate, and cefalexin in veterinary preparations was for the first time developed. Electrochemical behavior of these antibiotics on the mercury film electrode was studied, and the working conditions (background electrolyte, deposition potential) were defined for getting analytical signals using the voltammetry. The methods of real objects preparation for determination of tylosin tartrate, gentamicin sulfate, and cefalexin were offered. The techniques for the voltammetric determination of antibiotics in the veterinary preparations may be used in cefalexin ranging from 0.1 to 2.0 g/dm³, tylosin tartrate in the range from 0.1 to 1.7 g/dm³, and gentamicin sulfate from 0.1 to 1.5 g/dm³ (Sr is not more than 25 %)

1 Introduction

According to the monitoring Research Techart research company, about 19 % of antibiotics are used - as growth promoters, 58% - as a prophylactic agents, 23 % - for the treatment [1].

Gentamycini sulfas (GS), tylosinum tartrat (Tiltar) и cefalexinum (CF) - broad-spectrum antibiotics, which are used in the fields of agro-industrial complex of Russia. With a huge selection of products, practically solve the problem of counterfeit drugs, which is compounded by the fact that for a number of existing control methods of antibiotics is not completely meet modern requirements analysis. Therefore, the timely control of the content of antibacterial drugs - a problem that requires an immediate solution.

One of the first methods for quantifying antibiotic steel immunological and microbiological express - Tests [2, 3]. Among the physical and chemical methods of analysis of antibiotics popular planar chromatography (modern TLC), high performance liquid chromatography and capillary electrophoresis [4, 5].

Each of these methods has its advantages and disadvantages. These disadvantages are the duration of the analysis, the definition of a small range of concentrations, the low sensitivity of the method and the high cost analysis or equipment. Alternatively, these methods may make variations voltammetry (VA). VA taking into account the low cost of analysis and equipment, combine high information capabilities with ease of signal measurement process that allows, along with the definition of the content of the relevant substances in veterinary medicines to carry out pharmacokinetic studies.

2 Experimental

Chemicals and solvents

The aim of the work was to develop a way of measuring the voltammetric tylosin tartrate, gentamicin sulfate, cephalexin and methods of their determination in pharmaceuticals. Substances of antibiotics were provided by the "Altay - vitamins", Russia.

A stock solution containing 1000.0 mg/dm³ appropriate antibiotic was prepared by dissolving dry powder sample of pharmaceutical preparation (containing the basic substance is not less than 99.5 %) in bidistilled water. Working solutions - certified mixtures were prepared by serial dilution in double distilled water, basic solutions.

In order to select the operating examined supporting electrolyte of 0.1 mol/dm³ solution: $(NH_4)_2SO_4$, KCl, $Na_3C_6H_5O_7$, $C_6H_8O_7{\cdot}2HN_3$, $C_4H_4O_6Na$, $C_4H_6O_6$, $C_6H_8O_7$, buffer and Britton Robinson buffer citrate-phosphate mixture. Background electrolyte solutions prepared by dissolving the corresponding salts in the sample with bidistilled water.

Instrumentation and voltammetric conditions

The work on the voltammetric analyzer «STA» (LLC «ITM», Tomsk Russia). All studies were carried out on a mercury-film electrode (RPE) with the working surface 15 mm^2. The auxiliary electrode and the reference electrode – silver/silverchloride in a saturated solution of potassium chloride. Mass concentration of the antibiotic was calculated by the method of the additive mixture in a certified solution to be analyzed. Delete oxygen carried out by bubbling of nitrogen.

3 Results and discussion

On the model solutions studied by choice of operating conditions (background electrolyte, potential (Ee) and the electrolysis time (tэ)) receiving antibiotics analytical signals. When determining drugs plays crucial pH because it affects the rate and mechanism of the electrode process. The dependence of the maximum potential of the cathode of the peaks of the investigated antibiotics pH is complex. The pH change resulted in a change of the peak shape, or to shift it. Offset potential towards more negative value indicates a loss recovery process antibiotic that may be related to the preceding reaction protolytic deprotonation of the protonated forms of antibiotics. May precede the deprotonation step an electron transfer from the electrode to the molecule depolarizer or occurs simultaneously with it. In this regard, we have conducted research on the effects of cation-anion composition of electrolytes and pH. It was found that the best background for Tiltar electrolyte is 0.1 mol/dm³ $C_6H_8O_7{\cdot}2HN_3$ for gentamicin - background 0.1 M $NaC_4H_4O_6$ (pH 5 – 6), and for cephalexin - background $Na_3C_6H_5O_7$ 0.1 M (pH 4 – 5) where voltammograms recorded with a distinct analytical signal and good reproducibility.

The dependence of the peak current on the potential electrolysis and concentration on RPE antibiotics. Selected potentials are accumulating in which there is maximum analytical signals antibiotics. Antibiotics studied peak current depending on the concentration of the solution.

The obtained calibration relation is stored in the concentration range of tylosin tartrate (CTiltar) (0.10 – 1.70) g/dm³, gentamicin (0.1 – 10.0) g/dm³ and cephalexin (0.1 – 2.0) g/dm³. At higher concentrations, there is a deviation from the linearity of the calibration curve, which, apparently, due to the saturation of the surface of the indicator electrode. Analytical signal tylosin tartrate first obtained under these conditions is shown voltammograms (Figure 1).

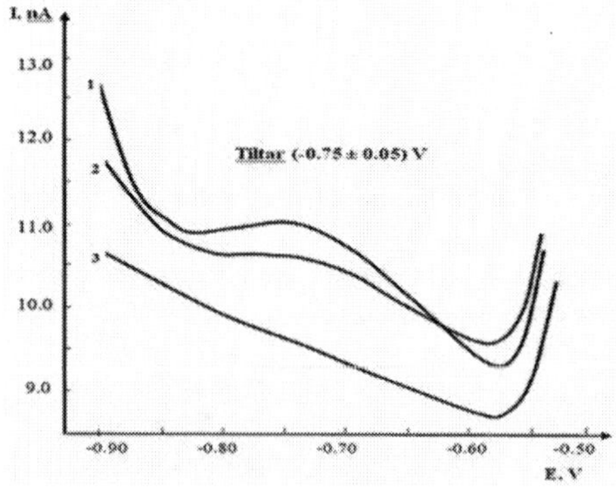

Fig. 1. Voltammograms Tiltar (tэ = 30 s): 1 – 0.1 mol/dm³ $C_6H_8O_7{\cdot}2HN_3$, pH 6; 2 – C_{Tiltar} = 0.2 g/dm³; 3 – C_{Tiltar} = 0.4 g/dm³

As can be seen (Figure 1) on the voltammogram observed a pronounced Tiltar signal at a potential maximum cathodic peak E = -0.75 V. The introduction of certified additives mixture, proportional to the concentration entered an antibiotic increases the analytical signal that indicate the possibility of quantification.

Voltammograms CF and GS for the first time obtained under these conditions are shown in Figure 2.

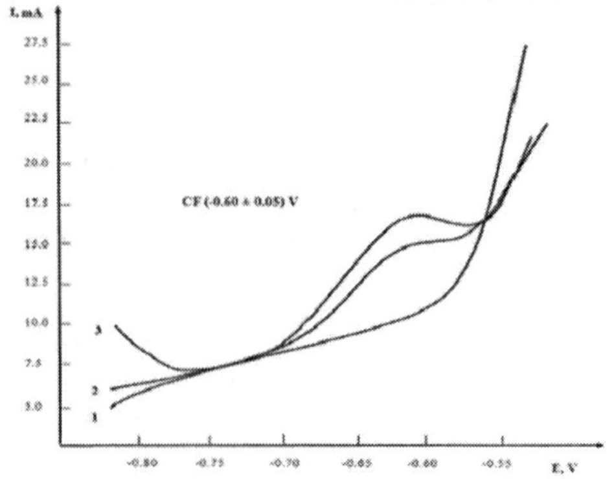

a)

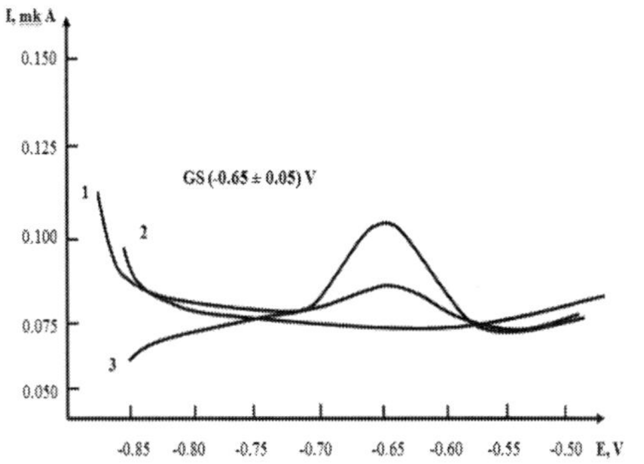

b)

Fig. 2. Voltammograms cephalexin (a) and gentamicin (b) on RPE: a) 1 - background solution - 0.1 M $Na_3C_6H_5O_7$, pH 4-5; $E_э = -0.35$ V, $t_э = 30$ s; 2 – $C_{CF} = 0.2$ g/dm³; 3- $C_{CF} = 0.4$ g/dm³; b) 1 - background solution 0.1 M $NaC_4H_4O_6 \cdot H_2O$, pH 5-6, $E_э = -0.3$ V, $t_э = 30$ s; 2 – $C_{GS} = 2$ g/dm³; 3 – $C_{CS} = 4$ g/dm³.

As can be seen from Figure 2 voltammograms there are well-defined signals CF at E = – 0.60 V, and in GS at E = – 0.65 V. The introduction of additive mixtures of certified analytical signals antibiotics increase proportionally to the introduced concentration, indicating that the possibility of quantification. An algorithm for the sample preparation of drugs, which is simple and reduces the sample to dissolve in water. In some cases, when the antibiotic is administered in a feed additive, to better carry out the deposition of the matrix requires the hydrolysis of 0.1 mol/dm³ HCl, and zinc sulfate salting (to saturation).

Using the proposed algorithm, the analysis of real-world objects with the declared content of antibiotics. The results and the method of checking the correctness of «added-found» in the table.

Table. The results of VA-determination of antibiotics and validation method of «added-found» (P = 0,95, n = 3)

Object	Antibiotic	Contents, g/g		
		On packing producer (in test)	Introduce	Find
«Farmazin» «HUVEPHAPMA», Bulgaria	Tylosinum tartrat	0.50	It isn't entered	0.48 ± 0.05
			0.40	0.93 ± 0.15
«Cephalex», «SP Veterinaria, SA», Spain	Gentamycin	0.50	It isn't entered	0.51± 0.05
			0.30	0.81± 0.13
«Gentamizin» OOO «KPK d.d.», Slovenia	Gentamycini sulfas	0.10	It isn't entered	0.11± 0,02
			0.10	0.20 ± 0.03

As the table shows, the results are consistent with those declared by the manufacturer.

4. CONCLUSION

Based on the studies developed methodology for the voltammetric determination of antibiotics in aqueous solutions of drugs used in veterinary cephalexin content ranging from 0.1 to 2.0 g/dm³, tylosine tartrate in the range from 0.1 to 1.7 g/dm³ and gentamicin sulfate from 0.1 to 1.5 g/dm³ (Sr is not more than 25 %). The methods are simple to implement, do not require costly and can be used for detection of counterfeit products.

References

1. Biomedia of the Russian Federation. Popular scientific portal. [Electronic resource]//access Mode: URL: http://биомедиа.рф

2. B. Peset *Praktikum according to the technical analysis and control in a priozvodstvo of chemical and pharmaceutical preparations and antibiotics* (1981).
3. GOST P 53912-2010 Foodstuff. An express method of definition of antibiotics
4. G. Hancu, H. Kelemen, A. Rusu, A. Gyéresi, J. Serb. Chem. Soc. **78**; 9 (2013).
5. T. Voyeykova, J. Biotechnology. **2** (2008).

The determination of sorption material properties for water softening in the dynamic conditions

Vladimir Somin [1], *Lubov* Kurtukova [1] and *Larisa* Komarova[1]

[1] Altay Technical University, Barnaul, Russia

Abstract. Hardness ions sorption properties of bentonite and paraffine-based material have been studied under dynamic conditions. The temporal characteristics of sorption process have been defined. The comparative analysis of obtained sorbent with cation exchange resin «Lewatit» has been implemented.

1 Introduction

Natural water softening is the urgent problem for many areas of Altai region, where there are almost no surface water sources and available ground water sources are characterized by heightened content of some elements, including the hardness salts. The content of hardness salts can fluctuate from 0,01mg-Eq/L to 7 mg-Eq/L depending on the category and function of water [2,3].

The traditional softening water methods are chemical and thermochemical methods, reverse osmosis and ion exchange. These methods are high-efficiency, but they demand significant costs of energy and resources. Recently the technologies based on using of sorption and ion exchanging materials from natural mineral raw material are introduced with the purpose of cost minimization on cleaning [4,5].

The most known group of mineral sorbents is the bentonite clay, that forms big deposits in nature and has an industrial value. At present there is a considerable amount of known bentonite deposits in the USA, France, India, Greece, Russia, Kazakhstan, Georgia, Armenia and other countries.

For effective usage of bentonite clay exchange capacity it is necessary to expose clay to preliminary activation for calcium and magnesium ions replacement by more mobile ions (sodium and hydrogen) in the crystal lattice structure [6]. And it is necessary to consider the nature of extractive substance and bentonite content. For metal ions removing from water solutions there are the following types of bentonite clay activations: heat, acid, salt and soda. The conducted research showed a higher efficiency of soda activation with the following extraction of hardness compounds for bentonites of all considered fields compared with other activations [7,8].We have detected, that the Millos bentonite (Greece) of soda activation has the highest static hardness ions sorption capacity (1,9 mg-Eq/L).

The sorption properties of this clay are comparable with sorption properties of ion exchange resin in the static conditions, named «Lewatit», using for softening.

* Corresponding author: vladimir_somin@mail.ru

In dynamic conditions it is impossible to use bentonite because of high hydraulic resistance and the infiltration carryover with flow. In connection with this research of bentonite clay fixation on different frames including the mineral fiber and waste wood are carried out in the Altai State Technical University at the department of Chemical technique and Engineering ecology [10,11]. The alternative version is paraffin, that can be used as a bonding agent. The obtained material has showed the high denotation of sorption capacity on hardness ions in static condition [9]. Determination sorption parameters in dynamic condition is more important for organization of technological process and so further research has been directed to of full dynamic exchange capacitance of this material studying.

2 Experimental

The research of filtration process through bentonite packing is complicated by high hydraulic resistance due to the ability to swell in the water and fill the loading channels. That is why it has been decided to use the «soft» filtration method, that has been conducted without water slick formation above the filtration material and the water is evenly distributed as a film on its surface.

The full dynamic capacity determination was realized by method GOST 20255.2-89 [12] by the way of water solution filtration of hardness salts through the module with an inside diameter of 35 millimeters, with packing material weight of 40 grams and height of 21 centimeters. Sorbent packing was made layer by layer with distributive grids using, that laid between material layers. The solution quantity flow was 26 ml/minute, the filtration speed was 1,6 meters/hour. The discharge under the material layer was made by vacuum-filter.

The model solution was applied with the content of hardness ions 10 mg-Eq/L, which was used for the preparation of calcium chloride analytical reagent grade and magnesium sulfate heptahydrate. Ions ratio of calcium and magnesium was 2:1, it is more appropriate for natural waters of Altai region. Sample picking of 15-20 ml volume was been realized after every 100 ml of filtrate, after that the sample was analyzed for the content of total hardness by titrometric method with chelating agent III with pH=10 using and indicator Black Eriochrome [13] presence. The process was carried out to equalize the filtrate and working solution.

The full dynamic exchanging capacity ($D_п$) was calculated in moles per gram (mg-Eq/g) by the formula (1):

$$D_п= ((V_ф \cdot C - \Sigma V_п \cdot C_п) \cdot 1000)/m, \quad (1)$$

where: $V_ф$ – the total filtrate volume, passing through the material to equalize the filtrate and working solution, cc;
C –concentration of working solution, mole/cdm;
$V_п$ –filtrate portion volume, cc;
$C_п$ –solution concentration in the filtrate portion, mole/cdm ;
m –sorbent weight, g.

The value of full exchange capacity of sorbent was being determined before and after regeneration. The regeneration of researched sorption and ion-exchange material was being implemented by 5% sodium hydrogen carbonate solution, after that there was the washing sorbent off by distilled water.

After the first cycle of saturation (the original material) the value $D_п$ was 0,31 mg-Eq/g, after the second one (the 1st regeneration) it was 0,26 mg-Eq/g, after the third one (the 2nd regeneration) it was 0,21 mg-Eq/g. As it can be seen regeneration reduces the full exchange capacity. The full dynamic exchange capacity of cation resin Lewatit was determined in comparison with the analogical condition, that came up to 0,47 mg-Eq/g after the first cycle of saturation, that is comparable with bentonite-paraffin materials.

In the research process of dynamic capacity the dependencies of ions hardness extraction effectiveness on the bentonite-paraffin materials and cation resin Lewatit from the solution with concentration of 10 mg-Eq/L on the time were determined, that is shown in figure 1.

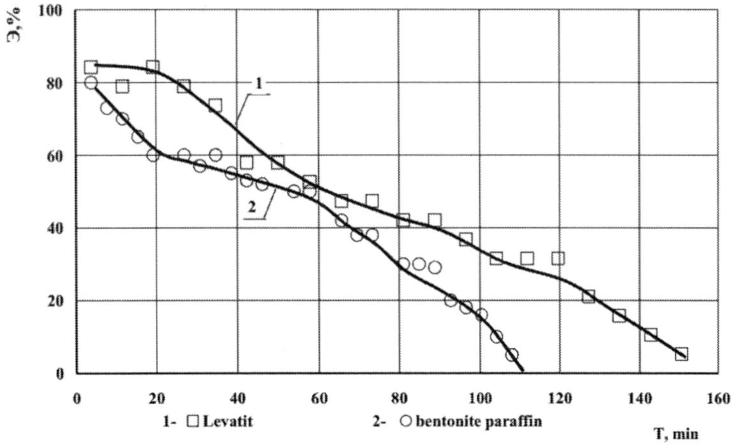

Fig. 1. The dependence of ions hardness extraction effectiveness (Э) on time (T) on the bentonite paraffin materials and cation resin Lewatit.

For both materials gradual decline of effectiveness with time can be marked. The maximum of softening was 84 % for Lewatit within 20 minutes with the following decrease by 5% within 160 minutes. For paraffin-bentonite material the maximum cleaning effect came up to 80% for the first portion of filtrate with the following decrease. The full saturation was achieved within 110 minutes.

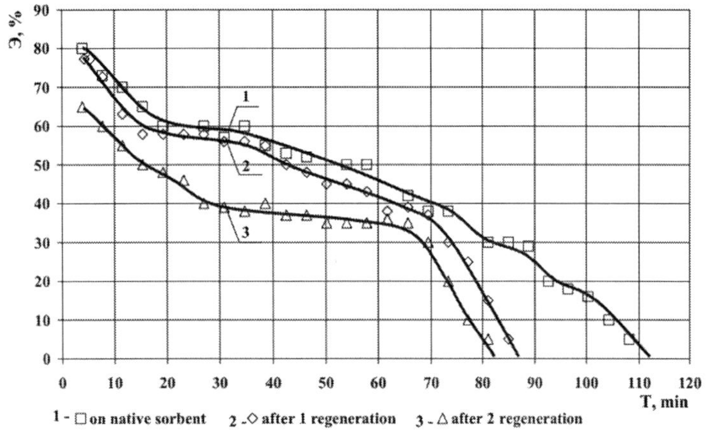

Fig. 2. The dependence of ions hardness extraction effectiveness (Э) and time (T) on the bentonite paraffin materials before and after regeneration.

On figure 2 the curves of effectiveness of ions hardness extraction from solution with concentration 10 mg-Eq/L and time depending on regeneration quantity are shown. It can be seen on picture 2, that after regeneration there are two sites on a diagram: a gradual loss of effectiveness in the interval from 5 to 70 min and a sharp drop after 70 min. It is noted, that the maximum effectiveness made up 77% after the first regeneration, and after the second one it reached 65%. By these curves time parameters of cleaning in dynamic conditions were determined: the time of

 DOI: 10.1051/matecconf/20168501034

protective action (to 50% of cleaning effect) and the time of full dynamic exchange capacity attainment.

On figure 3 there is the duration of the protective action time of filter depending on regeneration quantity (a) and dependence of time of the of full dynamic exchange capacity attainment depending on regeneration quantity (б).

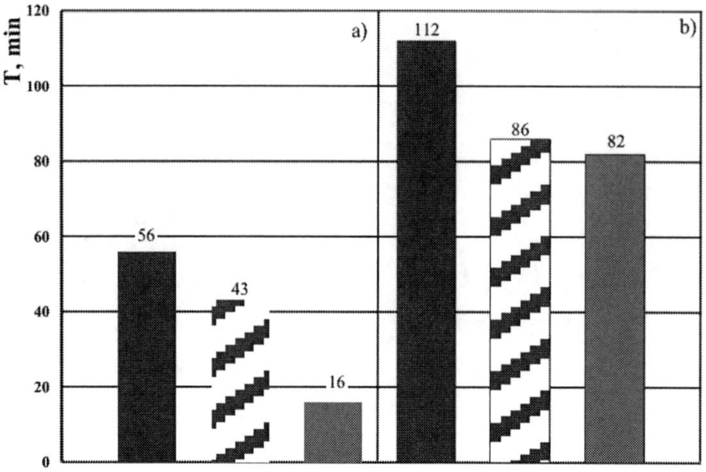

Fig. 3. The dependence of the protective action time of filter (a) and the dependence of the time of full dynamic exchange capacity attainment (б) on regeneration amount.

As it can be seen, after the first regeneration the protective action time falls by 23% (to 43 min), after the second regeneration it is more – by 64% (in comparison with the 1st regeneration).

The time of full dynamic exchange sorbent capacity attainment determination showed that regeneration reduces it (on average by 25%), but it is not reduced depending on the quantity of regeneration.

In spite of the fact that the protective action time reduces significantly after regeneration, the time of full dynamic exchange capacity attainment changes insignificantly, that means, that material can be used before its full saturation.

3 Conclusion

- Betonite and paraffine-based material can be used for water cleaning from the hardness ions. As it has been studied, the effectiveness of cleaning by this material is comparable with of cation resin Lewatit effectiveness.
- The full dynamic exchange capacity of the research material was determined, that came up to 0,31 mg-Eq/g after the first cycle of regeneration, after three regenerations it was 0,21 mg-Eq/g.
- The protective action time and the time of full dynamic exchange capacity attainment have been reduced from 56 to 16 min after the first regeneration and from 112 to 82 min after the second regeneration.

References

1. *The state report about the state and protection of the environment in the Altai region in 2010.* Barnaul, 175 p. (2011)
2. Belikov Ed. S.*Water Treatment: A Handbook.* Moscow: Aqua-Therm, p. 240. (2007)

3. *Sanitary rules and norms 2.1.4.1074-01 for drinking water. Hygienic requirements for water quality of centralized drinking water supply systems. Quality control.* 2nd ed. Moscow , p.127. (2002)
4. Nazarenko O., ZarubinaR., Veysgeym A.*Bulletin of the Tomsk Polytechnic University.* 3rd ed. Tomsk, p.28-32. (2011)
5. Vigdorovich V., Tsygankova L., Nikolenko D.,Aculov A., Rumyantsev F. *Sorption and chromatographic processes.* Tomsk, p 121-126. (2010)
6. KomarovV., Ratko A.*The Adsorbents: preparation, structure, properties*. Minsk: Belarus Science, 256 p. (2009)
7. Kurtukova L., Somin V., KomarovaL., BotsenkoA. *Polzunovsky Gazette.* 4th ed. Barnaul, p. 150-152. (2011)
8. Kurtukova L., Somin V., Komarova L.*The success of modern science.* 12th ed. Barnaul, p. 29-31. (2011)
9. Kurtukova L., Somin V., Komarova L. *Polzunovsky Gazette.* 3rd ed. Barnaul, p. 217-219. (2012)
10. Patent. RF №2345834 RU, MPK51 B01J20 / 16, B01D39 / 06.*A method of producing filter-sorption material.*
11. Patent RF №2394628 RU, MPK51 B01D39 / 14, B01J20 / 22. *A method of producing an ion-exchange sorption material.*
12. GOST 20255.2-89. *Ion exchangers.Methods for determination of dynamic exchange capacity.*
13. RD 52-24.395-2007. *Water hardness.Methods of measurement by titrimetric method with Trilon B.*

TUNGSTATE AND CARBONATE IONS SORPTION USING ANION EXCHANGERS AV-17-8 AND PUROLITE A400

S. Chegrintsev[1,2], R. Kraydenko[1,3], A. Dyachenko[1,4] and A. Kiselev[1,5]

[1]Tomsk Polytechnic University, Tomsk, Russia, [2]chesen@tpu.ru , [3]kraydenko@tpu.ru, [4]atom@tpu.ru, [5]hassem@tpu.ru

Abstract. The current paper shows the results of tungstate and carbonate ion sorption using strongly basic anion exchangers AV-17-8 and Purolite A400. It has been established that anion exchanger AV-17-8 in the chloride form with parameters of 168 g of tungstate ion and 157 g of carbonate ion per 1 kg of anion exchanger has the maximum capacity for the tungstate and carbonate ions.

1 Introduction

Traditional tungsten technology consists in obtaining tungsten concentrate by physical methods of enrichment and its processing by autoclave soda leaching [1, 2]. Tungsten is usually extracted from the solution by tungstic acid precipitation or by extraction [3, 4]. In both cases the processing reagent, used for the sodium carbonate leaching, is decomposed by acids to reduce the pH level of productive solution [5, 6]. Thus, it is impossible to regenerate the processing reagent and use it repeatedly for autoclave soda leaching.

Ion exchange is an alternative method of the direct solvent extraction of tungsten from soda solution (Na_2WO_4) [7]. When using strongly basic anion exchangers, there is no necessity to change pH of the solution, and, thus, the excess of sodium carbonate is not decomposed [8]. In the alkaline solution tungsten is stable in the form of tungstate ion WO_4 [2-9]. The productive solution containing sodium carbonate and sodium tungstate as macrocomponents as well as such impurities as silicate and sodium arsenate is supplied to the ion exchange.

The Russian strongly basic anion exchanger AV-17-8 and its foreign analogue Purolite A400 are proposed to be studied as a strongly basic anion exchanger (table 1). The most important characteristic of ion-exchange anionites is a total exchange capacity indicating the amount of adsorbed substance [10, 11]. This article presents the study of the total exchange capacity of anion exchangers in a chloride, nitrate, hydroxide and sulfate form for tungstate and carbonate ions.

Table 1. Characteristics of anion exchangers AV-17-8 and Purolite A-400 [12, 13].

	AV-17-8	Purolite A-400
Type	Strongly basic anion exchanger	Strongly basic anion exchanger
Matrix	Styrene-divinylbenzene gel	
Functional Group	Quaternary amine	
Physical appearance	Spherical granules from light yellow to dark yellow	Transparent golden spherical particles
Delivery form	OH^- / Cl^-	OH^- / Cl^-
Total capacity, no less, m-mole/cm^3	1.30	1.30
Dynamic exchange capacity, no less, meq./l	850 - 900	800
Moisture content, %	42 - 48	48 - 57
Bulk density, g/ml	0.67 - 0.73	0.69
True density, g/ml	1.07 - 1.10	1.06
Granule size, mm	0.32 - 1.25	0.30 - 1.20
Effective size, mm	0.40 - 0.70	
Uniformity index, ≤	1.6	1.6

2 Experimental

All reagents used in this study were of analytical grade (A.R). They are the ammonium paratungstate (A.R.), Na_2CO_3 (A.R.), NH_4Cl (A.R.), NH_4OH (A.R.), NH_4NO_3 (A.R.), $(NH_4)_2SO_4$ (A.R.), anion exchangers AV-17-8 and Purolite A400. Using ammonium paratungstate and sodium carbonate, stock solutions for the sorption were prepared. Sodium chloride, hydroxide, nitrate and sulfate solutions were used to recharge the anion exchanger (table 2).

Table 2. Solutions concentration to recharge anion exchangers.

Anion	*Compound*	Solution concentration
Cl^-	NH_4Cl	350 g/l
OH^-	NH_4OH	1700 g/l
NO_3^-	NH_4NO_3	1500 g/l
SO_4^{2-}	$(NH_4)_2SO_4$	600 g/l

The content of a tungstate ion in solution was determined by atomic emission spectrometry iCAP 6300 DUO. Analytical wavelengths 207.911 and 209.475 nm were used for the analysis. The analytical method was calibrated by SSRS (State Standard Reference Sample) of tungsten solution. The content of carbonate ion in solution was measured by acid-base titration. Methyl orange was used as an indicator, and 0.1 N HCl was used as a titrant.

To determine the total exchange capacity, a 25-ml burette filled with 10 g of the studied anion was used. The anion exchanger was recharged into the required counterion before the experiment. 200 ml of the experimental stock solution was passed through the recharged anion exchanger with the rate of 1 drop per second (3 ml/min). Sampling of 1 ml was selected after each 5-8 ml of stock solution to determine the concentration.

The study has shown that the anion exchanger AV-17-8 in the chloride form with a total exchange capacity for tungsten at least 168 g/kg of anionite is most appropriate to use for the sorption of tungstate ion from the solution (table 3, figure 1). In the case of a carbonate ion, the anion exchanger AV-17-8 in the chloride form has the best indicators; the capacity of the anionite is nearly 157 g of carbonate ions per 1 kg of the anionite (table 4).

Table 3. Capacity of the anion exchangers AV-17-8 and Purolite A-400 for tungstate ion.

Counterion	anion capacity, g/kg	
	AV-17-8	Purolite A-400
Cl^-	168.42	115.92
OH^-	111.91	125.06
NO_3^-	103.22	106.35
SO_4^{2-}	127.26	117.23

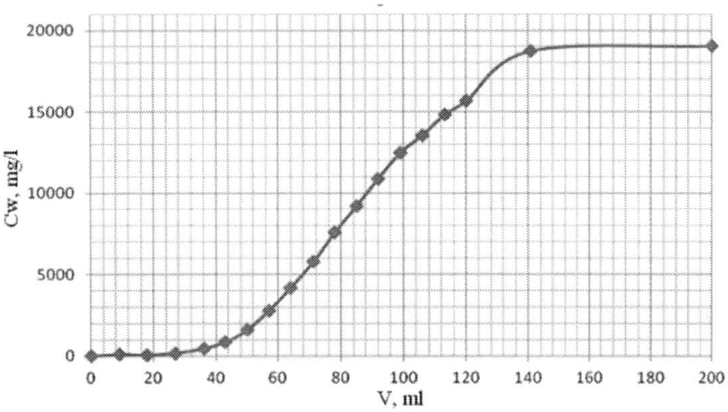

Figure 1. Dependence of the concentration of tungstate ion on the volume of solution passed through the ion-exchange column.

Table 4. Capacity of the anion exchangers AV-17-8 and Purolite A-400 for the carbonate ion.

Counterion	Anion exchange capacity, g/kg	
	AV-17-8	Purolite A-400
Cl^-	157.33	106.13
OH^-	118.32	122.74
NO_3^-	98.79	95.89
SO_4^{2-}	107.35	11.61

3 Conclusion

As anion exchangers AV-17-8 and Purolite A-400 are suitable for extracting tungsten from productive solution obtained after autoclave soda leaching.

The study has shown that the Russian anion exchanger AV-17-8 in the chloride form is the best option for the sorption of tungstate and carbonate ions.

Total exchange capacity of anion exchanger AV-17-8 in the chloride form for tungstate ion is 168 g/kg and for the carbonate ion is 157 g/kg.

Acknowledgements

This research is financially supported by the Government of the Russian Federation, contract № 02.G25.31.0118

References

1. Gedgagov E I, Besser A D, Yanakov V Ya and Smolyarchuk V P 2009 *Theoretical foundations of chemical engineering* **43** 529-538 (in Russian)
2. Naboichenko S S, Ni L P, Shneerson Ya M and Chugaev L V 2002 *The autoclave hydrometallurgy of nonferrous metals* p 940 (in Russian)
3. Zelikman A N and Korshunov B G 1991 *Metallurgy of Rare Metals* p 432 (in Russian)
4. Lassner E and Schubert W 1999 *Tungsten: properties, chemistry, technology of element, alloys, and chemical compounds* p 422
5. Paulino J F, Afonso J C, Mantovano J L and Vianna C A 2012 *Hydrometallurgy* **127** 121-124
6. Ivanov I M and Zaitsev V P 1995 *Nonferrous metals* **7** 4750 (in Russian)
7. Srinivas K, Sreenivas T, Natarajan R and Padamanabhan N P H 2000 *Hydrometallurgy* **58** 4350
8. Vol'dman G M and Zelikman A N 2003 *Theory of hydrometallurgical processes* p 464 (in Russian)
9. Gerhardt N I, Palant A A, Petrova V A and Tagirov R K *Hydrometallurgy* **60** 1-5
10. Figurovskaya N P, Zhuravleva L M and Cheremisova A M 1981 *Ion-exchange processes. Theoretical basis of the calculation and hardware design* p 88 (in Russian)
11. Bergaya F, Theng B K G and Lagaly G 2006 *Handbook of clay science* **1** 979–997
12. Faubert N 2001 Properties of Amberlite IRN78, Dowex 1-X8, and Purolite A400 strong anion exchange resins *Honours Essay/Dissertation (B.Sc.)* (Sudbury: Laurentian University) p 38
13. GOST 20301-74 *Ion-excahnge resins Anionites Technical specifications* (Moscow: State Standard of Russia) p 30 (in Russian)

MATEC Web of Conferences **85**, 01036 (2016) DOI: 10.1051/matecconf/20168501036
CCT 2016

ANALYSIS OF OPTIMAL PROCESS FLOW DIAGRAMS OF LIGHT NAPHTHA ISOMERIZATION PROCESS BY MATHEMATIC MODELLING METHOD

Vjacheslav Chuzlov[1,*], *Konstantin Molotov*[2]

[1]Tomsk Polytechnic University, Lenin av. 30, Tomsk, 634050, Russia
[2]OOO "Kinef", Enthusiastov Highway 1, Kirishi, Russia;

Abstract. An approach to simulation of hydrocarbons refining processes catalytic reactors. The kinetic and thermodynamic research of light naphtha isomerization process was conducted. The kinetic parameters of hydrocarbon feedstock chemical conversion on different types of platinum-content catalysts was established. The estimation of efficiency of including different types of isomerization technologies in oil refinery flow diagram was performed.

1 Introduction

The increase of straight run normal alkanes C_5-C_6 fraction research octane rating to 92 points is an effective decision of important applied problem of changing the structure of native gasoline pool. The high octane flow of isomerate does not content a sulfur, aromatic and olefin hydrocarbon that is why it is an excellent component for finished gasoline compounding. The selection of optimal process flow diagram (PFD) of light naphtha isomerization is an important research and practice task which can be solved more efficiently with a help of mathematical modelling method.

2 Research Object

In this paper the research object is catalytic isomerization of light gasoline fraction unit on sulfated catalyst with different recycle flows direction (figure 1) and isomerization unit with recycle of unconverted hexanes on chlorinated Pt-content catalyst (figure. 2).

* Corresponding author: chuva@tpu.ru

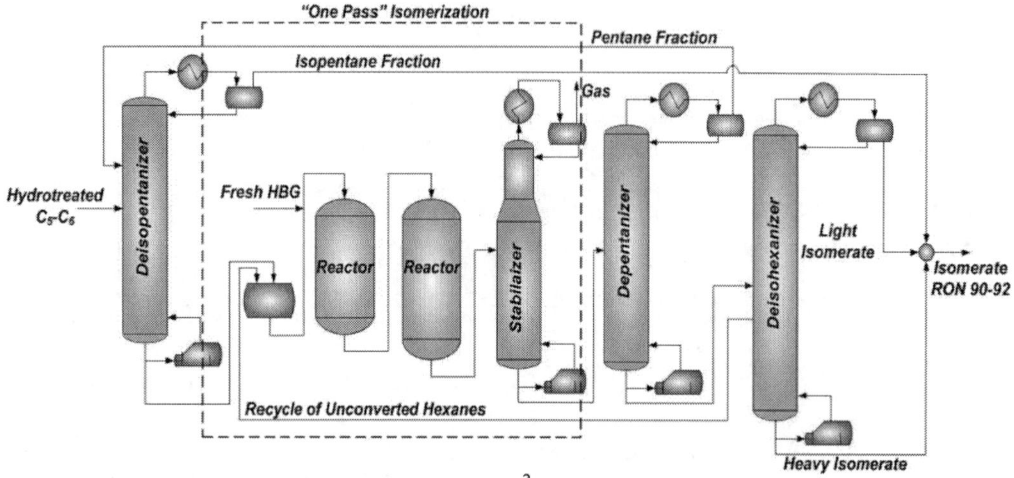

Fig. 1. Light naphtha isomerization PFD on $Pt/SO_4^{2-}/ZrO_2$ catalyst "One pass" and with recycles of n - C_5 and unconverted hexanes

Isomerization catalysts based on sulfated metal oxides combines high activity and resistance to poisons and have ability to regeneration [1-3]. "One pass" isomerization technology is the simplest, has minimal capital and operation costs and provides the production of isomerate with 80-82 RON. Including in "one pass" isomerization recycles of unconverted $C_5 - C_6$ hydrocarbons takes more difficult conditions of its operation but helps to reach the increase in RON about 10-12 points.

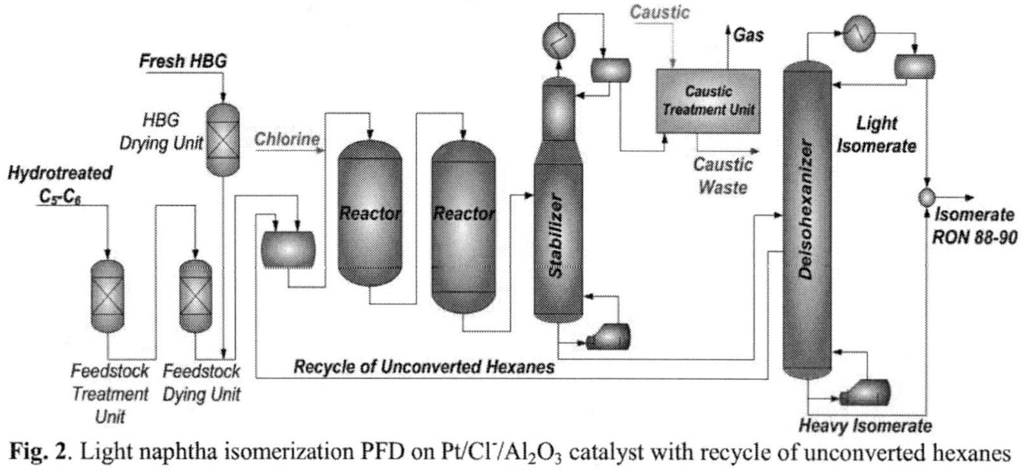

Fig. 2. Light naphtha isomerization PFD on $Pt/Cl^-/Al_2O_3$ catalyst with recycle of unconverted hexanes

Catalysts based on chlorinated alumina oxide due to longtime operation lose chlorine. In result its activity is decreased. In addition it requires a caustic treatment unit to utilize organic chlorine. Chlorinated catalyst can't be regenerated and its operation cycle is 3-5 years.

3 Methods

In this paper for description of unsteady catalytic processes of hydrocarbon feedstock isomerization the method of mathematical modelling is used. According to characterized methodology [4-15] the construction of mathematical model of complicated chemical-engineering process consist in sequential implementation of following stages: conducting of thermodynamic analysis, estimation of kinetic parameters of thermodynamically possible reactions, construction of reactor model and chemical-engineering system at all.

The full mathematical description of hydrocarbon feedstock catalytic isomerization process is formatted on components material and heat balances:

$$\begin{cases} G\cdot\dfrac{\partial C_i}{\partial z}+G\cdot\dfrac{\partial C_i}{\partial V}=\sum\limits_{j=1}^{m}a_j\cdot r_j \\ \dfrac{\partial T}{\partial z}+G\cdot\dfrac{\partial T}{\partial V}=\sum\limits_{j=1}^{m}a_j\cdot r_j\cdot\Delta H_j \end{cases}$$

if Z=0, C=0, where r=0, $C=C_0$, if Z=0, $T=T_0$, where r=0, T=Tin;

where G is a feedstock flow rate, m^3/h; z = G*t (t is a overall time of catalyst work starting from the new catalyst load, h); Ci is a concentration of ith component, mol/m^3; V is a volume of the catalyst bed,m^3; a is a catalyst activity; ρ is density of hydrocarbon mixture, kg/m^3; Cmp is a specific heat capacity of hydrocarbon mixture, J/(kg*K); Q_j is jth reaction heat, J/mol, T is temperature, K; rj is jth reaction rate, mol/(m^3 h), a – activity of catalyst.

The program realization of this model is implemented in Delphi 7. The initial data for calculation are feedstock composition and operation parameters of the process. The adequacy of the model is estimated by comparison of experimental data from industrial isomerization plant with calculated values [10]. The activity of catalyst is decreased due to processes of coke formation, poisoning and ageing.

4 Results and Discussion

The research of kinetic parameters of purposeful reactions of light naphtha isomerization process for chlorinated and sulfated catalysts is presented in table 1.

Table 1. The comparison of major reactions kinetic parameters of isomerization process

Reactions	**Relative rate constants**	
	$Pt/SO_4^{2-}/ZrO_2$	$Pt/Cl^-/Al_2O_3$
$n\text{-}C_5 \rightarrow i\text{-}C_5$	1	0.31
$i\text{-}C_5 \rightarrow n\text{-}C_5$	1	0.96
$nC_6 \rightarrow$2Methylpentane	1	0.47
2Methylpentane →2.3 Dimethylbutane	1	0.16
2Methylpentane →3Methylpentane	1	1.72
2.2 Dimethylbutane →2.3 Dimethylbutane	1	0.85

Rate constants of major reactions for sulfated catalyst are taken as 1, rate constants for chlorinated catalyst are performed as a part of 1. In such a way major reactions of isomerization process pass faster on sulfated catalyst.

The isomerate composition for different light straight run cuts isomerization technologies is presented in table 2.

Table 2. Calculated isomerate composition

Component	Isomerization Technology

	$Pt/SO_4^{2-}/ZrO_2$ System		$Pt/Cl^-/Al_2O_3$ System
	"One Pass"	With n-C_5 and C_6 Recycle	With Unconverted C_6 Recycle
n-butane	0.16	0.42	0.44
i-butane	0.13	0.00	0.06
n-pentane	14.32	0.00	11.47
i-pentane	33.38	2.32	30.72
n-hexane	6.01	6.87	0.54
2-methylpentane	15.13	0.00	5.20
3-methylpentane	7.70	9.87	0.75
2,2-dimethylbutane	11.74	61.75	36.12
2,3-dimethylbutane	4.41	14.90	5.20
n-heptane	0.04	0.13	0.00
Sum i-C_7	3.28	3.74	0.00
Sum C_8	1.79	0.00	0.00
Cyclopentane	0.84	0.00	9.11
Methylcyclohexane	1.06	0.00	0.40
Isomerate RON	***80.20***	***91.58***	***82.85***

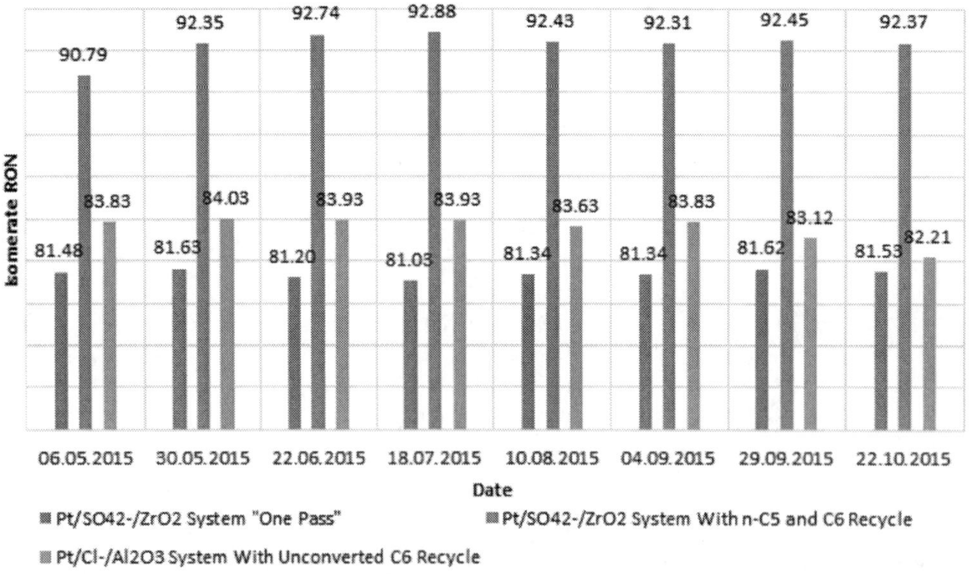

Fig. 3. Isomerate RON vs different types of catalyst and PFD

Application of catalyst based on sulfated zirconia in industrial isomerization process with recycles of unconverted hydrocarbons C5-C6 allows to produce isomerate with RON 10-12 points higher than traditional "one pass" isomerization technology and with RON 8-9 points higher than technology based on chlorinated catalyst with unconverted C_6 hydrocarbons recycle (figure 3).

References

1. Shakun A.N., Fedorova, M.L. Isomerization of light gasoline fractions: The efficiency of different catalysts and technologies, Catal. Ind., **6(4)**, 298-306, (2014)
2. Shakun, A., Aranovich, I., Reis, E. Isomalk-2: A low-temperature, light naphtha isomerization process, 2012 12AIChE - 2012 AIChE Spring Meeting and 8th Global Congress on Process Safety, Conference Proceedings
3. Shakun, A.N., Yas'yan, Yu.P., Litvinova, S.M. Some stability aspects of new zeolite containing catalysts for selective hydrocracking, Khimiya i Tekhnologiya Topliv i Masel, **2**, 39-40, (2001)
4. Frantsina, E.V., Ivashkina, E.N., Ivanchina, E.D., Romanovskii, R.V. Developing of the mathematical model for controlling the operation of alkane dehydrogenation catalyst in production of linear alkyl benzene, Chem. Eng. J., **238**, 129-139, (2014)
5. Kravtsov A.V., Ivanchina E.D., Ivashkina E.N., Frantsina E.V., Kiselyova S.V., Romanovsky R.V. Thermodynamic Stability of Coke-Generating Compounds Formed on the Surface of Platinum Dehydrogenation Catalysts in Their Oxidation with Water, Pet. Chem., **53**, 267-275, (2013)
6. Dolganova I.O., Dolganov I.M., Ivashkina E.N., Ivanchina E. D., Romanovsky R. V. Development of Approach to modelling and optimization of non-stationary catalytic processes in oil refining and petrochemistry, Pol. J. Chem. Technol., **14(4)**, 22-29 (2012)
7. Ivashkina E.N., Frantsina E.V., Romanovsky R.V., Dolganov I.M., Ivanchina E. D., Kravtsov A.V. Developing a Method for Increasing the Service Life of a Higher Paraffin Dehydrogenation Catalyst, Based on the Nonstationary Kinetic Model of a Reactor, Catal. Ind., **4(2)**, 110-120, (2012)
8. Dolganova I.O., Dolganov I.M., Ivashkina E.N., Ivanchina E.D. Development of computer modeling system as a tool for improvement of linear alkylbenzene production, Pet. Coal, **53**, 244-250, (2011)
9. Ivashkina E.N., Youriev E.M., Ivanchina E.D., Kravtsov A.V., Frantsina E.V., Romanovskiy R.V. Development of an intelligent system for controlling paraffin dehydrogenation catalyst operation in production of linear alkyl benzenes, Catal. Ind., **2(2)**, 137-144, (2010)
10. Chekantsev N.V., Gyngazova M.S., Ivanchina E.D., Mathematical modeling of light naphtha (C5, C6) isomerization process, Chem. Eng. J., **238**, 120-128, (2014)
11. Ivanchina, E., Ivashkina, E.,Frantsina, E., Silko, G., Kiseleva, S.Determination of the optimal operation mode of the platinum dehydrogenation catalysts // Advanced Materials Research, **880**, 25-31, (2014)
12. Nazarova, G.Y., Ivashkina, E.N.,Ivanchina, E.D., Stebeneva, V.I.,Seytenova, G.Z.Effect of Group Composition of the Vacuum Distillate from Heavy Kazakhstan and West Siberian Oil on the Yield of Light Fractions during the Catalytic Cracking // Procedia Engineering, **152**, 18-24, (2016)
13. .S. Belinskaya, E.D. Ivanchina, E.N. Ivashkina, V.A. Chuzlov, S.A. Faleev. Mathematical modeling of the process of catalytic hydrodewaxing of atmosheric gasoil considering the interconnection of the technological scheme devices // Procedia Engineering, **113**, 68-72, (2015)
14. Belinskaya N.S., Ivanchina E.D., Ivashkina E.N., Silko G.Y. Effect of feed composition changing at naphtha catalytic reforming unit due to involvement of gasoline fraction obtained by diesel fuels hydrodewaxing into the processing // Procedia

Chemistry, **10**, 267-270, (2014)
15. N.S. Belinskaya Mathematical modelling and optimization of catalytic dewaxing of diesel oil cuts and atmosheric gas oil // Peteroleum and Coal, **58(1)**, 126-134, (2016)

AUTHOR INDEX

Abakumov Alexander 93
Abramova R. 130
Andropov M. O. 103
Antonova Svetlana 74
Apalkov Gleb 110
Astashkina Anna 116
Bakibayev Abdigali 116
Belinskaya Nataliya 51, 125
Bobrovnikova A. 9
Bushmanov A. 130
Chaves-Rivas F. 69
Chegrintsev S. 190
Cherdantseva Daria 150
Cherkasova T. 9
Chernyakova Ekaterina 98
Chernykh Tatyana 159
Chernyshov A. 130
Chukhareva Natalya 15
Churkin R. A. 103
Chuzlov Vjacheslav 193
Deryabina Valentine 181
Dolganov Igor 23
Dolganova Irena 23
Duchko Maria 29
Dyachenko A. 190
Dyachenko A.N. 1
Dyachenko Alexander 45, 110
Dyachenko E.N. 33
Fakhrislamova R. S. 103
Frantsina Evgeniya 125
Gavrilova K.A. 177
Gindullina Tatyana 181
Govorova L.P. 137
Hernandez Espinoza M.A. 69
Ikonnikova Kseniya 143
Ikonnikova Lyubov 143
Ilyin Alexander 40
Ivashkina Elena 23, 51
Kabieva Saule 181
Kaduchagov Petr 29
Kantaev A.S 1
Karmanova Anna 63
Kiselev A. 190
Koksharov Anton 98
Kolbysheva Yuliya 116
Kolobova E. 69
Kolpakova N.A. 33, 60
Koltunova Ekaterina 143
Komarova Larisa 185
Korotchenko Tatiana 15
Korotkova Elena 74
Kotkova Elena 154
Krasnoshchekova Lyubov 150
Kratochvil Bohumil 74
Kraydenko R. 1, 190
Kraydenko Roman 45
Krivtcova Nadezhda 154
Kurganova Elena 159
Kurtukova Lubov 185
Lashtur A.L 1
Lesnikova Margarita 45
Makarov B.I. 79
Makarovskikh Alexandra 159
Malyutin Lev 45
Martemianov Dmitrii 74, 163
Martemianova Irina 74, 163
Mazurov A.K. 79
Mezentseva Olga 5
Mikheeva Elena 5
Minaev K. M. 103
Molotov Konstantin 193
Mostovshchikov Andrei 40
Murashkin Maxim 143
Nadeina Luiza 163
Nazarova Galina 51
Nikiforova Alena 116
Novoselov K.L. 177
Okanov Aydin 181
Ol'khovatenko V.E. 56
Oskina Y. A. 60
Panova S.M. 60
Paromov Vladimir 170
Pasechnik Elena Yu. 63
Peresedova Alexandra 29
Pestryakov A. 69
Petlin Ilya 45
Petranovskii V. 69
Pisarev Mikhail 23
Plotnikov Evgenii 74, 163
Popov Viktor K. 63
Popova Natalya 125
Potapova A.P. 137
Preis Yulia 29
Pryanishnikov E.G. 56
Pshenichkin A. Ya. 60
Reshetova A.A. 137
Reva I. 130
Reva I.V. 79
Root Liudmila 40
Rousskikh Irina 29
Rudmin M. 130
Rudmin M.A. 79
Sartakov Mikhail 15
Savichev Oleg 170
Serebrennikova Olga 29
Shafran Tatyana 51
Shakhova Nina 86
Shchukina Tatyana 181
Shirokolobova A. 9
Shvagrukova E.V. 137

AUTHOR INDEX

Slepchenko Galina 5, 181
Smatveenko Irina 170
Smirnov Sergey 110
Somin Vladimir 185
Strelnikova Eugenia 29
Syskina A. 9
Tataurshikov Anton 154
Toropkov Nikita 93
Tuan Hoang Tran 86
Ulyanova O. S. 103
Vakalova T.V. 137
Voronova Olesya 74
Voroshilov V. 130
Yakupova Inna 98
Yanovsky V. A. 103
Yurkova Maria 150
Yurmazova Tatyana 86
Zaytsev Nikolay 5
Zhabin Andrey 110
Zhuravkov Sergey 74